URBANIZATION, PLANNING, AND NATIONAL DEVELOPMENT

Urbanization, Planning, and National Development

JOHN FRIEDMANN

SAGE Publications Beverly Hills / London

For information address:

SAGE PUBLICATIONS, INC.
275 South Beverly Drive
Beverly Hills, California 90212

SAGE PUBLICATIONS LTD
St George's House / 44 Hatton Garden
London E C 1

Printed in the United States of America

International Standard Book Number 0-8039-0157-7

Library of Congress Catalog Card No. 72-84049

FIRST PRINTING

To my collaborators at the
Centro de Desarrollo Urbano y Regional
of the Catholic University of Chile

PREFACE

The study of urbanization in a context of national development is the result of both intellectual and practical concerns. On the intellectual side, studies of national systems of cities moved into the forefront of research in the early 1950's. At a conference held at the University of Chicago in 1954, Eric Lampard presented a study of urban history that radically departed from the manner, customary till then, of treating the subject as the biography of individual cities.[1] Concentrating on the United States, Lampard saw the evolution of cities tightly interwoven with the making of modern America. An equally germinal paper was contributed by Bert Hoselitz who had convened the conference and whose deep knowledge of economic history had suggested to him a causal connection between cities and the economic growth of nations. Hoselitz addressed a question that has preoccupied students of urbanization ever since: whether cities "generate" economic progress or "drain" an economy of its strength for development.[2]

Decisive as it was in refocusing scholarly attention, the Chicago conference was not immediately followed by massive research on the problems it had raised. Ten years later, a few of the relevant contributions were collected in a volume on regional development and planning edited by William Alonso and myself.[3] For several critics, however, the inclusion of these chapters was incomprehensible. They failed to see how urban studies might help explain the basic processes of regional (and, by implication, of national) development or, for that matter, what their policy significance might be.

Regional development had originally been identified with the harnessing of river basin resources for economic production. Under the impact of the dramatic Tennessee Valley development scheme, river basin planning had become virtually synonymous with regional planning. But as the limitations of this approach came to be recognized, a number of national governments began to emphasize urban development as a focus for subnational planning. The theoretical basis for this was provided by François Perroux who contributed the seminal concept of *pôle de croissance* to the discussion.[4] Although ambiguous and poorly understood, growth pole "theory" was quickly embraced by regional planners who applied it somewhat indiscriminately to their work. The results of these policies have recently been evaluated in a penetrating study by Lloyd Rodwin, himself one of the chief promoters of the concept.[5]

In the late 1960's, intellectual and practical concerns began to merge, giving impetus to a new and rapidly growing field of comparative urbanization.[6] Though an academic emphasis still predominated, it was difficult to write about cities without considering ways of guiding their future or of coming to grips with the problems occasioned by rapid urbanization.

My own interest in the subject dates back to my doctoral dissertation on the spatial impact of economic development in the Tennessee Valley.[7] I was persuaded at the time that urbanization had, indeed, a much more decisive role in regional development than planning doctrine allowed. I argued that economic growth not only influenced the spatial structure of production but also, that this structure had important effects on the future course of regional development.[8]

> "The growing emphasis on the city region as the center of economic development for an area encouraged a shift in the development process from natural to metropolitan resources, such as transportation, space, and community organization. The efficiency with which the city region functions as an integrated economic environment as well as its potential rate of economic advance—given a favorable national climate for expansion—depends ultimately on the degree to which these man-made resources are being developed and used."

My education on these matters continued in Chile where, for a number of years, I headed a program of advisory assistance to the central government. Half the chapters in the present book were originally written under the direct impact of this experience. As I reflected on the multiple influences—economic, social, and political—which a prolonged process of accelerated urbanization had exerted on the Chilean economy (chapter 5), I was made keenly aware of the subtle interplay of forces between urbanization and national development. Once aroused, my interest led to further theoretical work, the results of which are brought together as Part I of this volume. At the same time, my advisory work compelled me to think systematically about urban and regional development policies and the means of carrying them out. This led to chapters now included in Part II and the second chapter of Part III. The remainder of Part III deals with topics of interest chiefly to practitioners.

Although the essays included here were written at various times during a five-year period (the earliest article appeared in 1966), I thought that they displayed sufficient coherence to warrant their publication as a book. Some of the articles had to be substantially rewritten for the present edition; others were close to their final form and required only minimal editing. As a result, the major divisions of the book display considerable consistency, though the volume as a whole lacks that ultimate cohesion expected of a monograph-length study. But perhaps the time is not yet ripe for such perfection.

I wish to take this opportunity to thank the many people who contributed with their ideas and astute criticisms to individual chapters: Ricardo Jordan, Guillermo Geisse, René Eyhéralde, Ernesto Cohen, Thomás Lackington and Andres Necochea; Walter Stöhr, John Miller, Ralph Gakenheimer, Frank Earwaker, Charles Frankenhoff, Lawrence Mann, Francisco Vázquez and Michel Marié; Clinton Bourdon, Marguerite Diggs and Nancy Zimmerman; Barbara

Stuckey, Chung-Tong Wu and Eileen McGlynn; Ari Shachar; and the students of my seminar at UCLA in the Winter and Spring of 1971.

–*John Friedmann*

Brentwood
July 1971

NOTES

1. Eric E. Lampard, "The History of Cities in Economically Advanced Areas," Economic Development and Cultural Change, Vol. 3, No. 2 (January 1955), pp. 81-136.

2. Bert F. Hoselitz, "Generative and Parasitic Cities," Economic Development and Cultural Change, Vol. 3, No. 3 (April 1955), pp. 278-94.

3. John Friedmann and William Alonso, eds., *Regional Development and Planning.* A Reader. Cambridge, Mass.: The M.I.T. Press, 1964, Part III.

4. François Perroux, "Note sur la notion des pôles de croissance," Economie Apliquée, 1 and 2 (1955). For a review of growth pole literature, see David F. Darwent, "Growth Poles and Growth Centers in Regional Planning–A Review," Environment and Planning, Vol. 1, No. 1 (1969), pp. 5-32.

5. Lloyd Rodwin, *Nations and Cities: A Comparison of Strategies for Urban Growth.* New York: Houghton Mifflin Co., 1970.

6. Gerald Breese, *Urbanization in Newly Developing Countries.* Englewood Cliffs, N.J.: Prentice Hall, 1966; and *The City in Newly Developing Countries: Readings on Urbanism and Urbanization.* Englewood Cliffs, N.J.: Prentice Hall, 1969.

7. John Friedmann, *The Spatial Structure of Economic Development in the Tennessee Valley: A Study in Regional Planning.* Chicago: University of Chicago, Department of Geography, Research Paper No. 39, March 1955.

8. Ibid., p. 16.

ACKNOWLEDGMENTS

Chapter 2. Adapted from "The Role of Cities in National Development," American Behavioral Scientist, Vol. XII, No. 5 (May-June 1969), pp. 13-21. Reprinted by permission of Sage Publications, Inc.

Chapter 3. Slightly revised from "A General Theory of Polarized Development," in Niles Hansen, ed., *Growth Centers in Regional Economic Development.* New York: The Free Press, 1972, pp. 82-107. Reprinted by permission of The Free Press.

Chapter 5. Adapted from "Hyperurbanization and National Development in Chile: Some Hypotheses," Urban Affairs Quarterly, Vol. II, No. 4 (June 1970), pp. 3-29. (With Thomas Lackington.) Reprinted by permission of Sage Publications, Inc.

Chapter 6. Adapted from "The Urban-Regional Frame for National Development," International Development Review, Vol. VIII, No. 3 (September 1966), pp. 9-14. Reprinted by permission of the Society for International Development.

Chapter 9. Adapted from "Toward a National Urbanization Policy: Problems, Decisions and Consequences," presented at the United Nations Interregional Seminar of the Financing of Housing and Urban Development, Copenhagen, 25 May-10 June, 1970. Reprinted by permission of the United Nations.

Chapter 12. Adapted from "The Implementation of Regional Development Policies: Lessons of Experience," in International Social Development Review, No. 4 (1972).

Chapter 13. Revised from "The Uses of Regional Science: Policy Planning in Chile," Regional Science Association Paper, Vol. 18 (1966), pp. 207-222. (With Walter Stöhr.) Reprinted by permission of Regional Science Association.

Chapter 14. Adapted from "The Guayana Program in Regional Perspective," in Lloyd Rodwin and Associates, *Planning Urban Growth and Regional Development.* Cambridge, Mass.: The M.I.T. Press, 1969, pp. 147-162. Reprinted by permission of The M.I.T. Press.

Chapter 15. Revised from "Intention and Reality: the American Planner Overseas," Journal of the American Institute of Planners, Vol. 35, No. 3 (May 1969), pp. 187-194. Reprinted by permission of the American Institute of Planners.

Chapter 16. Revised from "Education for Regional Planning in Developing Countries," Regional Studies, Vol. II, No. 1 (September 1968), pp. 131-138. Reprinted by permission of Pergamon Publishing Co.

CONTENTS

Part One

THEORY

Chapter 1

TOWARDS A THEORY OF URBANIZATION

The possibility of a theory of urbanization is not yet granted universally. Its right to existence will be acknowledged only when a theory appears that students of urbanization will find illuminating and useful. Recent collective works, such as the monumental study sponsored by the Social Science Research Council, bring together a wealth of partial insights and specialized theoretical propositions.[1] They do not, however, delineate a general framework capable of providing the connecting links for more detailed observations of the phenomena of urbanization.

A theoretical framework of this sort must satisfy three formal conditions. First, it must be able to account for urbanization as a set of related social processes. This implies a shift in focus from the city as a cultural artifact to processes which (1) lead to the expansion of cities, and (2) generate and diffuse elements of urban life and culture. Second, it must be capable of being expressed in terms of a single unifying concept, similar to that of utility in economics, to which all urbanization processes may be explicitly related. Third, it must be able to encompass all relevant phenomena in a context of systemwide relationships.

In the theoretical chapters which follow, I have attempted to satisfy these conditions. It remains to be seen whether the resulting model of urbanization is both valid and productive in furthering research and clarifying the issues confronting public policy.

The first condition is met by identifying five major processes of urbanization, including the generation and diffusion of innovations, decision-making, migration, and investment, as the principal foci of theoretical analysis. These processes interact in a variety of ways, giving rise to major urbanization patterns that are themselves closely interconnected. The processes of innovation and decision-making describe urbanization in its socio-cultural dimension while the two remaining processes contribute primarily to the city-building activities of a society.

The second condition is met by adopting the unifying concept of *spatial* relationships within a simple core-periphery model in which peripheral areas are defined in terms of their dependency relationships to relevant core regions. Urbanization processes are consequently viewed as spatial processes connecting core and peripheral areas in ways agreeable to systematic study.

The third condition is met by adopting the nation as the relevant unit for the study of urbanization. Individual core areas (arranged in a hierarchy according to the degree of their autonomy in decisions affecting their development), together with their peripheries, are defined as the irreducible spatial subsystems of the nation.[2]

This shift in the perspective of urbanization research from individual cities to national systems and urban subsystems each of which, in turn, is composed of a dominant core and dependent periphery has two major implications. On the one hand, it permits the study of urbanization to be integrated with more encompassing theories of national development; on the other hand, it provides a solid foundation for national urbanization policies whose principal objective is the spatial integration of the society. These policies are growing concerns of governments throughout the world (see Part II).

The next five chapters illuminate these basic concepts. Chapter 2 provides a general overview of the subject. A number of hypotheses are advanced concerning the creative and integrative role of cities in national development. The chapter is written in a somewhat polemical tone to counteract the anti-urban bias of much sociological writing.

Chapter 3 pulls together theories in the social sciences into a body of propositions concerning urbanization viewed as a spatial process. It incorporates a series of interconnected definitions and hypotheses concerning (1) the origin and diffusion of innovations, (2) the relationship of innovation, power, and authority in spatial systems, (3) the resulting authority-dependency relations, and finally (4) polarized development in a hierarchy of spatial systems.

Chapter 4 carries the analysis to the level of empirical observation with a review of the literature on urbanization processes in a comparative setting. This chapter introduces the core-periphery model for empirical analysis and explicitly identifies the five major spatial processes with which the theory of urbanization is concerned. Decision-making as a spatial process is suggested as being of critical importance in influencing the behavior of the remaining processes.

Chapter 5 presents a detailed case study of Chile and advances the concept of hyperurbanization as one which merits further study. Hyperurbanization is defined as a prolonged condition of rapid urban growth producing strains and tensions in the social system. The successful resolution of these tensions, it is argued, permits an accelerated advance towards the goals of national development.

Chapter 6, finally, focuses on some social dimensions of urban development. It suggests that "local communities differ in their abilities to respond to opportunities in their environment in ways that will promote their economic growth, and that these differences vary systematically with certain characteristics of these communities." On the basis of an empirical study of two provincial cities in Chile, this hypothesis is expanded to include the national decision-making environment as an intervening variable for explaining the differential responses of local communities.

These theories are intended to provide a framework for the formulation of public policy. Three principles govern the relationship between theory and policy. The first states that the number and rigidity of controls stands in an inverse relationship to knowledge. As knowledge about a set of phenomena increases, policy intervention can be more selective and efficient. The second states that knowledge is a function of integrative conceptual frameworks capable of satisfactorily accounting for the diversity of empirical observations. From these two principles a third may be derived which states that policies will tend to be more selective and efficient in accomplishing their objectives to the extent they are based on a solid theoretical understanding of the relevant phenomena.

Whereas the relationship between theory and policy may still be a debatable issue, that of theory and empirical research is not. It is generally accepted that empirical study advances our understanding more rapidly when it is properly grounded in theory in a manner that may be approximately described as follows:

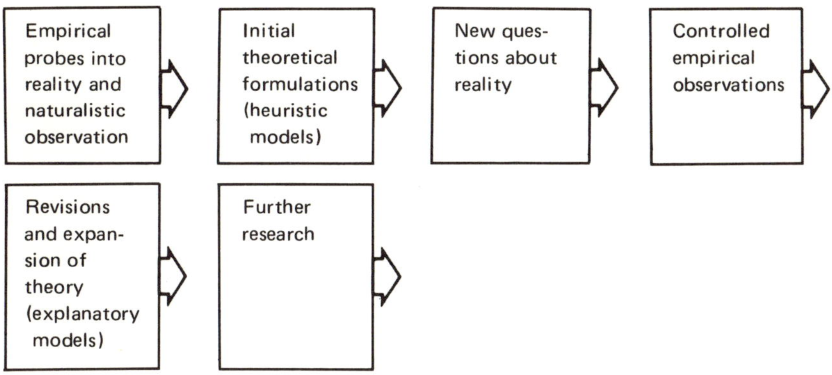

We may call this the theory-oriented model of research. A policy-oriented model, by contrast, is one in which questions for research are derived from problems of concern to policy makers and are constrained by the available means of intervention and control. The test of a theoretical proposition that is oriented to policy lies in its ability to generate *effective* policy interventions rather than in attempts to find empirical data that will contradict it. Policy-oriented research tends to be historically specific, normatively relevant, and primarily concerned with transformation paths, that is, with the various modes of achieving transitions from the present to a specified future. A policy-oriented model of research would take this form:

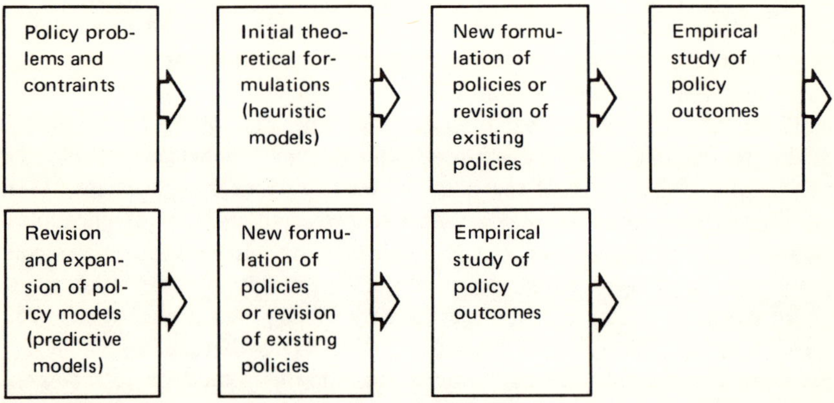

The chapters which follow are largely inspired by policy considerations. They are therefore concerned less with how given patterns are maintained and more with how they might be changed. It is clear, however, that they are also relevant for theory-oriented research, just as results inspired by the latter were used in formulating propositions of the former. In this manner, the two traditions of research are brought into conjunction.

NOTES

1. Philip M. Hauser and Leo F. Schnore, eds., *The Study of Urbanization.* New York: John Wiley and Sons, 1965.

2. Brian J. L. Berry, "Cities as Systems Within Systems of Cities," in John Friedmann and William Alonso, eds., *Regional Development and Planning. A Reader.* Cambridge, Mass.: The M.I.T. Press, 1964.

Chapter 2

THE ROLE OF CITIES IN NATIONAL DEVELOPMENT

THE MEANING OF NATIONAL DEVELOPMENT

The idea of national development is tantalizingly ambiguous; precisely for this reason, it has become one of the germinal ideas of our time. Its ambiguity arises from a value judgment: development is widely regarded as a social good. The onlookers—the students of development—may be of different minds when they are pressed for closer definitions. Each will envelope the term in an aura of ideology, but all are likely to agree that development, whatever its historically specific meaning, is a desirable achievement.

To be developing is good; stagnation is bad; indifference is inadmissible. There can be little doubt that national development displays this and many other features of a genuine social movement. It is taken for granted that national societies have a historical commitment to the idea of continuous and cumulative change.

This assumption is not quite exact, however. All national societies, it is true, may harbor elites that wish to enter such commitments, but effective power often rests with authorities that see development essentially as threatening their privilege and power. As if this were not enough, many societies are national in little more than name, lacking a genuine sovereignty, torn by antagonistic interests, and having yet to achieve a substantial measure of territorial integration. Development as a social movement may thus extend to something less than the entire world. But even nations that reject development must face prevailing international opinion that acts as if a moral obligation attached to the idea.

This essentially ethical view of development is not invoked for any particular set of policy prescriptions. It is agreed that nations differ in their historical conditions, and that every nation has a right autonomously to determine its own future.[1] Yet even granting this, no nation can entirely escape the moral imperative to "develop" and to continuously transform itself. These clashing views are somehow merged. The historically concrete is wedded to a universal expectation.[2]

There is, moreover, a growing conviction that the individual destinies of national societies are converging on a common path, and that the ultimate level of integration is a global one.[3] This belief has given rise to the distinction between underdeveloped (U) and developed (D) societies. The language suggests that the latter are passing through a phase in their historical unfolding that has not yet been reached by the former. Both types of society may be "developing," but, as late arrivals, the underdeveloped are obliged, in varying degree, to follow in the footsteps of the "developed" nations.

The argument can be properly understood only if we introduce two levels of abstraction. On the first, U-countries must, it is believed, acquire certain traits that are common to D-countries in order to survive in a developing world system. On a second, more specific level, each country must itself invent the concrete social forms and institutions corresponding to its actual condition.[4]

To give a simple illustration: all developing societies are faced with the necessity of steadily increasing the productivity of labor and of altering the composition of industrial employment. But it is left for each country to decide not only the pace at which these changes will be introduced, but also the rate at which different industrial sectors will increase their share of employment, the division of the means of production between public and private ownership, and the forms and methods of distributing the national product.

The bifurcation of the world into U- and D-countries is rudimentary. Nonetheless, it suggests that U-countries will hold in common certain comprehensive aims that, at an earlier period, were typical for the presently "developed" nations. These aims include:

(1) an increase in the autonomy of national society, by gaining effective sovereignty, increasing its military strength, enlarging the number and range of available choices, and acquiring a sense of national dignity;

(2) an increase in the levels of living of the population, by increasing production at faster rates than population growth, maintaining monetary stability, redistributing income in favor of the poorer segments of the population, and granting the rights of the people to certain basic claims of creature comfort;

(3) an increase in social integration, by promoting a wider and more effective participation of all the adult population in the decision processes of the society;

(4) an increase in modernization, by creating an institutional framework adapted to the requirements of generating and adjusting to continuous change and

promoting those activities, especially in science and artistic creation, which will permit the country to share more fully in an emerging global culture;

(5) *an increase in spatial integration,* by articulating the development process across the entire settled space of the nation through a balanced system of human settlements.

Although these aims are sufficiently general so that they will, to some extent, be present also in the developed countries, the idea of national development is not confined to them. After a critical point is reached in the evolution of a national society, the former aims may lose their urgency, while other social goals will rise to take their place. Contemporary D-countries appear to have shifted their attention to the uses of wealth and leisure, the problems of mass culture, the progress of science, and the conquest of space. These new aims of national development appear to be so remote from the perceived realities of U-countries, that they are thought to be of little relevance for them. The idea of development thus undergoes a radical change in content as we pass on to D-countries. The historical meaning is new; what remains is the driving ethical imperative.

Aims, however, are the governing criteria, and the performance of societies is judged in terms of them. Where it is satisfactory, no unusual measures need be taken. But implicit in the idea of development is the preparedness of a society to take responsibility for its performance and to intervene in the normal processes of change where the results expected from them fail to be produced. Development has the connotation of an *intended* good: it is a process endowed with moral purpose for which developing societies assume responsibility, and almost always this involves attempts at guided change or planning.[5]

THE SETTLEMENT SYSTEM

The main task of this chapter is to show, by way of introduction, the various modes of interaction between cities and national development.

As a form of human settlement, the city is a social system located in geographic space that occupies a precise position in a system of interconnected settlements, extending from hamlet to megalopolis.[6] National development occurs within this social interaction network stretched out over the landscape. But its occurrence in the spatial system is neither uniform nor simultaneous. Impulses for development originate at certain localities and are relayed to other localities in a definite sequence. The pattern of settlements creates a structure of potentials for development that will eventually be registered in indices of regional performance and will condition the evolving character of the society.

Since the goals of national development differ between U- and D-countries, each set of countries must be separately studied. In both, the spatial pattern of settlement will bear a close relation to the key events of national development, but the specific role of cities will be different in each.

These summary observations may be better understood if we look at a few salient features of the settlement system of a nation.

(1) The system of settlements is differentiated according to the functions performed by every subsystem within it.
(2) Each subsystem stands at the node of a communications network that originates, absorbs, and transmits impulses of change to the other subsystems.
(3) Each subsystem stands at the node of a field of forces that acts upon and influences the location of activities and population.
(4) Each subsystem serves as an agent of change for populations living in dependent areas.
(5) The development options of each subsystem are limited partly by its position within a complex system of dependency relations.
(6) The development options of each subsystem are partly determined by its internal capacity to respond to exogenous impulses for change.
(7) A change in the relative position of one of the subsystems in the hierarchy of dependency relations signifies a change in the structure of the system as a whole.
(8) A change in the relative position of one of the subsystems in the hierarchy of dependency relations can result from a disproportionate increase (or decrease) in subsystem size, from changes in its social and political organization, or from a change in its functions for the system as a whole.

If, as is claimed, the pattern of spatial dependency relations is a controlling one for national development, it is reasonable to focus attention on those subsystems that display a greater-than-average capacity for autonomous transformation and consequently for sustained innovation. The development capacity of the system as a whole will, therefore, come directly to depend on the performance of these centers or *core regions.* For each core region, one or more dependent *peripheral regions* may be defined, bearing in mind that lower-order cores may fall within the periphery of, and be dependent on, one or several higher-ranking cores. Core regions can thus be arranged into a hierarchy according to their relative autonomy in developing decisions. Peripheral areas, on the other hand, may be divided into upward- and downward-transitional regions in keeping with their estimated potentials for development. Finally, on core regions' peripheries, new settlement areas or *resource frontiers* may come into being.

The entire settlement space of the nation will thus be spanned by an array of regions that is articulated through a system of nodes and connecting channels of communication. This polarized space will display systematic variances in the capacity for development of given areas. Later comment will make the reasons for these differences more explicit. Here it is only necessary to point out that core regions possess the means for limiting and controlling the development of their peripheries and for extracting from them the resources that will contribute to their own accelerated growth. This imbalance of power between core and periphery, which displays a strong tendency to increase over substantial segments of the development path, will naturally lead to social and political tensions. Where these tensions cannot be held down to tolerable levels, they eventually undermine the stability of the system and its ability to generate further development.

Many appropriate examples come to mind. A striking account of the dangers inherent in this conflict between cores and their peripheries is given by Eric Wolfe in his description of the collapse of Theocratic society in Mesoamerica during the seventh and eighth centuries of our era.[7]

> "The web of Theocratic society contained another fatal flaw: a built-in imbalance between holy town and hinterland, between city and provinces. Ultimately, the towns grew wealthy and splendid, because the countryside labored and produced. Not that some of the wealth did not flow back into the rural area. Some benefits must be returned to the ruled in any complex society. Undoubtedly the Theocratic society forged its own chain of command with the final links in the villages and hamlets; some men in the countryside must have acted as messengers of the gods and received rewards commensurate with the utility of their services. The growing gap between center and hinterland was not based on an absolute enrichment of the center while the countryside remained absolutely impoverished. Both grew in their involvement with each other; but the centers grew more quickly, more opulently, and–more obviously.
>
> "Such disparity in growth may produce more conflicts than the absolute exploitation of one community by another. For a measure of prosperity which is yet confronted with the image of an ever more opulent prosperity in its midst breeds not only envy but also the desire to lay hands on the surplus for one's own purposes. Revolt occurs not when men's faces are ground into the dust; rather, it explodes during a period of rising hope, at the point of sudden realization that only the traditional controls of the social order stand between men and the achievement of still greater hopes. In complex societies, this confrontation of hope with the denial of hope pits rulers against ruled, rich against poor, and hinterland and periphery against core area and center. The periphery suffers by comparison while the center grows bloated with wealth and power. Yet it is also at the periphery that the controls of government and religion tend to be at their weakest; it is here that the forces of dissatisfaction can easily gain both strength and organization. Here the pull of the center and its ability to compel people to its will are at a minimum. Theocratic society witnessed this rebellion of the periphery against the center. The fissures opened up in three major regions; along the western margin of the Petén; along the southern border of the central highland; and in the northern marches that held the frontier against the hunters and gatherers beyond."

Spatial integration–which appears as a major development objective for U-countries–thus requires actions that will increase the autonomy of peripheral regions with respect to their controlling cores, diminish one of the important sources of conflict between them, and assure the stability of the system during its period of transition.

URBANIZATION AND THE DEVELOPMENT PROCESS IN U-COUNTRIES

City growth appears as an irrepressible accompaniment of modern development experience in U-countries. The statistical evidence leaves little question on this score.[8] The data for nearly all of the developing U-countries also show that large cities tend to grow at faster rates than smaller ones and that the national capital complex is particularly favored. But statistical associations constitute no proof of a causal relation. It is to this question of causality that we shall therefore turn by examining three fundamental forces in the process of development: the passage from tradition to innovation, from a limited culture to one of constantly expanding opportunities, and from an elitist to a mass political system striving for national integration.

Innovation and Urbanization

Invention must be distinguished from innovation. The former consists in the creation of something new out of a rearrangement of already existing elements, while the latter may be viewed as the transformation of inventions into historical fact. Both terms may be applied to tangible and intangible objects.[9]

> "The term 'invention' for anthropology does not mean only the deliberate creation of a machine, of any type of mechanism, or of any other material achievement of a culture, an achievement in the nature of a radical innovation, but also new ideas, new concepts or patterns in social, political or religious organization, as well as new economic systems which play such an important role in the entire historical evolution. Man's inventive mind and the cultural process which ensues from it do not act exclusively upon elements of culture that are in some way tangible, but certainly also upon its universal life."

Accepting this broad interpretation, we may then say that comprehensive development occurs when single innovations–large and small–are linked into innovative clusters and systems. Since innovations arise out of an established traditional matrix, their integration into larger systems occurs as an asynchronic process, pitting new forces against an already established order.

We can now assert a first hypothesis: *the frequency of inventions is positively correlated with a high potential for interaction, that is, a high probability of information exchange or communication.* The probabilities of communication over a given surface can be plotted on a map so as to yield a landscape of communication potentials or communication fields. Cities, especially large cities, appear as peaks in this landscape. It is the large city where the frequency of new idea combinations is generally greatest.[10]

A second and related hypothesis will say, first, *that urbanization will positively vary with the probability of communication at any given locality*–city growth is related to increases in potential communication, and, second, *as cities are joined into systems*–as the channels for transmitting information expand among cities–*the probability of information exchange among them will increase.* The growth of a single subsystem and expansion of intercity linkages are, to a large degree, substitutes for one another.

When we pass from invention to innovation, a third hypothesis may be stated, according to which *the frequency of innovation in a given locality is a function of its internal structure of social power.* A tightly controlled, rigid, hierarchical system of power will be less permissive of innovations than an open, horizontal, nonbureaucratic system of dispersed power. We should expect to find significant variations in the structure of social power among cities. But speaking generally, it is possible to assert that increasing size creates conditions that will bring the city's power structure into greater conformity with our second model.[11]

We are now in a position to ask whether a high probability for information exchange is causally related to certain forms of social and political organization. The answer, presumably, would be affirmative. If so, a fourth hypothesis may be stated: *increases in the level of urbanization will lead to an increasing openness of the system of social power and hence to an increase in the frequency of innovation.* This, then, may be regarded as the first positive contribution of urbanization to national development. It will transform traditional society into an innovative system.

Opportunity and Urbanization

George M. Foster describes peasant societies in terms of what he calls "an image of the limited good."[12] In such a society, the interests of one person can be advanced only, it is believed, by depriving someone else of the coveted benefits. The result is a brutally egocentric, defensive society whose members behave in accordance with this rule: "Maximize the material short-run advantage of the nuclear family; assume that all others will do likewise."[13]

The image of the limited good–a kind of zero-sum game–tends to be associated with isolated peasant societies. But contemporary urbanization is based on the opposite image of constantly expanding opportunities.[14] Cities undergoing development tend to transform the bitter little zero-sum games of peasant society into competitive nonzero-sum games. We should expect this image of the "unlimited" good to vary a good deal with both the rate of increase in employment and with city size. Both variables will increase economic choice and multiply the upward channels of mobility available to individuals.

Of course, image and experience do not always coincide. Expectations are not always bartered for the real thing. The older residents of the city, having pressed against the very real obstacles to opportunity that exist, may think society a fraud. Their political behavior will reflect a growing frustration with the social order. It is the recent migrant to the city–comparing his present condition to a dismal rural past–who looks upon his new environment with hope.[15] Return migration to the countryside is rare.[16]

What we may call the culture of the "unlimited" good is, on the whole, less tooth-and-claw, more given to accommodation, than societies organized around the opposite cultural image. Urban, in contrast to rural, society in U-countries tends to be more tolerant and generous towards the other, reinforcing the achievement impulses of large numbers of people. Continuous expansion of the urban system will tend to lower the level of interpersonal friction that might otherwise be expected.[17] The modern notion that development involves unceasing, positively valued change has grown out of urban experience. This, then, is the second contribution of urban to national development: *the city has not only invented the concept of unlimited opportunity, but also made it possible and finally internalized it as the basis for its own unique form of culture.*

Political Transformations and Urbanization

Industrialization and urbanization are not always parallel phenomena. Migration to cities in U-countries generally take place at faster rates than the absorption of new labor force into employment of high productivity. Accelerated urbanization in this sense will inevitably lead to the formation of a massive urban proletariat whose members, lacking special skills, are only partially within the labor market, if at all, and for this reason are excluded from most of the material—and spiritual—benefits of the developing, participant society. The political significance of this situation derives from a heightened visibility of the city proletariat that, being looked at and looking at the world from which it is excluded in any but the stictest functional sense, may acquire a keen sense of its own "marginal" condition and become potentially available as a potent political force to whatever leadership is capable of capturing its confidence.

The second result for political development is the emergence in cities of innovative groups that pose a challenge to the established powers in their attempt to legitimize new claims and create a social climate favorable to further innovation, insofar, at least, as this would be consistent with their own schedule of values. In this endeavor, innovative counter-elites may seek alliances among successively larger, more inclusive sectors of the population: the upper-middle class, the lower-middle strata, the workers, the urban proletariat, and eventually also the excluded rural proletariat on the periphery. As their base for political action expands and the masses become politicized, the politics of innovation first turn populist, and later integrationist, on a national scale.[18] Strongly doctrinaire and ideological during the struggle for power with the established elites, innovative groups will tend to become open to compromise once they have risen to power.[19] Political parties may consequently become transformed into giant bartering systems striving to aggregate ever widening circles of multiple and overlapping interests.

This outcome is not inevitable; it is merely possible. We may, therefore, wish to test the following hypothesis: *the resistance of the "established" powers to innovative counter-elites diminishes with the acceleration of the urbanization process.* This may be regarded as the third main contribution of the city to national development: *accelerated urbanization hastens the coming of a national mass politics based on bargaining and compromise among competing interest groups.*[20]

Summary

We have tried to suggest that the three basic processes of national development–innovation, social, and political transformation–are closely linked to yet another process, that of urbanization, which tends to reinforce the latent predispositions to developmental change through increasing communication potential and a change in the pattern of social organization from Euclidean hierarchies to Einsteinian systems existing in time. We have further tried to show that large cities are more effective in promoting these changes than small cities, and that the linkage of cities into urban systems has an effect analogous to increasing city size. It is also probably true that these variables–urbanization, city size, and expanding urban systems–are positively correlated with conflict. Urban society is a conflict society, but this conflict is only the external and highly visible accompaniment of more creative processes beneath the surface. Where conflict is absent, we can be certain that development does not occur.[21]

THE DEVELOPMENT PROCESS AND CHANGES IN SPATIAL ORGANIZATION

In the preceding sections, we have described the development process as if it extended uniformly across a nation, except for a sharp intensification–and acceleration–at a small number of core regions. Superimposed upon a uniform development "surface," core regions polarize surrounding space and set up fields of tensions between themselves and their peripheries. If we now examine this surface more closely, we discover that it is finely articulated by a system of settlements which, subject to change itself, ends by completely transforming the structure of spatial relations in the social economy. The form in which this spatial transformation occurs exerts a decisive effect on the ability of a nation to achieve a satisfactory development of its internal resources.

Two pervasive forces underlie a change in spatial organization. The first is a steady improvement in the overall accessibility of the system that comes as a direct result of the five universal characteristics of modern development:

(1) a progressive expansion of transport and communication capacity over larger areas,
(2) an increase in the number of possible interconnections within the system,
(3) a reduction in the unit cost of transportation and communication,
(4) a rise in real incomes per capita,
(5) an increase in the speed and efficiency of the transport and communication system.

Changes in the location propensities of firms and households is the second force determining the path of spatial transformation. The tendencies will differ with the experience of each country, but among the more significant changes in the location propensities of industrial and commercial firms–as development proceeds—will be shifts from a predominant resource to a market orientation, from single to multiple firm corporations, from an emphasis on the presence of economic to social infrastructure, from small to large firms (in employment), and from isolated firms to clusters of related enterprise.

During the early and intermediate phases of development, the location of households tends to be determined by the geographic pattern of emerging economic opportunities, but gradually this grim determinism will be relaxed as the classical formula of "labor follows industry" is reversed, and the natural and social amenities of different community environments come to exert a growing influence on where people decide to settle.

The relation between these changes in accessibility and location propensity, on the one hand, and spatial organization, on the other, is not a linear one, however, and must be studied by tracing the process along a series of feedback loops. Every change introduced into the spatial system will accordingly modify the conditions determining location decisions, and, in a second round, produce further changes in the system. With this important qualification in mind, it is now possible to describe—in the form of a model—the four types of transformation in the spatial organization of society that the development process engenders. It should be added—though it may seem obvious—that a model, which is a guide for research, can never be a substitute for it. Empirical departures from the model elicit questions inviting further and more detailed study.

The Transformation from Simple to Complex Structures[22]

The spatial structure of U-countries is, by comparison to later periods, a very simple one. Regional economies show relatively little specialization, since most of the population obtains its living from agriculture. Farming activities may be grouped into broad ecological or production zones, but the cities that organize the national economy show only minor functional distinctions. One or two of them, however, will eventually stand out from the rest, evolving to the point where they can be unambiguously identified as core regions that will reduce the rest of the country's economy to a peripheral status. As core-region development proceeds, this simple structure is gradually filled out: territorial specialization occurs and urban functional hierarchies—following a modified Christaller pattern for all but the largest centers—become established. The new system will be coordinated—it would be wrong to speak here of direction—through spatial aggregates at the top of the hierarchy whose internal complexity is so great that new terms have had to be invented to describe them: metropolis, urban field, development corridor, megalopolis. Development comes to be concentrated in these potent spatial forms through which the former national peripheries are reorganized into elaborate patterns of interdependency. As larger and larger areas are brought into the structure of core regions, the national periphery is trimmed to a vestigial status. Most of the population comes eventually to live within these new foci of development.

Transformation from Imbalanced to Balanced Structures

The initial spatial imbalance may be observed not only in the pronounced core-periphery structure of U-countries, but also in the relationship of population size to rank of cities in the system. According to a widely accepted rule, the size of each city in a fully developed system is determined by its rank, so that by dividing the size of the largest city, by the rank of any city, we obtain its population size.[23] There is a second rule that connects the size and function of cities with regional income. This states that the number of central functions of a city will increase with regional per capita income—holding size constant—but will also increase with an increase in city size, for constant income.[24] Combining these two rules with what has been said about growing functional specialization in the preceding paragraph, it may be asserted that, whatever the initial structure, continued development will be accompanied by an increase in the rank-size ordering of cities.[25] If we now extend the posited relationship horizontally across the settled space of a nation, the result of a process of increasing rank size will be a fully developed, hierarchically ordered spatial system, in which all places have approximately equal access to similar urban functions.[26] The system minimizes distances and is a stable one in its overall configuration, even though it may continue to evolve internally. Thus, lower-order functions may be recombined in larger centers, thereby eliminating the smaller places in the hierarchy, while, at the top, entirely new urban forms come into being.

Transformation from Narrow Impact of Urban Life-Styles to Total Immersion in Urbanism

The spread of urban life-styles is not coincident with the pattern of cities, though it is strongly influenced by it. Initially, the gradients of urban influence are very sharp: urbanism is typically a large-city phenomenon, but the population affected by it will tend to live in very close proximity and constant interaction with the center.[27] Development, however, acting through the forces of equalizing access and location propensities, will work to reduce these gradients and to introduce elements of urban life-styles into the remotest corners of the nation. Vast zones of an engulfing urbanism thus appear as the large city penetrates smaller towns and even the countryside with mass information media and modern forms of large-scale organization (e.g., commercialization of agriculture, unionization of farm labor, industrial production). Since urban life-styles may set up expectations that cannot always be fulfilled in the locality, massive population readjustments occur as migrant streams move from peripheral areas lightly touched by urbanism to the more thoroughly urbanized development regions of the country.[28]

Transformation from Partial to Complete Regional Integration

Each of the preceding three transformations of spatial systems—functional specialization of areas, increasing urban balance, and incorporation of most of the population into the culture of urbanism—contributes to the full regional integration of the nation. They do so not only by forging multiple and stronger linkages among all parts of the country, but also by encouraging greater geographical mobility of labor, capital, and commodities in expanding national and regional markets. At this point, the classical economic law of equal factor remuneration comes into full force, allocating resources in an efficient way according to the returns expected in each case.[29]

But regional integration has an even more basic meaning. It suggests the transformation of an entire nation into an innovative system with great capacity for self-renewal, the integration of most of the population into a social realm of constantly expanding opportunities, and the creation of a truly national polity whose individual members have roughly equal access to decision-making centers. These results will also tend to lead to less authoritarian patterns of social organization and to a society held together by force of innumerable strands of criss-crossing communications, producing a "new man."[30]

The system of cities thus appears as a dynamic agent of development, not only generating but also mediating development impulses—an agent subject to change no less than changing, whose structural elements may be arranged in optimal ways to facilitate the process of national transformation.

NOTES ON URBANIZATION AND NATIONAL DEVELOPMENT IN D-COUNTRIES

Sometime during the course of spatial transformation just described, a country will pass, quietly and unnoticed, from U to D, from a condition of underdevelopment to one of development. The exact timing of this changeover is somewhat arbitrary, for the process of transition is a relatively smooth one, and precisely where one elects to place the magical barrier that divides these different worlds is a matter on which experts may differ. Moreover, admission to the select company of D-countries signifies neither that all internal phenomena will automatically be consistent with each other—the oxcart coexists with the jet engine—nor that the process of development will increase. On the contrary, election of D-status means that a country has pushed forward into a period of self-reinforcing and continuous change, based, as in the past, on invention and innovation. Inevitably, therefore, discrepancies arise among the elements of which developing societies are made. Development appears as an asynchronic process.[31]

Nevertheless, in taking stock of the changes that have occurred, students of development will discover that the general situation of a D-country differs radically from earlier periods. For our purposes, the most important change to have occurred is the substitution of time for place. Expressed in different language, this means:

(1) increased freedom from economic constraints in location decisions,
(2) increased geographic mobility of population and productive resources,
(3) larger networks of social interchange,
(4) larger number of decision-making centers, and
(5) easier communication among centers.

The substitution of time for place also means a substantial reduction in the so-called friction of distance that will enlarge the options for locational choice and make the real time of reaching other places in the system a more decisive fact than out-of-pocket costs of movement. Eventually, even time may cease to be important. As the time for covering a certain distance shrinks, a situation of approximately equal access is eventually approached. This will extend to the entire space of the nation the salient traits of a core region.

Modern development tends to push these changes farther in the same direction. This does not mean, however, that location decisions will henceforward be treated with whimsy. For even in postindustrial societies, the past bears heavily upon the present. Old centers persist in having strong attractive power, roughly in the same proportions as over decades past, though activities may spill out into urban fields, enlarging the concept of city and constituting a new unit of spatial integration.[32] On the other hand, the reconcentration of communication-oriented services and industries into broad zones will tend to influence the location choices of those firms that still perceive an advantage in being physically near to the means of interchange.

If the substitution of time for place reduces economic constraints on location decisions, the city ceases to be a propulsive force of development. Spatial patterns of development were primordial national concerns of an underdeveloped society seeking greater autonomy, higher living levels, more social integration, greater world participation, and a more complete integration of its territory. They cease to be so for countries that are rapidly slipping into the postindustrial era.[33] The new salient interests in leisure, mass consumption, and science are no longer dependent on an urban focus—they may be called trans-urban. Nor is the city—in its traditional meaning—any longer an important instrument for their fulfillment.

This is not to claim total irrelevance of the settlement pattern for development in these societies. Problems of transition will continue to command the attention of policy makers.[34] Pockets of peripheral poverty remain, dramatic changes in the internal ecology of meta-urban units call for difficult adjustments; the poor are too handicapped to take advantage of the new location freedom; massive transportation problems demand urgent solutions. In addition, the policy of linking national economies into supranational systems, such as a common market, will pose problems of spatial reintegration similar to those familiar to U-countries and amenable to a similar treatment. In both cases, the urban pattern will play a major role.

The problem of urban form, however, remains an open one where the dominant social concerns fall more in the realm of culture than of politics or economics. For locational freedom imposes a special kind of responsibility: freedom for what? The forms of urban settlement are therefore not indifferent to postindustrial societies.

Which patterns are optimal for this new phase of development? In attempting to answer this question, we trespass on unknown territory. U-countries, after all, have the example of the more developed societies as a guide, and we are able to study the regularities and phases of succession in spatial organization. But on what model may D-countries pattern *their* development? We are only now beginning to know postindustrial society and are still missing a full vision of its potentialities.

THE QUESTION FOR POLICY

Given the open-ended nature of urban development in D-countries, and without denying a need for urban policies to guide the evolution of their complex spatial systems, it is primarily for U-countries that a scientific basis for policy may be devised.

But to what kinds of question is urban policy in U-countries to be addressed? Without attempting to be complete in listing them, nor deciding whether urban development policy had not better be called regional, we may pose six leading issues that need to be considered in any serious thinking on the subject. The questions are:[35]

(1) What is the optimal rural-urban balance for successive phases in a process of national development?
(2) What is the optimal rate of urbanization at successive phases of the development process?
(3) What are the optimal patterns of spatial organization at successive phases in the development process?
(4) What are the optimal transformation paths in shifting from one type of spatial organization to another?
(5) What should be the criteria to determine the proper timing in shifts from one kind of urban development strategy to another?
(6) What criteria may be applied to determine the optimal balance between centralization and decentralization in administrative and political decisions for urban development?

Along with these specific issues of urban development policy, more general questions that apply to any attempt in guiding system change ought to be considered.

(1) What are the key points of intervention in a system?
(2) How can systemwide balances be maintained during the transition period while making maximum use of imbalance for promoting change?
(3) How can the asynchronic elements of a system undergoing transformation be reintegrated around emerging focal points of innovative change?

The study of urban development has never been approached in quite this way. The perspective is normative; the ultimate aim is prescription. The analysis appropriate to this objective is cast in terms of a dynamic systems model in which the relevant aspects of economics, sociology, political science, and geography are brought together. Combined with this is an unfamiliar emphasis on the spatial aspects of development. We are therefore conscious of making a first reconnaissance in difficult terrain.

NOTES

1. This is the ideal. In practice, large-scale intervention is often found, spheres of influence are carved out by contending powers, and national options are limited by the criteria governing international assistance.

2. This probably explains why many American observers admire Communist China's staggering effort to lead a population of 800 million into the twentieth century (despite possible disagreement with the methods used), and why they hold right-wing (but pro-American) dictatorships, from Portugal to Paraguay, in contempt for failing to do likewise. Moral imperatives are stronger than political philosophy.

3. According to Gideon Sjöberg, "industrial cities over the world are becoming alike in many aspects of their social structure. . . . Implicit in our analysis is another hypothesis: as technology becomes increasingly complex, a significant number of structural imperatives become more narrowly defined. . . . Cultural values induce stylistic differences among industrial cities that cannot be ignored. However, a value system cannot modify structural arrangements in an infinite variety of ways. Industrial cities share certain values because of their dependence on scientific method and modern technology. Overemphasis of cultural values as an independent variable leads to historicism and a denial of the possibility of making cross-cultural generalizations" ("Cities in Developing and Industrial Societies: a Cross-Cultural Analysis," in Philip M. Hauser and Leo F. Schnore, eds., *The Study of Urbanization.* New York: John Wiley and Sons, 1965, pp. 249-250). To the extent that development implies the spread of industrial culture, the thesis of cultural convergence may be maintained with a certain logic.

4. See Reinhart Bendix, "Tradition and Modernity Reconsidered," Comparative Studies in Society and History, Vol. 9 (April 1967), pp. 292-346, for a subtle elaboration of this argument.

5. The literature on development planning has grown impressively in recent years. For a comprehensive account of national development planning in practice, see Albert Waterson, *Development Planning: Lessons of Experience.* Baltimore: Johns Hopkins Press, 1965.

6. Differences in urban social systems do not concern us for the moment. Later, when we return to consider this question, it will be shown that only a few, but important features are relevant for national development.

7. Eric Wolfe, *Sons of the Shaking Earth: The People of Mexico and Guatemala.* Chicago: University of Chicago Press, 1962, pp. 168-169.

8. A good recent summary of data is found in Gerald Breese, *Urbanization in Newly Developing Countries.* Englewood Cliffs, N.J.: Prentice Hall, 1966. For a sophisticated comparative analysis of urban population trends in one of the world's major underdeveloped regions, see Lowdon Wingo, Jr., "Recent Patterns of Urbanization Among Latin American Countries," Urban Affairs Quarterly, Vol. 2 (March 1967), pp. 81-109.

9. Gonzalo Aguirre Beltrán, "Confluence of Culture in Anthropology," Diogenes, No. 47 (Fall 1964), pp. 4-5.

10. An early but, nonetheless, excellent reference on the origins of innovation is H. G. Barnett, *Innovation: The Basis of Culture Change.* New York: McGraw-Hill, 1953. Barnett lists nine conditions favoring the initiation of innovation: the accumulation of ideas, the concentration of ideas, the collaboration of effort, the conjunction of differences, the expectation of change, the lack of dependence on authority, the competition of rivals, the deprivation of essentials, and the modification of a dominant correlate (chain reactions, as change in dominant elements sets up propitious, and even compelling, conditions for change in other elements). Even superficial analysis, however, will show that the large city tends to present the optimal combination of these conditions.

11. The two alternatives describe extreme situations. Optimal innovative systems may result from some combination of authoritarian and democratic power structures. As ideal types, however, the two polar models of social organization may still be useful as a heuristic device. See T. Burns and G. M. Stahler, *The Management of Innovation.* Chicago: Quadrangle Books, 1962, pp. 119-125.

12. George M. Foster, "Peasant Society and the Image of the Limited Good," American Anthropologist, Vol. 67 (April 1965), pp. 293-315.

13. Edward C. Banfield and Laura Pasano Banfield, *The Moral Basis of a Backward Society.* New York: The Free Press, 1958, p. 85. Oscar Lewis confirms this finding. He writes: "recent comparative analysis of the quality of interpersonal relations in small peasant societies, based on anthropological monographs, shows that they are characterized by distrust, suspicion, envy, violence, reserve, and withdrawal. . . . In some villages, peasants can live out their lives without any deep knowledge or understanding of the people whom they 'know' in face-to-face relationships. By contrast, in modern Western cities, there may be more give and take about one's private, intimate life at a single 'sophisticated' cocktail party than would occur in years in a peasant village. I suspect that there are deeper, more mature relationships among sympathetic, highly educated, cosmopolitan individuals who have chosen each other in friendship, than are possible among sorcery-ridden, superstitious, ignorant peasants, who are daily thrown together because of kinship or residential proximity." ("Further Observations on the Folk-Urban Continuum and Urbanization with Special Reference to Mexico City," in Philip M. Hauser and Leo F. Schnore, eds., op. cit., p. 498.)

14. This is illustrated by abundant data in a study carried out by the Centro de Estudios del Desarrollo of the Universidad Central de Venezuela *(Estudio del Conflicto y Consenso.* Serie de Resultados Parciales, Vols. 1-3. Caracas, 1965). This Venezuelan survey shows a surprisingly high incidence of hope among urban population samples. In this connection, it is worth noting that Oscar Lewis considers the "wide range of alternatives for individuals in most aspects of life . . . one of the most distinctive characteristics of cities, whether industrial or pre-industrial" (op. cit., p. 499). It is therefore those population sectors that fail to gain access to some or all of these alternatives for improving their condition that are "marginal" to urban society. This definition may be useful for reinterpreting the frequently cited fact that large numbers of urban dwellers in U-countries partake of village culture (see, for instance, Richard D. Lambert, "The Impact of Urban Society Upon Village Life," in Roy Turner, ed., *India's Urban Future.* Berkeley, Calif.: University of California Press, 1961). These "urban villagers" are, in fact, "marginated" from the dominant urban culture and may be usefully regarded as being oriented towards, but excluded from, the urban world of constantly expanding opportunities. Their persistance in village folkways is therefore not surprising. It is the only world they know and can rely upon.

15. Myron Weiner, "Urbanization and Political Protest," Civilisations, Vol. 17, Nos. 1-2 (1967), pp. 44-52. Weiner's findings concerning voting behavior, that recent migrants to the city vote more conservatively than those who have established longer residence, may be interpreted in terms of the theory of "marginality" mentioned in note 14. The recent rural migrant to the city is initially absorbed into the "marginal" population sector where he can follow his customary life-style with only minimal adjustments. But for rural population, this style includes a profoundly conservative outlook. It is, therefore, not especially surprising to discover that the voting behavior of recent migrants to the city is also conservative. It is only as the migrant gains a substantial foothold in the dominant urban society, breaks through the barrier of marginality in at least some respects, and thus becomes a "resident"–for instance, a minor bureaucrat in a public service–that his political behavior may tend to switch to a more radical position. As Weiner and others have pointed out, urban areas tend to be more "progressive" in their politics than rural areas and peasant farming districts in particular.

16. Richard D. Lambert, op. cit., pp. 135-136.

17. This, of course, is still a relatively unorthodox view. More commonly, rural arcadia is contrasted with urban violence and anomie. There are no definitive studies on this subject, but the evidence on rural violence is steadily accumulating. Irving Louis Horowitz, for instance, cites data on the positive rank correlation of deaths from group violence and numbers of people in farming. ("Electoral Politics, Urbanization, and Social Development in Latin America," Urban Affairs Quarterly, Vol. 2 (March 1967), Table 7.) At the same time, he notes an inverse correlation between the urbanization process and these causes of death. There are, however, significant exceptions at very high levels of urbanization. This suggests that one should perhaps distinguish between two forms of violent conflict: the urban form, tending to be goal-oriented and limited in time and purpose (e.g., a strike, a street demonstration), even though the frequency of such incidents in the total population may be high; and the rural variety, geared to a culture of the "limited good," which is characterized by "bitterness and hostility between factions." According to G. M. Foster *(Traditional Cultures and the Impact of Technological Change.* New York: Harper, 1962, p. 102), "the length to which people will go to humiliate their rivals (in small peasant societies) is difficult to believe." Because of environmental circumstances, then, rural conflict tends to be endemic and generational. More recently, however, urban forms of violence have taken root also in rural areas. Revolutionary guerilla activity in the countryside is an urban invention (see Aníbal Quijano Obregón, "Contemporary Peasant Movements," in Seymour Martin Lipset and Aldo Solari, eds., *Elites in Latin America.* New York: Oxford University Press, 1967).

18. Torcuato Di Tella, "Populism and Reform in Latin America," in Claudio Veliz, ed., *Obstacles to Change in Latin America.* New York: Oxford University Press, 1965, pp. 47-74; and Kalman H. Silvert, ed., *Expectant Peoples: Nationalism and Development. New York: Random House, 1963.*

19. John Friedmann, *Venezuela: From Doctrine to Dialogue.* Syracuse: Syracuse University Press, 1965; and Fred D. Levy, Jr., "Economic Planning in Venezuela," Yale Economic Essays Vol. 7 (Spring 1967), pp. 273-321.

20. Francine F. Rabinovitz, *Urban Development and Political Development in Latin America.* Occasional Paper. Bloomington, Ill.: Comparative Administration Group, 1967. An opposite interpretation of the same set of facts is found in Shanti Tangri, "Urbanization, Political Stability, and Economic Growth," in Roy Turner, op. cit. The reason for this difference in interpretation is given by Tangri's implicit belief that economic development

can be successfully pursued without major structural change in the political system. Our own position is that conservative, rural-based political structures are incapable of bringing development about and must be changed to achieve development objectives in U-countries. In this connection, it may be observed that rural societies tend to be more prone to totalitarian appeals than highly urbanized societies, though the evidence on this point is admittedly still inconclusive (see S. M. Lipset, *Political Man.* Garden City, N.Y.: Doubleday, 1963).

21. The problem of controlling urban conflict is to guide it towards constructive ends while keeping rapidly changing social systems from disintegrating. See Lewis Coser, *The Functions of Social Conflict.* New York: The Free Press, 1956.

22. The following paragraph is based on Brian J. L. Berry, *Geography of Market Centers and Retail Distribution.* Englewood Cliffs, N.J.: Prentice Hall, 1967; John Friedmann and John Miller, "The Urban Field," Journal of the American Institute of Planners, Vol. 31 (November 1965), pp. 312-320; and J. R. Lasuén, "Urbanization Hypotheses and Spain's Cities System Evolution." Paper presented at the Workshop on Regional Development. Institute of Social Studies, The Hague, October 1967 (mimeo).

23. Brian J. L. Berry, "City Size Distributions and Economic Development," in John Friedmann and William Alonso, eds., *Regional Development and Planning. A Reader.* Cambridge, Mass.: The M.I.T. Press, 1964.

24. J. R. Lasuén, op. cit.

25. J. R. Lasuén et al., "City Size Distribution and Economic Growth: Spain," Ekistics, Vol. 24 (August 1967), pp. 221-226.

26. Brian J. L. Berry, *The Geography of Market Centers and Retail Distribution,* op. cit., pp. 76-79.

27. Richard A. Ellefsen, "City-Hinterland Relations in India," in Roy Turner, op cit., pp. 94-116.

28. Aníbal Quijano Obregón, "La Urbanización dela Sociedad in Latinoamerica," Santiago: United Nations Economic Commission for Latin America, Division of Social Affairs, 1967 (mimeo).

29. In calculating investment returns, social considerations may, of course, lead one to assign values that are different from those of individual enterprise operating in a freely competitive market.

30. Karl Mannheim was among the first to realize the importance of psychological adaptation to the new conditions of an industrializing society. *(Freedom, Power, and Democratic Planning.* New York: Oxford University Press, 1950. Part 3: "New Man–New Values.") David Riesman's concern with "other-directed" personality types is a more recent example.

31. The following analysis of Japan's development is to the point: "The attempt to evaluate Japan's place in the scale of economic development reveals a paradox. In terms of a set of variables for which data have been presented by Ginsburg, Japan is in many respects among the most developed countries in the world. On the other hand, in terms of the criteria of backwardness described by Leibenstein, Japan may be classified as a backward country. Specifically, Japan is backward as measured by the criterion of per capita national income" (Leon Hollerman, "Japan's Place in the Scale of Economic Development," Economic Development and Cultural Change Vol. 12 (January 1964), p. 139).

32. John Friedmann and John Miller, op. cit.

33. Daniel Bell, "Notes on the Post-industrial Society," Public Interest, Vol. 6 (Winter 1967), pp. 24-35 and Vol. 7 (Spring 1968), pp. 102-118.

34. U.S. Department of Commerce, Area Redevelopment Administration, *Area Redevelopment Policies in Britain and the Countries of the Common Market.* Washington, D.C. Government Printing Office, 1965.

35. These questions were first posed—though with considerably less precision—during an international conference on India's urbanization about ten years ago (see the contributions of Chandhuri, Brado, Harris, Wurster, and Meier, as well as the editor's Postscript in Roy Turner, op. cit.). In general, the American contributors to this symposium were more urban-oriented than their Indian colleagues.

Chapter 3

A THEORY OF POLARIZED DEVELOPMENT

I. INTRODUCTION

In many parts of the world today, regional development planning, along with national planning, is considered an important aspect of guided change in societal systems. Regional planning is characterized by its focus on the attainment of objectives for territorially organized–or spatial–subsystems of national societies and, with growing frequency, also of multi-national systems. Although the practice of regional planning in this sense is becoming increasingly common, it still needs to be put on an adequate theoretical foundation. And the most appropriate theory, it would seem, is one that would set forth and explain systematic interrelations between development and space, a theory, in other words, of the development process in its spatial dimension.[1]

Before presenting this subject, existing theories will be reviewed.[2] Despite their relative shortcomings, these theories complement each other in a number of ways and may each be relevant for specific kinds of planning analysis. In several important respects, however, none of them can be accepted as an adequate framework for regional development planning in its more general connotation.

Classical location theory is concerned with the optimum location of the firm.[3] More recently, this theory has been expanded to include industrial complex analysis which, in turn, has led to the theory of "growth centers" as formulated by François Perroux and others.[4] In a perspective of regional planning, however, classical location theory must be considered deficient because its concern is essentially with point locations rather than regional (or spatial) systems. The link between the location of economic activities (micro-theory) and the development of a system of regions (macro-theory) is not yet clearly established.

Spatial organization theory has been specifically advanced as a basis for regional planning.[5] This theory focuses on the structural characteristics of a system of point locations. To the extent that it adopts a systems approach, spatial theory represents an advance over classical location theory. It remains unsatisfactory for our purposes, however, because it is formulated in terms of a general equilibrium model: the theory can explain the patterns of point locations at two separate periods of time; it contributes little to our understanding of the processes by which the historical transformation of spatial systems occurs.

In regional growth theory an attempt is made to overcome this difficulty and to treat specifically the dynamics of spatial systems. Separate theories have been evolved for individual regions[6] and systems of regions.[7] In the case of the former, the theory presents some of the same difficulties that restrict the applicability of classical location theory to regional policy formation: its analysis falls short by concentrating exclusively on the case of single regions. In the more elaborate attempts to deal with the dynamics of a system of regions, the theory tends once more to revert to a general equilibrium formulation, where interregional flows of labor and capital are looked upon as the principal mechanisms for reestablishing a balance that might have been disturbed.[8] Although superior to all preceding theories as a foundation for regional development planning, it suffers the additional limitation—contrary to national growth theory—of being formulated entirely in economic terms. Albert Hirschman and Gunnar Myrdal have attempted to advance beyond these limitations, but their ideas have not been full elaborated.[9]

Perhaps the most comprehensive theoretical text on regional growth published until now is that of Horst Siebert who proceeds to formulate models for both an individual region and a system of regions.[10] Although an economist, Siebert deals imaginatively with significant topics such as technical knowledge, diffusion of innovations, and communication which brings his formulation close to the present one.

The avowed purpose of this chapter, then, is to formulate a theory of the development process in its spatial dimension. Society is spatially organized in the sense that human activities and social interactions are space-forming as well as space-contingent. It follows that as a society undergoes development, its spatial structure will be transformed, but the development process will also be influenced by the existing patterns of spatial relation and the dynamic tensions that result from them. Space is used here in the non-physical sense of a field of forces (e.g., energy levels, decision-making power, communications) that obeys its own laws of transformation.[11] In order to state a spatial theory of the development process, therefore, it is necessary to establish a linkage between the separate but correlative theories of social change and spatial organization. There are two main contenders for a relevant theory of social change—Everett Hagen and Ralph Dahrendorf.[12] Hagen's theory is grounded in individual psychology and is consequently difficult to integrate with a theory of spatial organization.

Dahrendorf works with a conflict model of social change in which the principal variable is the pattern of authority-dependency relationships that characterizes any organized social system. Since spatial systems are territorially organized systems of social relation, the Dahrendorf model was selected as a promising beginning for formulating a general theory of polarized development.[13]

The theory is stated in the five sections that follow. In Section 1, "development" is distinguished from "growth" and is characterized as an innovative process leading to the structural transformation of social systems. Societies that fail to achieve such transformations either become "arrested" in their growth or start to degenerate as their internal order breaks down into increasingly unstable and random relations.

In Section 2, six conditions favorable to innovation are discussed. These conditions, it is claimed, are generally present in large and rapidly growing urban systems or, more generally, at the points of highest potential interaction in a "communication field."

In Section 3, innovation is linked to the concepts of power and authority in territorially organized social (spatial) systems. Innovations enhance the potential power of the innovating agents over a given environment, but the full possibilities for increased environmental control can be extracted from an innovation only if the powers inherent in it are accepted as socially legitimate. Socially legitimate power is called authority. The conflict that may result from the efforts of innovating agents to legitimate their power is seen as frequently involving a wider conflict over the legitimacy of some or all authority-dependency relations within a spatial system. It consequently bears directly on the social bases for integration of that system and on its ability to achieve development.

In Section 4, the preceding argument is carried further by a discussion of authority-dependency relations in a spatial system. If innovations coincide in large measure with the points of highest interaction potential in a communications field, their origin will be found in a relatively small number of centers of change from which innovations tend to diffuse downwards (and outwards) to areas of lower potential interaction. Centers of innovative change are called core regions. With respect to any core region, a periphery may be defined by its relation of dependency to the core. This relationship is further analyzed in terms of the self-reinforcing effects of core region growth and the rising opposition of peripheral elites to being dominated by the core, a condition that, to a large extent, is based on the core's superior capacities for innovation. The ensuing conflict may lead, among other possible outcomes, to a sharing of core region authority with peripheral areas or the replacement of core region elites by others whose political strength lies essentially with peripheral populations.

In Section 5, the core-periphery relation is generalized to an hierarchical order of spatial systems. A nested hierarchy of five spatial systems is posited, including the world, the multi-nation region, the nation, the sub-national region, and the province. Specific mechanisms of core region dominance are discussed, and the ability to challenge core region elites successfully is stated as a basic condition for the structural transformation of spatial systems. This can occur through the establishment of new core regions in the periphery and the gradual linkage of core regions, old as well as new, into more encompassing spatial systems. One result of this process will be the exponential growth of innovations and the consequent acceleration of developmental change in society.

The chapter concludes with a brief discussion concerning the relation of the general theory to existing theories of regional growth. The latter are shown to be special cases of the general theory of polarized development.

II. THE THEORY

1. Development as Innovation

1.1. (D)* Historical progress may be understood as the temporal succession of one socio-cultural paradigm by another.[14] Simon Kuznets attributes this succession of paradigms to the appearance of what he calls epochal innovations.[15] He writes: "The pattern of life during the epoch must be seen as the realization of the potentialities involved in the single complex identified as the epochal innovation. . . . To put it in technical terms, the epoch must be defined so that a single trend line can be effectively fitted to it."[16] These innovations have the capacity of transforming social values in such a way that all pre-succession history will tend to be rewritten from the vantage point of new perspectives. Epochal innovations are self-validating as initial opposition to them is changed into support, and a new world view, expressing the new norms, is born.[17]

According to Kuznets, the major innovation of our own epoch is the "extended application of science to problems of economic production."[18] This innovation—in reality a vast complex of interlocking institutional, technical, and cultural innovations—has long antecedents in the past.[19] It began as a series of isolated innovations that occurred at different moments in history and gradually became linked to each other, inducing a structural transformation of the traditional social system from which they had emerged. The transformation towards which contemporary—or modern—development is moving is a social system that has a high capacity continuously to generate and adapt to innovative change.[20]

*D stands for a statement of definition. All other statements are intended as formal hypotheses.

1.2. (D) In accordance with this interpretation, development may be studied as a discontinuous, cumulative, process that occurs as a series of elementary innovations which become organized into innovative clusters and finally into large-scale systems of innovation.[21] The innovations may be technical or institutional and, if the latter, may be usefully subsumed under the customary categories of social, economic, political, and cultural.

1.2.1. (D) With reference to societal systems, "growth" should be conceptually distinguished from "development." The former refers to an expansion of the system in one or more dimensions without a change in its structure. Such growth normally follows the general form of an S-curve so that if system-wide growth is to continue, a transformation of the constraining structural relations becomes necessary.[22]

1.2.2. (D) "Development" refers to the unfolding of the creative possibilities inherent in society. But this can only occur if growth is allowed to pass through a series of successive structural transformations of the system.

1.2.3. (D) Societies which fail to achieve such transformations either become "arrested" in their growth or start to degenerate as their internal order breaks down into increasingly unstable and random relations.

1.3. Development always occurs through asynchronic processes in which leading, innovative forces arise from or are injected into an existing mould of traditional structures.

1.3.1. (D) No particular model of traditional society has to be assumed; traditional is simply that which, at any given moment, is established and with respect to which an innovation is defined.

1.3.2. The cumulative effect of successive innovations is to transform the established structure of society by attracting creative or innovative personalities into the enclaves of accelerated change;[23] by encouraging the formation of new values, attitudes, and behavior traits consistent with the innovation;[24] by fomenting a social environment favorable to innovative activity; and by bringing into existence yet further innovations.[25]

2. The Conditions of Innovation

2.1. (D) Invention is the act of conceiving and presenting new structural combinations of preexisting cultural elements or ideas; it may also refer to the result of an inventive act; invention is an idea or artefact discovered.

2.2. (D) Innovation is the successful introduction of ideas or artefacts perceived as new into a given social system.[26]

2.2.1. (D) Innovations may be based on ideas or prototypes that are invented, borrowed, or imitated. What is already established in one place may, by borrowing or imitation, become an innovation in another. It is its perception as something "new" that is the decisive criterion for innovation.

2.2.2. Every innovation requires a measure of organization and adaptation to the conditions and functional requirements of the medium into which it is introduced or, alternatively, a change in certain aspects of the medium itself. Innovation and medium must be structurally compatible.[27]

2.2.3. Innovations require individuals or institutions that will organize the necessary resources and assume the risks of failure; every innovation requires an innovative agent.[28]

2.3. The likelihood that innovative activities will be generated at a given location and be ultimately successful depends on six basic conditions, including:

2.3.1. First, the number of problems resistant to solution by traditional means.[29] This may be said to represent the effective demand for innovation. The greater this demand, the higher will be the probability of innovation.

2.3.2. Second, the probability that two or more previously unconnected mental frames of reference have for mutual confrontation.[30] Such confrontations raise the likelihood that innovations will occur.

2.3.2.1. (D) This probability may be expressed in terms of a concept of information exchange or communication in open systems.

2.3.2.1.1.(D) For every territorially organized social system—henceforth to be called spatial system—a communication field may be defined in which the probability of information exchange may vary from 0 to 100, forming a spatial frequency distribution.[31]

2.3.2.1.2. Innovations are more likely than elsewhere to occur at those points in a given communication field where the probability of information exchange is relatively high (see 2.3.2.).

2.3.3. Third, the capacity of the existing social system to absorb proposed innovations without itself undergoing major structural changes.[32] The greater this capacity, the greater will be the probability of innovation.

2.3.3.1. Hierarchical, centrally controlled social systems (Type I) tend to have a relatively low capacity for continuously generating and absorbing innovations.[33]

2.3.3.2. Non-hierarchical, multi-centric, horizontally integrated social systems (Type II) tend to have a high capacity for generating and absorbing innovations.[34]

2.3.3.2.1. Where power approximates an even distribution among multiple centers of decision, however, and agreement on objectives is lacking among them, Type II systems may also display a low capacity for innovation, because multiple interests in a horizontally integrated social system are able to exercise effective veto power over the initiatives of any one of them.[35]

2.3.3.3. From the standpoint of innovative capacity, the optimal system of social organization appears to be a combination of Types I and II where leadership, central information, and certain conflict-resolving functions are hierarchically superimposed upon Type II systems.[36]

2.3.4. Fourth, the frequency of innovative personality traits in the population of a given society.[37] The greater the innovative propensity of a population, the higher is likely to be the frequency of innovation in that population.

2.3.4.1. Socio-cultural systems will significantly differ in the proportion of innovative personalities they produce or attract from the outside.

2.3.4.2. The propensity to innovation may also result from the social position occupied by certain collectivities. Foreign immigrant groups or deviant sub-cultures often account for a substantial share of total innovative activity in a society.[38]

2.3.5. Fifth, the ability to marshal the human and material resources sufficient for effective innovation, from the initial introduction of a new idea to its ultimate adoption and diffusion through the social system.

2.3.6. Sixth, the social rewards offered for innovative activity. The higher these rewards, the more are innovations likely to occur.

2.4. Historically, conditions especially favorable to innovation are generally present in large and rapidly growing urban systems.[39] Ten conditions appear to favor innovative activity in the urban complex (see 1.3.2.).

2.4.1. The pressure of unprecedented problems resulting from rapid growth, high densities, and the presence of culturally heterogeneous population groups in large cities generates urgent demands for solution (see 2.3.1.).

2.4.2. The traditional frames of reference are incapable of solving many of the new problems arising from accelerated urbanization (see 2.3.1.).

2.4.3. The bankruptcy of traditional mental frames leads to an intense search for new and adequate solutions (see 2.3.2.).

2.4.4. A relatively heavy volume of information from outside the urban center transmits new knowledge together with new ways of looking at the world. Elements of "exotic" culture encounter in the city a pattern of established beliefs and practices from which they are set apart and with which they often come into conflict (see 2.3.2.1.2. and 2.3.4.2.).

2.4.5. The resulting clash of new and existing culture elements in an environment of high potential interaction leads to substantial exposure of certain population groups to different mental frames of reference (see 2.3.2.).

2.4.6. The large city or urban region tends to have a relatively loose if complex social structure as well as a relatively diffused structure of power; both conditions will tend to encourage and facilitate experiment and innovation (see 2.3.3.).

2.4.7. The large city or urban region tends to attract creative personality types in a proportion greater than their frequency in the population as a whole (see 2.3.4.).

2.4.8. The large city or urban region provides exceptional opportunities for marshalling the necessary human and material resources for innovation (see 2.3.5.).

2.4.9. The large city or urban region reinforces creative responses to new situations by extending economic, social, and political rewards for innovative activity (see 2.3.6.).

2.4.10. A process of innovation sustained over long periods of time is likely to lead to the institutionalization of this process.

2.5. These ten conditions, however, are not equally satisfied in all large-scale and expanding urban systems. Holding size constant, urban systems may differ, for example, with respect to their rates of growth (and consequently the urgency of their demands), cultural heterogeneity, degree of openness to external information, and social organization conducive to innovation.[40] Theocratic cities, for instance, will tend to be less innovative than large commercial-industrial centers.

2.5.1. The historical correlation between large city size and rapid urbanization on the one hand, and innovation on the other, seems nevertheless to be well established, especially if it is borne in mind that innovation may result from borrowing and imitation no less than from invention.

3. Innovation, Power, and Authority in Spatial Systems

3.1. (D) To have power is to exercise a measure of autonomy in decisions over a given environment and to have the ability to carry out these decisions.[41]

3.1.1. Successful innovation gives to both innovators and initial adopters a relative advantage over possible competitors in controlling a given environment. Successful innovation tends to increase the potential power of innovators.

3.2. The full possibilities for increased control over an environment can be extracted from an innovation only when the potential power inherent in it is institutionalized; in other words, when the full exercise of its power is accepted as socially legitimate.

3.2.1. (D) Socially legitimate power will be called authority.[42]

3.2.2. Innovators generally seek to translate a prospective gain in power into some form of (legitimate) authority.[43]

3.2.3. Where the innovation is socially compatible (see 2.2.2.), this process is relatively automatic. Society will reward the innovator with esteem and invest him with a measure of authority (see 2.3.6.).

3.3. Where the innovation is incompatible with the social medium, however, efforts to render a prospective gain in power legitimate will put innovators in conflict with the holders of established authority (see 1.3.).

3.3.1. Individuals or groups seeking to legitimize prospective gains in power frequently establish alliances among themselves in order to confront established authorities from a position of collective and, consequently, greater strength.

3.3.2. (D) Such alliances lead to the formation of counter-elites to established authority.

3.4. (D) The presence of groups exercising authority within a territorially organized social (spatial) system implies the existence of other groups dependent on them for vital decisions.[44]

3.4.1. Spatial systems are integrated through a given structure of authority-dependency relations that is maintained partly by a belief in the legitimacy of the relation itself and partly by coercion.[45]

3.5. The conflict between innovating counter-elites and already-established authorities (see 3.3.) is often posed as a wider conflict over the legitimacy of some or all authority-dependency relations within a given spatial system.[46] It is consequently a conflict that may bear directly on the social bases for integration of that system (see 3.4.1.).

3.5.1. This conflict can have four possible outcomes:

3.5.1.1. *Suppression:* counter-elites are prevented from gaining access to positions of authority (their attempt to legitimize prospective gains in power is frustrated).

3.5.1.2. *Neutralization:* established authorities oscillate between acceptance and rejection of the proposed innovations, frequently adopting their external forms but managing them so as to neutralize their effects.[47] Counter-elites fail in gaining full access to authority.

3.5.1.3. *Coöptation:* counter-elites are drawn into the established structure of authority.[48] Their innovations may be adopted, but in a highly restricted form. Counter-elites are pacified by being allowed to share to a limited extent in the exercize of established authority.

3.5.1.4. *Replacement:* counter-elites are successful in replacing the established authorities and substituting their own authority for them.

3.5.2. Conflict over authority may take either legitimate or illegitimate forms.

3.5.2.1. (D) Conflict is legitimate when it (a) occurs within a framework of rules established for regulating social conflict, and/or (b) does not challenge the fundamental bases for social integration of the spatial system (see 3.2.3. and 3. 4.1.).

3.5.2.2. (D) Conflict is illegitimate when it fails to meet either of these two criteria.

3.6. Where the outcome of conflict over authority is resolved in favor of the counter-elites (see 3.5.1.3. and 3.5.1.4.), the existing structure of authority-dependency relations and, consequently, the social bases for integration of the spatial system, will be transformed (see 3.5.).

3.6.1. The magnitude of this transformation will depend on the extent of the claims made on authority.

3.6.1.1. Since authority in complex spatial systems tends to be fragmented, it may be possible for the system to absorb relatively minor and compatible innovations without a change in its fundamental structure; in this case, the effects of individual innovations are contained although their cumulative impact may be very large.

3.6.1.2. Substantial transformation occurs when authority-dependency relations important for the social integration of the system are successfully challenged. In this case it is possible to speak of the "development" of that system (see 1.2.1.).

3.6.2. If successive counter-elites are gradually absorbed into the established ruling stratum (see 3.5.1.3.), the transformation of the social bases for integration will tend to occur as a gradual, evolutionary process.

3.6.3. If the conflict is resolved with innovating groups replacing the established authorities (see 3.5.1.4.), the consequent changes in the social bases for integration will tend to be violent, rapid, and far-reaching.

3.6.4. Having once gained access to positions of authority, former counter-elites may either foreclose further innovative efforts or create a social environment capable of absorbing successive innovating groups into an adaptive system of authority-dependency relations (see 3.6.1.).

4. Authority-Dependency Relations in a Spatial System

4.1. Development, viewed as occurring through a discontinuous but cumulative process of innovation (see 1.2., and 1.3.), will tend to have its origin in a relatively small number of *centers of change* located at the points of highest potential interaction within a communication field (see 2.3.2.1.2.); innovations will tend to spread downward and outward from these centers to areas where the probability of potential interaction is lower.[49]

4.1.1. (D) Major centers of innovative change will be called *core regions;* all other areas within a given spatial system will be defined as *peripheral.*[50] More precisely, core regions are territorially organized subsystems of society which have a high capacity for innovative change; peripheral regions are subsystems whose development path is determined chiefly by core region institutions with respect to which they stand in a relation of substantial dependency.

4.1.2. (D) Peripheral regions may be identified by their relations of dependency to a core area.

4.1.3. (D) Core and periphery together constitute a complete spatial system.

4.1.4. (D) A spatial system is integrated through a pattern of authority-dependency relationships that is focused on the dominant core regions (see 3.4. and 3.4.1.).

4.1.4.1. The domination of the periphery by its core is the result of earlier and critical innovations that were incorporated into the central authority structure of the system (see 3.5.1.3. and 3.5.1.4.).[51]

4.1.5. The conflict over the legitimacy of established authority-dependency relationships resulting from continuing core region growth may undermine the bases for social integration of the spatial system and give rise to prolonged conflict between core and peripheral elites.

4.1.5.1. This conflict will be especially severe where core region elites belong to a cultural tradition different from that of peripheral populations.

4.2. Four major propositions may be advanced concerning the relations of core regions to their peripheries:

4.2.1. *First,* core regions impose a condition of organized dependency on their peripheries.

4.2.1.1. This dependency relation results from a penetration of the periphery by institutions that are effectively controlled by core region authorities (see 5.2.1.).

4.2.1.1.1. In the perspective of the periphery, many of these institutions will be perceived as innovations (see 2.2.1. and 4. 1.4.1.).

4.2.1.1.2. Peripheral elites will attempt to resist those innovations they regard as threatening their own authority positions and as imposing a lasting dependency on external authorities (see 3.3.).

4.2.1.1.3. From the standpoint of the periphery, core region elites may be viewed as counter-elites intent on usurping peripheral authority positions (see 3.3.2.). However, where the structural conditions are roughly compatible, certain core region innovations may be integrated into the periphery with relatively little difficulty (see 2.2.2.).

4.2.1.1.4. Massive psychological, material, and coercive resources at the disposal of core region elites give them a significant advantage in their struggle to reduce peripheral areas to a status of dependency. Local elites may be suppressed, neutralized, coopted, or replaced (see 3.5.1.).

4.2.1.2. (D) Successful penetration of the periphery by core region institutions implies that decisions vitally affecting local populations will henceforward by made by relevant core region authorities.

4.2.2. *Second,* the process by which core regions consolidate their dominance over the periphery tends to be self-reinforcing.[52] This may be ascribed to six major feedback effects of core region growth:

4.2.2.1. *Dominance effect,* or the steady weakening of the peripheral economy by a net-transfer of natural, human, and capital resources to the core.[53]

4.2.2.2. *Information effect,* or the increase in potential interaction within a given core region resulting from its own growth in population, production, and income. Among other things, this effect will tend to induce a higher rate of innovation (see 2.4.4. and 2.4.5.).[54]

4.2.2.3. *Psychological effect,* or the creation of conditions favorable to continued innovation at the core, such as rendering the opportunities for innovation more visible, reducing the risks of innovation through imitation, and creating expectations for further innovation (see 2.4.5., 2.4.7. and 2.4.9.).[55]

.2.2.4. *Modernization effect,* or the transformation of existing social values, behavior, and institutions in the direction of greater acceptance of and conformity with rapid cumulative change through innovation (see 1.3.2.).[56]

.2.2.5. *Linkage effects,* or the tendency of innovations to breed other innovations by creating new service demands as well as new markets for the services the core region is itself able to supply to other areas. However, not all innovations will have the same capacity in this respect; the multiplier or linkage effects of certain innovations will be greater than for others.[57]

.2.2.6. *Production effects,* or the creation of an attractive reward structure for innovative activity operating through the exploitation by innovators of their temporary monopoly position, the appearance of linked systems of innovations, and growing specialization. All of these will tend to increase economic returns, while increasing external economies to scale and urbanization economies (e.g., greater efficiency in the use of social overhead facilities) will tend to reduce the costs of innovation (see 2.4.10.).[58]

4.2.3. *Third,* introducing core region innovations into the periphery will augment the flow of information to the dependent region from the core (see 4.2.1.1.). This will tend to be an unanticipated side effect of core region dominance.

4.2.3.1. Sustained contact with the core region will tend to arouse portions of the peripheral population, not only to possible new ways of life, but also to their own comparative disadvantage in gaining access to them. New desires and frustrations will encourage demands for greater regional autonomy in areas of vital decision and may lead to a prolonged conflict with the core (see 4.1.5.).[59]

4.2.3.2. Individuals and groups most directly exposed to information originating in core regions will gradually awaken to the periphery's and their own dependency (and powerlessness) and will be among the first to insist upon greater autonomy for the periphery. Alternatively, they will emigrate to the core to be drawn there into the established structures of authority.

4.2.3.3. To stem off the pressures for greater autonomy on part of the periphery, and possibly to avoid violent conflict, core region elites may move to implement a policy of limited "decentralization" which, among other things, may project the build-up in the periphery of new, subsidiary core regions.[60]

4.2.4. If the outcome of this conflict should favor the interests of core region elites, the subsequent acceleration of spread effects to the peripheral region will lead to a substantial sharing of decision power between the old and newly implanted core areas (see 3.5.1.3.).[61]

4.2.4.1. Such sharing of authority is not likely to be fully equal, however. In many important respects, the spatial system will continue to be dominated by the older core regions; the relation will be asymmetrical.

4.2.4.2. But such an outcome is not inevitable. In the process of extending their institutions into the periphery, older core regions may at some point "exhaust" their capacities for further creative response and decline to a subordinate, peripheral position relative to the cores they had helped to bring into existence.

4.2.4.2.1. Specifically, former peripheral elites successful in their endeavors to rise to positions of core region authority may act to foreclose further innovative activity (see 3.6.4.), while the positive feedback effects of growth in the old core regions (see 4.2.2.) may be weakened as control over their former peripheries is lost.[62]

5. Polarized Development in a Hierarchy of Spatial Systems

5.1. (D) Core regions are located within a nested hierarchy of spatial systems (see 4.1.1.).

5.1.1. (D) Possible spatial systems are the world, the multi-nation region, the nation, the subnational region, and the province.

5.1.1.1. (D) Whether or not a given area constitutes a spatial system depends on the pattern of its internal relations. Where a core can be shown to dominate some of the vital decisions of populations in areas external to itself, a spatial system will be said to exist.

5.1.2. A given spatial system may have more than one core region.

5.1.3. The territorial extent of a given core region will tend to vary with the physical size or scope of the relevant spatial system of which it forms a part. For a major world region, the appropriate core may be a vast and complex area such as Megalopolis in the northeastern part of the United States; at the level of provincial systems, the relevant core may be a single city region of moderate size.

5.1.4. Core regions generally perform a wide variety of functions for their dependent areas; some specialization may occur, however, especially in the cores of lower-order spatial systems.[63]

5.2. Five major propositions concerning the role of core regions in the development of spatial systems may be suggested.

5.2.1. Core regions organize the dependence of their respective peripheries through systems of supply, market, and administrative areas (see 4.2.1.).[64]

5.2.1.1. As centers of production and consumption, core regions organize their peripheries into sets of supply areas for furnishing them the raw materials, food stuffs, and semiprocessed commodities they need.

5.2.1.2. As supply centers, core regions organize their peripheries into sets of market areas (e.g., through the establishment of regional sales and service offices or a "colonial" pattern of transport and communication services which generally serves the needs of the core region more than those of the periphery).

5.2.1.3. As centers of legitimate decision-making power, core regions organize their peripheries into sets of administrative areas for the purpose of securing central bureaucratic domination over them.[65]

5.2.2. For any spatial system, a loose hierarchy of core regions may be identified in accord with the functional importance of each core for specified characteristics of system performance. This hierarchy will tend to be asymmetrical, however, with some functions placing a core region at one hierarchical level, while others may indicate a different position.[66]

5.2.2.1. For the same spatial system, any two core regions of approximately the same level in the hierarchy will tend to have a greater and more balanced volume of mutual interaction, modified by distance, then either will have with individual lower-order cores.

5.2.2.2. Within a given spatial system, the competition for the exploitation of specific innovations will be greatest among core regions of approximately the same hierarchical level.

5.2.2.3. A core region may, at the same time, belong to the hierarchies of two spatial systems. Thus, a second-order core region for a higher spatial system may also appear as a first-order core for a lower-ranking system. The inverse of this situation is not possible.

5.2.3. Core regions systematically transmit impulses of innovation to the peripheries they dominate (see 4.2.1.1.).[67] Three further propositions may help to clarify this process.

5.2.3.1. The rate of innovation tends to be greatest in core regions corresponding to high-level systems. Innovations tend to be propagated from high-order cores downward to lower-order cores of the same spatial system or laterally to similarly ranking core regions in parallel systems.[68]

5.2.3.2. The successful diffusion of innovations from given core regions will partly depend on the structural and behavioral characteristics of the receiving areas which must, in at least some respects, be consistent with those of the sending area;[69] on the ability of the innovative forces to overcome the resistance of local authorities likely to be encountered (see 4.2.1.1.2.); and on the general conditions prevailing in the receiving areas which may favor or not a sustained process of innovation (see 2.3.3.).[70]

5.2.3.3. Because of their greater and more effective control over both external and policy variables, and because of the larger volume of resources at their command, higher-order core regions are more autonomous than lower-order cores with regard to their ability to make critical development decisions.[71]

5.2.3.3.1. Because of the asymmetrical nature of their relationship to higher-order cores and their partial dependency on them, all lower-order cores are also peripheral to a set of dominant core regions (see 4.1.2). All propositions concerning core-periphery relations stated in Section 4 above, therefore apply also to the hierarchical relations *among* core regions.

5.2.4. Up to a certain point in time, the self-reinforcing character of core region growth (see 4.2.2.) will tend to have positive results for the development process of the relevant spatial system; eventually, however, it will become dysfunctional, unless the spread effects of core region development to the periphery can be accelerated, and the periphery's dependence on the core region reduced (see 4.2.3.).[72] The approach of this critical turning point will be registered in the growing political and social tensions between core and periphery (see 4.1.5.) that are likely to drain core region strength and reduce its capacity for further development (see 1.2.3.).

5.2.5. The probability of innovation will increase over the surface of a given spatial system with increases in the probability of information exchange over that system.[73] This tendency—the result of a successful challenge of core region autonomy by their peripheries—will induce the physical spread of existing core regions, a weakening of their hierarchical order, the emergence of new core regions on the periphery, and the emergence of new core regions on the periphery, and the gradual incorporation of large parts of the periphery into one or more system cores.[74]

5.2.5.1. Enclaves of economic backwardness will appear in the areas lying between expanding core regions and/or in limited urban sectors of the cores themselves.[75]

5.2.5.2. The theoretical upper limit of core region spread is given by the extension of a single core region over the inhabited parts of the entire globe.

III. RELATION OF THE GENERAL THEORY TO CURRENT THEORIES OF REGIONAL GROWTH

As noted in Section I, existing growth theory for regional economies is fundamentally a variant of classical location theory.[76] The economic growth of a single region is usually traced to some natural and consequently immobile resource for which an external demand exists. Investments to exploit this resource are attracted and lead to the further expansion of productive activity in the region through the stimulation of local demand and the creation of substantial external economies.[77] The region's growth relative to that of competitive subnational economies is in the long run made to depend on the extent of its participation in the rapidly expanding sectors of the total economy. But this participation can similarly be treated as a question in the economies of location. Finally, it is customary to emphasize that any welfare disequilibria that may result from the differential location behavior of economic activity will ultimately be restored to balance by compensatory flows of labor (and capital) from "surplus" to "deficit" regions.[78] That, in brief, is the substance of contemporary regional growth theory; the theory of polarized development goes considerably beyond it.

This theory treats economic growth as a function of changes in the structures that inevitably limit a system's capacity for expansion and, in the specific case of growth based on the application of science to problems of economic production, also of the system's capacity for the continuous generation and absorption of innovations. This formulation assigns a decisive influence to the institutional and organizational framework of society and, specifically, to the patterns of authority and dependency that result from the unusual capacity of certain areas to serve as cradles of innovation. In this view, it is by no means indifferent what regions organize the new investments and make the relevant decisions on location, or what purposes are to be served by these activities. Dominant core and dominated periphery together constitute a relatively stable spatial system in which the latter is successfully "colonized" chiefly to sustain the continued growth of the former.

The further growth of core regions, however, is in the final analysis constrained by the tensions that tend to build up from the ever more visible discrepancies in the rates of expansion and modernization between core and periphery. The increasing flow of information from core to periphery, together with an aroused awareness of potentially modernizing elites in the periphery of the conditions of their own dependency, produce conflict with core region authorities over the extent of permissible autonomy. This conflict, capable of seriously affecting the social bases of integration in a spatial system, may have four possible outcomes: first, the violent repression of peripheral elites by the core so that the existing spatial system is maintained even though its overall potential for growth may decline; second, the neutralization of peripheral elites, a process that may occasionally lead to a gradual, if inadequate, modification of the system's authority-dependency structure; third, the replacement of core region by peripheral elites followed by either the acceleration of system-wide growth (where the new elites are dedicated modernizers) or stagnation and even retrogression (where the new elites follow a traditional orientation); and fourth, the coöptation of peripheral by core-region elites, leading to a more equitable sharing of powers that may be accompanied by a process of political and economic decentralization and the creation of new, or the expansion of already existing, core regions in the periphery. Under coöptation, the authority-dependency relations between cores and their peripheries are likely to diminish and may eventually disappear altogether except for relatively minor enclaves of economic backwardness located in "interstitial" areas or limited urban sectors in the cores themselves. The actual working out of core-periphery relations, however, is made exceedingly complex by the hierarchical superposition of spatial systems ranging from sub-national province to a (theoretical) world system.

In the context of the theory, contemporary regional growth theory may therefore be treated as a special case, applicable only to situations where the dualism of core and periphery is of relatively little consequence, such as the reasonably advanced and integrated spatial systems of the United States, West

Germany, or Sweden. It is applicable neither to the industrializing countries nor to multi-national regions where core-periphery dependency relations are still predominant influences. In their case, the transformation of authority-dependency patterns in spatial systems is a fundamental condition of development and, consequently, also of sustained economic growth.

NOTES

1. Between theory and practice there is inevitable tension. Nevertheless, they mutually enrich and reinforce each other. For further discussion of this point, see Chapter 13.

2. Horst Siebert has recently published an excellent survey of the literature: "Regionalwirtschaftslehre in den USA: Ein Uberblick," Jahrbuch für Sozialwissenschaft, Vol. 20, No. 1 (1969), pp. 51-83. For an earlier but still excellent review in English, see John R. Meyer, "Regional Economics: A Survey," American Economic Review, Vol. 53, No. 1 (March 1963), pp. 19-54.

3. An excellent summary statement of classical location theory is William Alonso, "Location Theory," in Friedmann and Alonso, eds., *Regional Development and Planning: A Reader.* Cambridge, Mass.: The M.I.T. Press, 1964, pp. 78-106. For a more extended treatment, see Martin Beckmann, *Location Theory.* New York: Random House, 1968, and L. H. Klaassen's *Methods of Selecting Industries for Depressed Areas.* Paris: OECD, 1967.

4. On the subject of industrial complex analysis, chapter 9 of Walter Isard's *Methods of Regional Analysis.* Cambridge, Mass.: M.I.T. Press, 1960, still serves as a useful introduction to the subject. On the concept of *pôle de croissance,* see François Perroux, *L'économie du XX*[e] *siècle.* Paris: Presses Universitaires de France, 1961, Part II. Other French literature on the subject is cited in *Cuadernos de la Sociedad Venezolana de Planificación,* Vol. II, Nos. 5-6 (August-September 1963), pp. 74-77, and in Niles M. Hansen, "Development Pole Theory in a Regional Context," Kyklos, Vol. XX, No. 3 (1967), pp. 709-726. An excellent recent contribution to growth pole theory is Heiko Körner, "Industrielle Entwicklungspole als Instrumente der Regionalpolitik in Entwicklungsländern," Kyklos, Vol. XX, No. 3 (1967), pp. 684-708, which also contains comprehensive references. The entire body of literature is critically reviewed by David F. Darwent, "Growth Pole and Growth Center Concepts: A Review." *Environment and Planning,* Vol. I, No. 1 (1969), pp. 5-31.

5. Edwin von Böventer, "Spatial Organization Theory as a Basis for Regional Planning," Journal of the American Institute of Planners. Vol. XXX, No. 2 (May 1964), pp. 90-99; and by the same author, "Toward a Unified Theory of Spatial Economic Structure," Regional Science Association, Papers, Vol. X (Zurich Congress, 1962), pp. 163-18. Other major theoretical statements are: Walter Isard, *Location and Space-Economy.* Cambridge, Mass.: The M.I.T. Press, 1956; L. Lefeber, *Allocation in Space: Production, Transport and Industrial Location.* Amsterdam: North Holland Publishing Co., 1968; H. C. Bos, *Spatial Dispersion of Economic Activity.* Rotterdam: University Press, 1965; and Brian J. L. Berry, *Geography of Market Centers and Retail Distribution: Foundations of Economic Geography Series.* Englewood Cliffs, N.J.: Prentice-Hall, Inc. 1967.

6. Harvey S. Perloff with Vera W. Dodds, *How a Region Grows: Area Development in the U.S. Economy,* Supplementary Paper No. 17. New York: Committee for Economic Development, 1963. See also Charles M. Tiebout, *The Community Economic Base Study.* Supplementary Paper No. 16. New York: Committee for Economic Development, 1962.

7. Harvey S. Perloff, Edgar S. Dunn, Jr., Eric E. Lampard, and Richard F. Muth, Regions, *Resources and Economic Growth.* Baltimore: The John Hopkins Press, 1960; George H. Borts and Jerome L. Stein, *Economic Growth in a Free Market.* New York and London: Columbia University Press, 1964; John Friedmann, *Regional Economic Policy: A Case Study of Venezuela.* Cambridge, Mass.: The M.I.T. Press, 1966, and L.B.M. Mennes, Jan Tinbergen, and J. George Waardenburg, *The Elements of Space in Development Planning.* Amsterdam: North Holland Publishing Co., 1969.

8. For a critique of equilibrium models in regional analysis, see John Friedmann, op. cit., chapter I.

9. Albert O. Hirschman, *The Strategy of Economic Development,* New Haven: Yale University Press, 1958, and Gunnar Myrdal, *Rich Lands and Poor: The Road to World Prosperity.* New York: Harper and Brothers, 1957.

10. Horst Siebert, *Regional Economic Growth: Theory and Policy.* Scranton, Pa.: International Textbook Co., 1969.

11. General systems theory is relevant here. See, for example, James G. Miller, "Living Systems: Basic Concepts," Behavioral Science, Vol. 10, No. 3 (July 1965), pp. 193-237; "Living Systems: Structure and Process," Ibid., Vol. 10, No. 4 (October 1965), pp. 337-379; and "Living Systems: Cross-Level Hypotheses," Ibid., pp. 380-411. On the concept of nonphysical space, see François Perroux, "Economic Space: Theory and Applications," in John Friedmann and William Alonso, op. cit., pp. 21-36.

12. Everett E. Hagen, *On the Theory of Social Change.* Homewood, Illinois: The Dorsey Press, 1962; Ralph Dahrendorf, *Class Conflict in Industrial Society,* Stanford: Calif. Stanford University Press, 1959.

13. This is not the place for a full critique of Dahrendorf's model. Nevertheless, some important qualifications must be mentioned because they are relevant to what follows. According to Dahrendorf, authority-dependency relations are maintained only by restraint and coercion (pp. 158ff). Without denying the fundamental importance of coercion in system maintenance, it is necessary to point out that social values also play a significant role. This is implicit in the legitimacy which is accorded authority, and more specifically, the authority-dependency relation itself. Any conflict arising out of a given authority-dependency relation may therefore be either legitimate or illegitimate. It is illegitimate only when the legitimacy of the authority is itself put in question. Not all social conflict, however, turns on this issue.

A second objection to Dahrendorf is his insistence that for any given social system there is only one authority when, in fact, authority is rarely, if ever, either unitary or absolute (pp. 296ff). Any complex organization is structured as a system of polyarchy in which there are many leaders whose authority is limited with regard to both function and the control of the leaders themselves and by those who are dependent on them. On this point, see the excellent discussion in Robert A. Dahl and Charles E. Lindblom, *Politics, Economics and Welfare.* New York: Harper and Brothers, 1953, Part IV.

14. This is the basic thesis of Thomas S. Kuhn's excellent study, *The Structure of Scientific Revolutions.* Chicago: The University of Chicago Press, 1962. A closely related idea is Joseph A. Schumpeter's concept of "creative destruction" explained in *Capitalism, Socialism and Democracy.* New York and London: Harper and Brothers, 1947, chapter 6.

15. Simon Kuznets, *Modern Economic Growth.* New Haven and London: Yale University Press, 1966, chapter 1.

16. Ibid., p. 4.

17. Thomas S. Kuhn, op. cit., chapter 10.

18. Ibid., p. 9.

19. Charles C. Gillespie, *The Edge of Objectivity: An Essay in the History of Scientific Ideas.* Princeton: Princeton University Press, 1960.

20. This view of the development process is elaborated in a brillant essay by Manfred Halpern, "The Revolution of Modernization in National and International Society," in Carl J. Friedrich, ed., *Revolution.* Year book of the American Society for Political and Legal Philosophy, Vol. VIII. New York: Atherton Press, 1966. According to S. N. Eisenstadt, "the structural propensity to continuous change" is a crucial aspect of modernization. See "Breakdowns of Modernization," Economic Development and Cultural Change, Vol. XII, No. 4 (July 1964), p. 347. For a more extended treatment, see Eisenstadt's *Modernization: Protest and Change,* Englewood Cliffs, N.J.: Prentice-Hall, 1966.

21. The view of development as innovation is fundamental to Everett E. Hagen's analysis, *On the Theory of Social Change,* op. cit. Less explicitly, innovation enters as a primary datum in Albert O. Hirschman's well-known study, *The Strategy of Economic Development,* op. cit.

22. This extremely important principle is stated in K. E. Boulding, "Toward a General Theory of Growth," in Joseph J. Spengler and Otis Dudley Duncan, eds., *Population Theory and Policy: Selected Readings.* New York: The Free Press, 1956, pp. 118-119. See also Chandler Morse, "Becoming versus Being Modern: An Essay on Institutional Change and Economic Development," in Chandler Morse et al., *Modernization by Design: Social Change in the Twentieth Century.* Ithaca and London: Cornell University Press, 1969. The need for structural change lies, of course, also at the root of the developmental theories of Kuznets and Kuhn.

23. For detailed discussion of the concept of creative personality and its possible role in the process of social change, see Everett E. Hagen, op. cit., chapters 5-7.

24. A basic postulate of the theory of "cognitive dissonance" is that perceptions, attitudes, and ultimately values tend to become consistent with experienced reality and thus to maintain (or re-establish) the psychological equilibrium of a person. See Leon Festinger, *Theory of Cognitive Dissonance.* New York: Harper and Row, 1957.

25. For an extended discussion of this idea, see especially Hirschman, op. cit., chapters 4 and 6, and Gunnar Myrdal, *Rich Lands and Poor,* op. cit., chapter 3.

26. The definitions of "invention" and "innovation" are generally consistent with the usage in Everett M. Rogers, *Diffusion of Innovations.* New York: The Free Press, 1962, pp. 195 and 13 respectively. Rogers, however, further distinguishes between innovation as a new idea and its successful introduction or adoption, a distinction not made in this paper. Unfortunately, the empirical measurement of invention and innovative activity presents countless problems for which no adequate solution exists at the moment. For a penetrating discussion, see the papers in Part 1, National Bureau of Economic Research, *The Rate and Direction of Inventive Activity: Economic and Social Factors.* Princeton: Princeton University Press, 1962.

27. Rogers, op. cit., pp. 127-130.

28. In usage, the term innovative agent here is roughly analogous to Schumpeter's image of the "entrepreneur." See Joseph Schumpeter, *The Theory of Economic Development.* New York: Oxford University Press, 1961.

29. Kuhn, op. cit., chapters 6-9. See also Rogers, op cit., p. 125.

30. Arthur Koestler, *The Act of Creation.* New York: The Macmillan Co., 1964, passim. Koestler refers to this basic mechanism as "bi-sociation."

31. The concept of a "communication field" in relation to urbanization is discussed in John Friedmann, "An Information Model of Urbanization," Urban Affairs Quarterly, Vol.

IV, No. 2 (December 1968), pp. 235-244. See also the related treatment of the "mean information field" in Torsten Hägerstrand, *Innovation Diffusion as a Spatial Process.* Chicago: University of Chicago Press, 1967, chapter VII.

32. The relation between organizational structure and innovation is discussed in James G. March and Herbert A. Simon, *Organizations.* New York and London: John Wiley and Sons, Inc., chapter 7.

33. According to Young and Moreno, historical experience in the United States suggests the following conclusion: "Social rigidity and political rigidity, have been shown to be related to attributes of stagnant economic structure; they are inversely correlated with innovation, risk-taking, technical proficiency, organizational complexity, and distributive economic development, over and above the expected value of these attributes based on the general level of industrialization of these areas." Ruth C. Young and José A. Moreno, "Economic Development and Social Rigidity: A Comparative Study of the Forty-Eight States," Economic Development and Cultural Change, Vol. XIII, No. 4, Pt. I (July 1965), p. 449. See also Michel Crozier, *The Bureaucratic Phenomenon.* Chicago: The University of Chicago Press, 1964.

34. This hypothesis is derived from Warren Bennis, "Beyond Bureaucracy" Trans-action, July-August 1965 (reprint). See also Paul R. Lawrence and Jay W. Lorsch, *Organization and Environment.* Boston: Harvard University, Graduate School of Business Administration, 1967.

35. This thesis is brilliantly demonstrated in Edward C. Banfield's study of interest group politics in Chicago, *Political Influence.* New York: The Free Press, 1961. Its theoretical statement is found in James Q. Wilson, "Innovation in Organization: Notes Toward a Theory," in James D. Thompson, ed., *Approaches to Organizational Design.* Pittsburgh: University of Pittsburgh Press, 1966.

36. This appears to be the main conclusion of Emmette Redford, "Centralized and Decentralized Political Impacts on a Developing Economy: Interpretation of American Experience," CAG Occasional Papers, February 1967.

37. The concept of innovative personality is discussed by Everett Hagen, op. cit., pp. 88-97.

38. This point is stressed, for example, by Seymour Martin Lipset, "Values, Education, and Entrepreneurship." in Lipset and Solari, eds., *Elites in Latin America.* New York: Oxford University Press, 1967, pp. 3-60. It is a basic point also in Hagen, op. cit., chapter 12, and in Rogers, op. cit., pp. 196-198.

39. Wilbur R. Thompson, "Locational Differences in Inventive Effort and their Determinants," in National Bureau of Economic Research op. cit., pp. 253-272. Many fo the hypotheses in this section are also consistant with the empirical findings of Alan R. Pred, *The Spatial Dynamics of U.S. Urban-Industrial Growth, 1800-1914.* Cambridge, Mass.: The M.I.T. Press, 1966, chapter 3.

40. On the variety of urban structures relative to economic growth and development, see Bert F. Hoselitz, *Sociological Aspects of Economic Growth.* New York: The Free Press, 1960, chapter 8, "Generative and Parasitic Cities." An especially intersting study, relevant in this context, is A. L. Epstein, "Urbanization and Social Change in Africa," Current Anthropology. Vol. 8, No. 4 (October 1967), pp. 275-284.

41. In this definition, I follow Dahrendorf (op. cit.) who, in turn, bases his concept on Max Weber. For Weber, power is the "probability that one actor within a social relationship will be in position to carry out his own will despite resistance, regardless of the basis on which this probability rests." (Dahrendorf, op. cit., p. 166). I have found it convenient,

however to alter the wording of this definition. See also Hans Gerth and C. Wright Mills, "Power and Authority: A Summary," in Lewis A. Coser and Bernard Rosenberg, *Sociological Theory: A Book of Readings.* New York: The Macmillan Co., 1957 and 1964, pp. 156-158.

42. Dahrendorf, op. cit., p. 166.

43. "All that is creativity, innovation and development . . . is due to no small extent, to the operation of conflicts" Dahrendorf, op. cit., p. 208.

44. "Authority relations exist whenever there are people whose actions are subject to legitimate and sanctioned presecriptions that originate outside them within social structure." Dahrendorf, op. cit., p. 168, See also pp. 171ff.

45. Dahrendorf, op. cit., pp. 162-163. How this integration is achieved specifically through systems of communication is explored in Philip E. Jacob and James V. Toscano, *The Integration of Political Communities.* Philadelphia: J. B. Lippincott, 1964, especially in the contributions of Karl W. Deutsch.

46. Ibid., p. 134. See also Gino Germani, "Social Change and Intergroup Conflict," in I. L. Horowitz, *The New Sociology.* New York: Oxford University Press, 1964, chapter 23.

47. S. N. Eisenstadt, "The Development of Socio-Political Centers at the Second Stage of Modernization–A Comparative Analysis of Two Types," International Journal of Comparative Sociology. Vol. VII, No. 1-2 (March 1966), pp. 119-137.

48. According to the inventor of the concept, "The process of absorbing new elements into the leadership or policy-determining strata of an organization as a means of averting threats to its existence or stability is called coöptation." Philip Selznick, *TVA and the Grass Roots.* Berkeley, Calif.: University of California Press, 1949, p. 259.

49. Kuznets, op. cit., chapters 1 and 9. Of the six variables identified as having a decisive influence on the probability of interaction (see 2.3), the communication variable is here singled out as having probably greater importance than the remaining ones. The surface of the communication field will subsequently have to be modified to take account of the other variables, including differences in the demand for innovation (see 2.3.1.), pattern of social organization (see 2.3.3.), frequency of innovative personality traits (see 2.3.4.), initial resource availabilities (see 2.3.5.), and the pattern of social rewards for innovation (see 2.3.6.), if it is to give a faithful expression of localized capacity for generating innovations. Though relative values may change topographical details of the communication field as a result of this operation, its general morphology will probably remain the same (see 2.4).

Concerning the mechanisms of innovative spread, see Lawrence A. Brown, *Diffusion Processes and Location.* Philadelphia: Regional Science Research Institute, 1968.

50. The concept of core region and periphery were first introduced to systematic regional analysis in John Friedmann, *Regional Development Policy,* Cambridge, Mass.: The M.I.T. Press, 1966, especially chapter 2. The present definitions, however, go considerably beyond the first conceptualization.

51. No separate explanation for the origins of core regions is here offered (but see note 50). They are posited to arise at points of relative initial advantage in terms of the structure of a given communication field. This initial structure may be assumed to be the result of randomly operating forces. For a very interesting but more deterministic study of the economic origin of core regions, see Eugene Smolensky and Donald Ratajczak, "The Conception of Cities," Explorations in Entrepreneurial History. Second series. Vol. 2 (Winter 1965), pp. 90-131.

52. The implication of a relatively stable system of cities and regions has been frequently noted. See, for instance, the discussion by Eric E. Lampard, "The Evolving

System of Cities in the United States: Urbanization and Economic Development," in Harvey S. Perloff and Lowdon Wingo, Jr., eds., *Issues in Urban Economics.* Baltimore: The Johns Hopkins Press, 1968, pp. 126-130.

53. Horst Siebert, op. cit., pp. 57-76. The phenomenon is widely noted and politically one of the most explosive in regional relations.

54. Ibid., pp. 70-74.

55. For a penetrating discussion of this effect, see Heiko Körner, op. cit.

56. Ibid.

57. Ibid.

58. Tibor Scitovsky, *Papers on Welfare and Growth.* Stanford, Calif.: Stanford University Press, 1964, chapter 3, "Two Concepts of External Economies."

59. The results for political development of a rising want/get ratio are most recently discussed by Daniel Lerner, "Conflict and Consensus in Guayana," in Frank Bonilla and Jose A. Silva Michelena, *Studying the Venezuela Polity.* Cambridge, Mass. and Caracas, Venezuela: Center for International Studies, Massachusetts Institute of Technology and Centro de Estudios de Desarrollo, Universidad Central de Venezuela, May 1966, pp. 479-512.

60. See, for example, Robert C. Fried, "Administrative Pluralism and Italian Regional Planning," Public Administration (London), Winter 1968, pp. 375-391. Also Niles Hansen, *French Regional Planning.* Bloomington: Indiana University Press, 1968 and John Friedmann, *Venezuela: From Doctrine to Dialogue.* Syracuse: Syracuse University Press, 1964. A full-fledged study of the politics of regional development policy is still lacking, however.

61. Some empirical evidence for this hypothesis is found in the best documented study to date of core region development: Lloyd Rodwin et al., *Planning Urban Growth and Regional Development. The Experience of the Guayana Program in Venezuela.* Cambridge, Mass.: The M.I.T. Press, 1969, especially in chapter 24 by Lisa Peattie.

62. The former case may be illustrated with the example of post-liberation Rangoon, where authority was transferred to a non-innovative, religious elite; the latter case may be exemplified by post-World War I Vienna and post-World War II London, where the loss of empire led to a decline in innovative capacity.

63. Evidence on this point is presented in Otis Dudley Duncan et al., *Metropolis and Region.* Baltimore: Johns Hopkins Press, 1960, chapter 11.

64. Studies of regional dependency relations are rare, partly, I suppose, because "dependency" is a value-laden term and, according to the traditional academic view, social science should be neutral with respect to values (Alvin W. Goulder, "Anti-Minotaur: The Myth of a Value-Free Sociology," in I. L. Horowitz, op. cit., chapter 13.) For contributions to the study of regional dependency, however, see Chandler Morse, "Potentials and Hazards of Direct International Investment in Raw Materials," in Marion Clawson, ed., *Natural Resources and International Development.* Baltimore: The Johns Hopkins Press, 1964, pp. 397-414; Espartaco, "La 'crisis Latinoamericana' y su marco externo," Desarrollo Económico (Buenos Aires), Vol. No. 6, No. 22-23 (July-December 1966). pp. 319-354; and Oswaldo Sunkel, "Política Nacional del Desarrollo y Dependencia Externa," Estudios Internacionales (Santiago), Vol. I, No. 1 (April 1967), pp. 43-75.

65. James W. Fesler, *The Political Role of Field Administration.* Papers in Comparative Public Administration. Ann Arbor, Michigan, 1962, and "Approaches to the Understanding of Decentralization," Journal of Politics, Vol. 27 (1965), pp. 536-566.

66. This hierarchy may be identified empirically by studying the organization of decision-making in the political-administrative space of the system and by the spatial

linkages of the different cores to their markets and supply areas. Much empirical work remains to be done in this field. For the concept of a loose, asymmetrical hierarchy of core regions, see Otis Dudley Duncan et al., *Metropolis and Region.* Baltimore: The Johns Hopkins Press, 1960, chapters 3 and 11.

67. For a fascinating discussion of this subject, suggesting a number of lines of inquiry, see Kuznets, op. cit, chapter 9. Also Torsten Hägerstrand, "The Propagation of Innovation Waves," Lund Studies in Geography, Series B. Human Geography, 1952. And, by the same author, "Aspects of the Spatial Structure of Social Communication and the Diffusion of Information," Regional Science Association, Papers, Vol. XVI, 1966, (European Congress, 1965), pp. 27-42.

68. Poul Ove Pedersen, "Innovation Diffusion Within and Between National Urban Systems," *Geographic Analysis,* Vol. II, (1970), pp. 203-254. This relation holds especially where the distance decay function is relatively high, i.e. during the early phases of industrialization. It may not hold for post-industrial systems like the United States.

69. For evidence, see Norman E. Whitten, "Power Structure and Socio-Cultural Change in Latin American Communities," Social Forces, Vol. 43, No. 3 (March 1965), pp. 320-329, and John H. Kunkel, "Economic Autonomy and Social Change in Mexican Villages," Economic Development and Cultural Change, Vol. 10, No. 1 (1965).

70. See Chapter 6.

71. See Kuznets, op. cit., chapter 6, for evidence with respect to world patterns of international trade and their bearing on political relations.

72. This hypothesis is consistent with the empirical finding that "rising regional income disparities and increasing North-South dualism is typical of early development stages, while regional convergence and a disappearance of severe North-South problems is typical of the more mature stages of national growth and development." Jeffrey G. Williamson, "Regional Inequality and the Process of National Development," Economic Development and Cultural Change, Vol. XIII, No. 4, Part II (July 1965), p. 44.

73. On the close correlation and interaction between communications and urbanization indices and the use of these indices for predicting political change, see Phillips Cutright, "National Political Development: Measurement and Analysis," American Sociological Review, Vol. 28, No. 2 (April 1963), pp. 253-264.

74. Poul Ove Pedersen and Walter Stöhr, "Economic Integration and the Spatial Development of South America," American Behavioral Scientist, Vol. XII, No. 5 (May-June 1969), pp. 2-12.

75. John Friedmann and John Miller, "The Urban Field," Journal of the American Institute of Planners, Vol. XXXI, No. 4 (November 1965), pp. 312-319.

76. The one notable exception to this generalization is Horst Siebert, op. cit. (see note 10).

77. The process measured by the regional multiplier is normally assumed to be an automatic result of initial investment. That this is by no means so is demonstrated and made the basis of important modifications of regional growth theory by Robert E. Baldwin, "Patterns of Development in Newly Settled Regions," in John Friedmann and William Alonso, eds., *Regional Development and Planning.* Cambridge, Mass.: The M.I.T. Press, 1964.

78. See, for instance, Charles M. Tiebout, *The Community Economic Base Study.* Supplementary Paper No. 16. New York: Committee for Economic Development, 1962; Harvey S. Perloff, with Vera Dodds, *How a Region Grows.* Supplementary Paper No. 17. New York: Committee for Economic Development, 1963; and Hugh O. Nourse, *Regional Economics.* New York: McGraw-Hill Co., 1968.

Chapter 4

URBANIZATION AND NATIONAL DEVELOPMENT: A COMPARATIVE ANALYSIS*

I. THE POSSIBILITY OF A THEORY OF URBANIZATION

Is a theory of urbanization possible? Some years ago, Leonard Reissman raised this question in his excellent book, *The Urban Process.*[1] But Reissman was ambiguous: at times he spoke of a theory of the city or more specifically, a theory of urban society; at other times, he seemed to have in mind a city-forming process. Lack of an adequate conceptual framework has hindered theorizing about one of the most compelling societal forces operating in history.

Let us define two forms of urbanization. The first refers to settlement and economic location, the second to what we shall designate as modernization:

Urbanization$_1$: The geographic concentration of population and nonagricultural activities in urban environments of varying size and form

Urbanization$_2$: The geographic diffusion of urban values, behavior, organizations, and institutions

In both meanings, urbanization refers to complex processes that occur in spatially organized social systems. The most convenient unit for the study of these processes we shall take to be the nation.

Clearly, urbanization$_1$ can reach a point of saturation when all of a country's population is living and working in urban settlements. But any settlement

*This chapter was written in collaboration with Eileen McGlynn, Barbara Stuckey, and Chung-Tong Wu.

pattern may continue to evolve in its spatial dimension even at "saturation" levels.

A minimum of spatial integration with respect to $urbanization_2$ is reached when a national population has come to share in at least some elements of the urban culture specific to that nation. This urban culture, however, undergoes further transformations, giving rise to regions that are "traditional" and "transitional" relative to those parts of the country where the most current forms of $urbanization_2$ are evolving. This complex pattern of regional variations will, in turn, influence settlement patterns and activity locations. Thus, the pre-industrial city was eventually superseded by the industrial city which is now beginning to give way to the expansive urban environments of post-industrial society.

But urbanization as a process of cultural diffusion is not confined to the industrializing countries alone; it is a universal process that can evolve indefinitely into the future in always newly elaborated forms. The "modernity" of every period is generated in an urban setting and diffused in various ways throughout the urban system.

Both processes of urbanization continuously reshape the spatial organization of the social system composed of (1) the spatial organization of the economy (economic activities + settlement networks), and (2) the spatial organization of modernization (socio-cultural pattern + the territorial organization of power). Process thus terminates in structural organization but is also affected by it. Process and structure co-vary in time.

$Urbanization_{1+2}$ can be further disaggregated into five basic processes which contribute directly to spatial organization. They are: innovation generation (including innovation by imitation), innovation diffusion, control of decisions, population migration, and investment.

These several definitions and concepts enable us to consider *the spatial organization of national societies* as the principal subject of a theory of urbanization. To be of value, such a theory would have to explain both processes and structures of urbanization and the systematic interrelations between them.

In a recent paper, Professor Richard Morse of Yale University argues that history can provide a "lens correction" for a theory of urbanization. The historian, he asserts, "is concerned more with morphology than with theory, more with cross-cultural comparison than with cross-cultural generalization, more with enduring occurrences and ironies of the human condition than with its ultimate purification."[2] The disciplines of social science and history can, indeed, be mutually enriching. Ideographic analysis can lead to a deeper understanding of human phenomena by way of a study of lawful behavior, just as nomothetic study can benefit from a knowledge of the historical uniqueness of social behavior. A comparison to musical composition suggests itself: the expressiveness of a fugue derives in large measure from the general form which underlies it. Structure gives clarity to invention: invention endows form with an enduring meaning.

Where Morse emphasizes cross-cultural comparisons, we will stress cross-cultural generalizations.[3] The specifics of cultural history will thus be fused with their recurrent patterns.

Comparative urbanization studies are still in their infancy. Little systematic testing of hypotheses has been attempted across cultural boundaries. Studies of the urbanization experience of Latin America, Africa, Asia, Oceania, Europe, and North America are carried on with few cross-references to other areas. There are a few notable exceptions, but comparative studies have, on the whole, been seriously hampered by the lack of a coherent paradigm of urbanization and the precise formulation of hypothesis for empirical testing.[4] In this chapter, such a paradigm will be presented and illustrated with suggestive findings from a review of recent literature pertaining to Latin America, Africa, and South Asia. These findings may be useful at a next stage in the formulation of testable hypotheses for cross-cultural study.

The principal use of theory is to provide efficient explanations and deeper understanding of important phenomena. But in the case of urbanization, verified theory can also serve as a foundation for public policies to guide the course of urbanization and regional development. Such policies, adopted in increasing measure throughout the world, have as their principal aim a more satisfactory spatial organization of society.[5] The efficient and productive use of resources, the spatial integration of national societies, and the enhancement of the physical environment for living are three recurrent objectives of national policies for urbanization. These policies are directed at different facets of urbanization, among them the spatial organization of political and administrative power, the volume and direction of population migration, the composition and location of investments, and the spread of social innovations. A number of policy scientists have endeavored to provide rationales for major alternatives, but their writings are not always grounded in urbanization theory as well as one might wish.[6] If this deficiency can be remedied, public policy for urban-regional development might be formulated with greater hope for relevance and success.

II. A PARADIGM FOR THE STUDY OF URBANIZATION

A graphic representation of our paradigm is shown in Figure 1. Absent from the paradigm is any reference to either cities or urbanization. Core and periphery are shown as the basic structural components instead. Core regions are defined as territorially organized subsystems of society which have a high capacity for generating and absorbing innovative change; peripheral regions are subsystems whose development path is determined chiefly by core region institutions with respect to which they stand in a relation of substantial dependency.[7] Core and periphery together constitute a complete spatial system.

The paradigm is presented here in a highly simplified form: in actuality, a country is likely to have more than one core area and a rather complex network

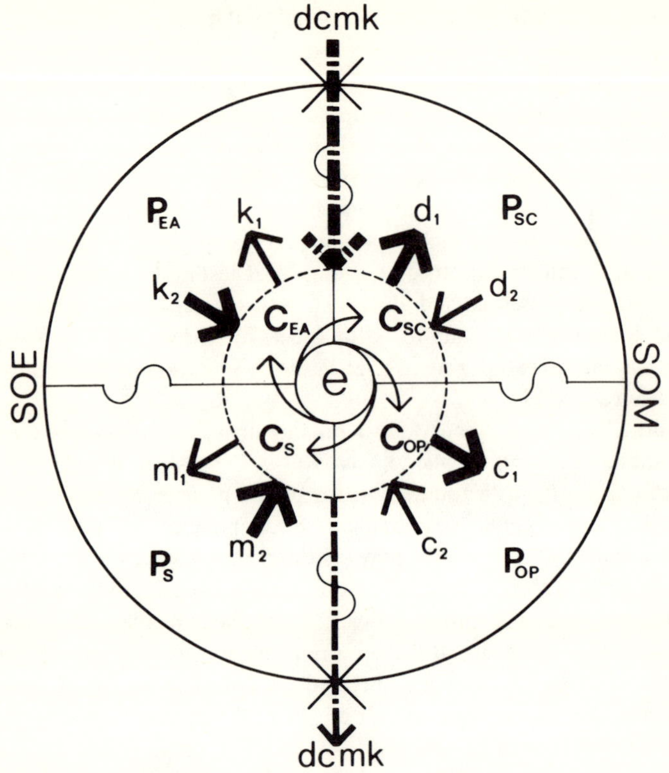

Figure 1: A Paradigm for the Study of Urbanization

of peripheries determined by patterns of dependency relations. The autonomy-dependency relation between core and periphery, however, is fundamental to the paradigm and lies at the root of its internal (spatial) dynamics. Because urban settlements can occur in the periphery and rural settlements in the core, it is possible for the periphery to become urbanized in either of the two meanings we have discussed without losing its dependency status. The core, on the other hand, though dominating the spatial system, may include sizable populations with only a low degree of urbanization_2.

The driving force in the development of a spatial system is a set of "emergent innovations" (material, technological, spiritual, institutional). This is symbolized in Figure 1 by the letter "e" centered in the core region. Diffusing outwards from the core, these innovations affect economic activity, socio-cultural structures, the organization of power, and settlement patterns in the periphery. It is thus assumed that continued innovation will force the spatial system through successive structural transformations.[8]

In addition to "emergent innovations," the paradigm includes four basic urbanization processes that relate core and periphery in the spatial system: innovation diffusion (e), control (c), migration (m), and investment (k). These processes, indicated by arrows of constrasting width, are shown to be asymmetrical, a condition that ultimately traces back to the fundamental imbalance in authority-dependency relationships between core and periphery. This asymmetry in the behavior of urbanization processes produces a series of disequilibria in the organization of the spatial system. A simplified reading of the paradigm might be as follows: the volume of controlling decisions that emanates from the core is greater than the reciprocal volume of controls from periphery to core. This causes a net-flow of capital from the periphery which, in turn, gives rise to a net-flow of migrants into the core area. At the same time a continuous stream of innovations diffuses from the core to the periphery where it ultimately helps to create conditions that lead to demands for at least a partial restructuring of the fundamental dependency relation.

Unless the social-political tensions produced by these four sets of disequilibrium can be successfully contained (or creatively resolved), development of the spatial system will either stagnate or break down.

Each of the four basic urbanization processes affects a subsystem of the spatial system: innovation diffusion alters the socio-cultural pattern (SC) of the spatial system; control processes yield a pattern of power relationships (OP–organization of power); migration leads to changes in the settlement pattern (S); and investment processes affect the spatial pattern of economic activities in both core and periphery (EA).

These spatial patterns are closely interrelated. The partial integration of power and socio-cultural patterns may be defined as the spatial organization of modernization (SOM), while the joint effects of economic activity and settlement patterns produce the unique spatial organization of the economy (SOE). Finally, the two sub-organizations–SOM and SOE–combine to form the spatial organization of society.

Spatial systems–even at the level of nations–are never completely closed to external influence. The paradigm therefore provides for the influence of exogenous core regions on urbanization processes and integration patterns. With respect to exogenous core areas, the spatial system under consideration is shown to be in a condition of (partial) dependency and, consequently, in a peripheral relation. On the other hand, it may have backward linkages with the controlling core areas and forward linkages to dependent, lower-order, peripheral systems. These relationships may be crucial, especially for newly industrializing countries, where a colonial past or a neo-colonial present may exert substantial influence on patterns and processes of urbanization.

This paradigm of urbanization is helpful, we believe, in visualizing national development in its widest terms as occurring through the operation of five basic processes that lead–if successful–through a series of structural transformations to successively higher levels of spatial integration. And the integration of national territories must be regarded as a major and critical variable in the historical development of new nations.

Spatial integration may be thought of as having essentially two meanings: (1) an increase in the volume of transactions among urban places and regions which leads, in turn, to a complex territorial division of labor, and (2) the extension over a given territory of a common basis for social life or, more accurately, a shared frame of socio-cultural expectations, including language, cultural values, political-legal-bureaucratic institutions, and a market economy. Integration in the latter meaning may be regarded as a precondition of integration in the former, inasmuch as stable transactions presume a measure of commonality among the parties concerned: in the marketplace, traders are bound to respect the rules by which a fair exchange may be accomplished.

In the paradigm of urbanization presented above, the spatial expansion of shared expectations is summed up on the right-hand side of the diagram by SOM; the result of a growing territorial division of labor appears on the left-hand side as SOE. The integration of SOM and SOE will yield the nation as a spatially organized social system.[9]

III. THE COMPARATIVE ANALYSIS OF URBANIZATION

The following review of recent urbanization$_{1+2}$ literature for Latin America, Africa, and South Asia reveals a surprising poverty of significant findings.[10] In part, this may be attributed to the extreme reluctance of American social scientists to turn to a study of macrophenomena such as urbanization, and ultimately to a failure to develop the concepts and investigative tools that would be needed for this task.

Most social scientists still turn instinctively away from the analysis of issues of public policy, especially when these are raised in a national context. Only economists are working with heuristic models that lend themselves to the analysis of national systems in ways that are relevant to policy, but few economists have applied themselves to the study of urbanization. Geographers, on the other hand, for whom spatial systems are of central interest, have only recently begun to turn to a consideration of urbanization policy. The great majority of them are still content with straightforward descriptive analysis.

For the most part, the studies reviewed focus on social behavior with respect to urban subsystems. Although repeated expressions of dissatisfaction with the rural-urban dichotomy were encountered, no useful suggestions for replacing this classical distinction were found. The principal subject of social inquiry appears to be internal migration and its consequences. Other urbanization processes, such as innovation diffusion, decision controls, and investment though admittedly more difficult to investigate, are widely neglected. Yet, unless migration studies are related to these other processes, little of any consequence can be inferred from the survey data on which the great majority of migration studies is based.

In the absence of a common conceptual framework, comparative studies of urbanization$_{1+2}$ have not been flourishing, though of late an upsurge of interest may be noted. Scientists working on Africa have ignored Latin American evidence; Latin Americanists have remained blissfully unaware of Asian studies; Orientalists have worked in isolation from Africanists. And so the circle of mutual ignorance is joined. Though notable in themselves, the few efforts in the direction of comparative, cross-cultural studies have so far failed to inspire the academic community to more ambitious undertakings.[11]

This failure may be seen as the result of the prevailing intellectual style of social analysis which is suspicious of theoretical formulations, is dryly empirical, bound by disciplinary traditions and primarily concerned with the study of structural relations and microphenomena. As a result, the field of urbanization$_{1+2}$ studies has remained one of the most disappointing and underdeveloped areas of social science research.

Rather than belabor the obvious, this thumbnail critique or urbanization$_{1+2}$ studies may be summarized in the form of a table in which the salient characteristics of these studies are set forth in contrast to what we consider to be a more appropriate approach.

Because of this deplorable state of affairs, analyses of urbanization$_{1+2}$ have made few worthwhile contributions to the construction of public policy.[12] Recommendations tend to be justified on only the flimsiest of evidence, with gaps in knowledge being filled by naturalistic observation, intuition, and moral conviction. The soundest predictions to date have asserted only that urbanization$_1$ will continue to be of great importance in the future of the industrializing countries,[13] that it will continue to cause numerous problems of social adjustment,[14] and that the resulting spatial structures of settlement and regional economy will exhibit considerable stability over time.[15] Is it surprising that policy makers have found such statements somewhat less than compelling?

TABLE 1

CHARACTERISTICS OF URBANIZATION$_{1+2}$ STUDIES

Current	*Desirable*
disciplinary	interdisciplinary
focus on closed urban subsystems (cities)	focus on open national systems
little explicit use of theoretical models, except for urban-rural dichotomy	extensive use of models based on core-periphery paradigm
no explicit comparative focus	explicit comparative, especially cross-cultural, focus
structures emphasized (comparative statics, at best)	dynamic process interaction emphasized, with structures as the limiting condition
policy questions not central to analysis	policy analysis central
urbanization studies without relation to more comprehensive processes such as national and spatial integration	urbanization studies tied back into national development and spatial integration: a systems approach

In the following pages, some of the major conclusions reached by current urbanization$_{1+2}$ research will be highlighted. Where nothing was found, it will be duly noted. Hopefully, this chapter will encourage the social science establishment to redirect its efforts and reach out to more socially relevant problem formulations.

Since there has been little research done on the interplay of urbanization$_{1+2}$ processes and structures, we have decided to emphasize the processes by arranging the materials we have studied under the headings suggested by the paradigm in Figure 1.

A. Emergent Innovations and Innovation Diffusion

The importance that innovations have for a country's modernization is widely acknowledged; indeed, the relationship would appear to be almost tautological. By definition, important changes in a social system cannot occur without innovation. It is all the more astounding therefore, how little attention has been given to the processes by which innovations are generated, diffused, and absorbed into a culture. Most innovation research has been conducted in Western Europe and the United States.[16] For industrializing countries, innovation processes have not been extensively investigated.

It may be assumed that the majority of modernizing innovations in these countries, which stand in a dependent relationship to external core areas, are imported, usually into the largest urban center. Brian Berry has described this process succinctly when he writes with reference to South and Southeast Asia:[17]

> "impulses of economic change are transmitted . . . simultaneously along three planes:
>
> a. Outwards from heartland metropoli to those of regional hinterlands.
>
> b. From centers of higher to centers of lower level in the hierarchy, in a pattern of 'hierarchical diffusion.'
>
> c. Outwards from urban centers into their surrounding fields."

A more detailed look at this process will follow later. What is of interest here is that the dependency relations which result from this process have been completely overlooked.

Three kinds of innovation may be identified: household, technological, and institutional. Household innovations tend to assimilate the tastes and fashions in peripheral areas to those of the originating core regions: the dependency here is a cultural one. Technological innovations are frequently imported under conditions of external control with the result that their use is usually subject to some form of non-local decision. Further, technological innovations arrive in the periphery with a certain time lag: this gives the originating core areas at least a temporary monopoly over their use and consequently a competitive advantage in marketing the commodities produced. Finally, institutional innovations are often poorly adapted to local conditions, causing internal adjustments in the culture that are as much involuntary as they are unanticipated. To the extent that expectations in the periphery are set for a continued flow of innovations from core areas, the incentive for self-generating innovations is reduced. As a result, the initial dependency relation is strengthened and the power of the dominant core region over the periphery consolidated.

All of these processes operate at different levels in a hierarchy of cores and the respective peripheries, from international to local systems.

The responses of peripheral areas to imported innovations is an important aspect of this process. Where these responses can be described with that most elusive of adjectives, creative, the imbedding of exotic innovations in a given cultural matrix may give rise to hybrid creations. These are of exceptional importance for the development of a country and warn us to avoid any simplistic notions of cultural dualism.[18] For instance, James F. Guyot in his study .of urbanization in Malaysia and William and Judith Hanna (Nigeria, Uganda) emphasize the merging of modern and traditional patterns in the new urbanization$_2$.[19] The Hannas, in particular, identify an emerging group of "community influentials" whose role is that of broker between the modernizing elites and peasant masses.

From Latin America comes similar evidence. In Peru and Bolivia the Spanish-speaking Mestizos corresponding to the Hanna's group of "influentials" are the *cholo* who mediate between the Indian peasants and the Spanish ruling castes and are, in effect, the bearers of a new cultural synthesis.[20] In the spread of innovations, their role would seem to be a particularly crucial one.

Innovation diffusion processes have been studied in considerable detail by Poul Ove Pedersen for Chile.[21] His is the first such study undertaken with explicit reference to the role of innovation diffusion in national development. Some of his principal conclusions may be restated in the form of testable hypotheses.

(1) At higher levels of economic development, innovations diffuse down the urban hierarchy, from large to small central places; at lower levels, however, the friction of distance appears to have greater influence on diffusion rates, so that cities further away from innovation-generating centers will tend to receive a smaller number of innovations during a given time period than more accessible places, regardless of their hierarchical position.

(2) The time required for the complete diffusion of an innovation over a given spatial system tends to decrease with an increase in the level of economic development of that system.

(3) Innovations which have a low threshold population size (and income) below which their introduction to an urban center becomes economically infeasible, will tend to diffuse more rapidly throughout the spatial system than high-threshold innovations.

(4) Higher rates of participation in communication processes linked to innovations permit the rate of diffusion to increase; but if the rate of participation in the

communication process systematically decreases with town size, the diffusion process will be retarded.

(5) A small number of entrepreneurs capable of translating relevant information into actual innovations will tend to retard the diffusion process. Since entrepreneurship is a scarce resource in all industrializing countries, and especially in their peripheries, the process of innovation diffusion will tend to occur more gradually during the early periods of industrialization than later.

The data used by Pedersen are highly aggregated. Considerably less is known about specific mechanisms of diffusion in various cultural settings, though the complementary roles of mass communication and face-to-face contacts are fairly well established.[22] Guyot has shown how rural development programs initiated by government introduce a wide variety of novel features into rural life in a process he calls "creeping urbanism."[23] Friedmann has drawn attention to a similar phenomenon in Chile where land reform programs, rural education, and the syndicalization of rural labor have contributed to the social mobilization and heightened political awareness of formerly passive rural populations.[24] In both cases, institutional innovations resulted from centrally directed efforts.

B. The Control of Decisions

In studying the ways in which core regions control their peripheries and the peripheries, in turn, try to minimize this control, the broader question of the relationship between urbanization$_{1+2}$ and political development is set aside. This relationship is a relatively new research focus, but it should not be confused with the study of the flow of controlling decisions between core and periphery. The latter process is critical to the spatial integration of society.

Although no interregional studies of control processes exist, the phenomenon has not gone entirely unnoticed. Ideologues call it colonialism–external, where it refers to the control of a weak country by one that is stronger, and internal, where it concerns the control of peripheral regions by a national core. In recent years, writers on Latin American have come to speak of the relationship as one of "dependency," and at least one author has attempted to show that urbanization$_1$ patterns in countries such as Peru are the result of economic and political domination by the United States. But Aníbal Quijano's argument is not supported by persuasive evidence.[25] Though unrelated to urbanization$_{1+2}$, a much more suggestive study of economic dependency has been published by Chandler Morse.[26] In all likelihood, the basic thesis which attempts to relate national urbanization$_{1+2}$ processes to the phenomenon of external dependency is correct; it simply has not been satisfactorily demonstrated.

Still, some general facts concerning interregional control processes are known. The railway networks of Argentina, Ghana, and India, for example, were built by English companies interested in facilitating the export of raw materials from the interiors of these countries to European markets. Rail lines fanned out from the principal coastal cities without providing for lateral ties that would facilitate interprovincial communications.[27] In this way, external control was manifested in internal relationships. Port cities grew rapidly as a result. But many of the new industries, banks, and large commercial houses remained in foreign hands and extended their branches deep into the country's periphery.[28] Dominance of the emerging primate city was strengthened by the military power of the central government (often with foreign advisers), and rebellious regions could always be beaten into submission. It should be added, however, that this forceful penetration also permitted the seepage of innovations into the periphery; some provincial centers eventually grew to respectable size; and the beginnings of a spatially integrated economy were laid down.

In one way or another, this process was repeated many times throughout the world. Core-region institutions organized their tributary areas through a complex network of supply, market, and administrative areas. Their bureaucratic hierarchies created a corresponding hierarchy of central places that was vertically connected by communication routes to the primate city. Institutional control—by influencing the direction of investment flows and migration—thus contributed heavily to the maintenance of the core-periphery structure and to the continuing (and even growing) preeminence of the primate city. The clustering of high-level control centers in the primate city acted itself as a major location factor. Moreover, the surplus wealth of the country was reinvested in core areas to beautify and rapidly modernize these most visible symbols of national power.[29] Few people realized that such power is hollow when the controlling national institutions are themsevles dependent on more powerful foreign interests.[30]

Sooner or later, however, thresholds of tolerance will be reached in the periphery. Demands will be made for redirecting investments away from the core to some of the more advanced of the peripheral centers. These demands may be met by repression or by conciliatory gestures. But where peripheral status coincides with feelings of cultural separateness, the resulting conflict may lead to civil strife.[31] Even in less extreme cases, however, the political life of the country will tend to be strongly colored by interregional conflict and may generate movements in the periphery capable of assuming control over core institutions by relatively peaceful means. This happened, for instance, in Venezuela and Uganda.[32] In both countries, stepped-up policies for encouraging development in the peripheral regions followed. The choice here was clearly in favor of national integration, even if it came at the expense of short-term increments of economic growth. This dilemma has been pointed out, with reference to Nigeria, by Akin Mabogunje. It is a classical issue in the formation of developing nations.[33]

The foregoing constitutes a general description of a process that appears to exist not only in many of the industrializing, but also in Western European countries, such as Holland, Belgium, Austria, Italy, France, Yugoslavia, and Spain. It is astounding that a process so essential to the construction and maintenance of national systems has scarcely been touched by students of urbanization$_{1+2}$. With extensive empirical evidence, it should be possible to elaborate the theory substantially by identifying the conditions under which the propositions offered would be falsified, modifying the nature of some of the relationships, and tracing through in painstaking detail the spatial consequences of control processes that join core and periphery into a single national system.

C. Investment Flows

Published studies of interregional investment flows are non-existent probably because of the extraordinary difficulty in collecting the necessary data. However, the general draining of rural surpluses to enhance the splendor of core regions is a widely accepted hypothesis. Evidence can be inferred most readily from sectoral analysis. This has been done for Chile and Argentina in a series of brilliant essays by Mamalakis and Merkx.[34] Unfortunately, comparative data for South Asia and Africa are lacking.

The immediate mechanisms for accomplishing interregional capital transfers are the traditional ones of taxation, budgeting, and banking. Countries conforming to the classical core-periphery structure draw more revenues from the periphery than they return in capital improvements, and branch banks—whether public or private—manage to transfer substantial peripheral savings to the core. These actions are governed by laws (e.g., restrictive loan policies for branch banks and budget legislation) and therefore reflect the political process that characterizes and maintains traditional forms of core-region dominance.

The direct result of differential capital flows is interregional differences in income. Such income differentials have been studied more intensively than the flow process itself. Jeffrey G. Williamson, for instance, has outlined the relationship between national development and changes in regional inequality which accompany this growth.[35] His conclusions may be summarized as follows:

> "Regional inequality passes through three distinct stages as an economy moves from early development to maturity. In the early stages of development, regional inequality widens. This is followed by a period of stable, high-level inequality, and finally a marked secular trend towards regional equality sets in as development proceeds."

Detailed historical studies bear out Williamson's thesis only in part. Although not strictly related to our study of the developing countries, data for Canada suggest only a small amount of convergence between 1929 and 1956, while earlier periods led to a gradual widening of the income gap.[36] And a recent study of regional income differences in Brazil, though showing a slight convergence in the decade of the fifties, is inconclusive on the long-term direction of the trend.[37] The best that can be said from the available information is that regional income convergence is an extremely drawn-out process. It occurs, when it does, at a relatively advanced stage in the industrialization process, and it rarely changes the rank-ordering of regions by income.[38] Core-region dominance, we may conclude, is not easily challenged. And when it is, the reversal of earlier trends may require a number of generations. The continuous outward expansion of core areas may occur when subsidiary centers are drawn into the political-economic subsystem of the core, but this process is one of gradual accretion; major interregional differences in income respond considerably more slowly.

Governments frequently propose to use transport investments as a development tool for reducing income differences among areas.[39] The theory underlying these programs is that by increasing the accessibility of a peripheral area to major market centers, economic development in the periphery will be accelerated. This should, in principle, lead to the closer integration of peripheries with core region economies.[40] The evidence, however, does not support such a simple generalization.

An increase in the periphery's accessibility from the core is equivalent to an increase in the core's accessibility from the periphery. As result, transport investments often strengthen the periphery's dependency instead of encouraging self-generated development. This outcome is demonstrated for São Paulo by Gauthier.[41] There, the failure to invest in transport routes between provincial centers helped to encourage primarily the growth of the state capital. Similar findings are reported by McNulty for Ghana and by Sendut for Malaysia.[42]

Peter Gould, in his analysis of Tanzania's spatial structure, suggests an additional reason why transport development fails to improve the condition of peripheral areas. He postulates that in the early phases of development, roads and rails serve a "space-bridging" rather than "space-organizing" function. Thus, at the start of industrialization, interregional transport systems simply connect "islands of early development," each of which is surrounded by non-participating peripheries.[43]

An interesting study by Pedersen and Stöhr for South America leads to similar conclusions. In an effort to test the spatial impact of economic integration policies, the authors conclude that "integration in the short-run can be expected to reinforce the existing trend towards geographic concentration. . . . Only when high levels of economic development are reached (in the existing core regions), can the effect of integration reverse itself . . . and support other policies for achieving a wider spread of economic activities."[44]

D. Interregional Migration Flows

In contrast to other urbanization processes, internal migration has received the most attention. This is so probably for a number of reasons: the long tradition of demographic studies that reaches back to Thomas Malthus (1798) and the work of the *physique social* group of French scholarship that followed in the early part of the nineteenth century; the long-standing concern, particularly among intellectuals and government bureaucrats, with the "flight" of the population from the land and the subsequent multiplication of social problems in the large cities; the greater visibility of population change compared to innovation diffusion and decision-making processes; and the ready availability of quantitative data from census studies and the direct usefulness of standard social survey techniques for the analysis of migration.

The large number of migration studies reviewed for this paper show an exclusive behavioral-descriptive orientation. Except for the work of Jorge Balán, none of the papers used theoretical models for predicting aggregate migrant streams.[45]

Following our paradigm, interregional migration may be treated as a resultant of other urbanization$_{1+2}$ processes, i.e., of emergent innovations, innovation diffusion, decision-making, and interregional investment flows. A more carefully structured model which has recently been elaborated by Paul Drewe includes the following major variables: (1) employment opportunities (which are highly correlated with other urban attributes, such as opportunities for education and social advancement), (2) accessibility (expressed as a function of linear distance, modified by the probabilities of information exchange), and (3) the number and intensity of prior contacts with the migrant's place of ultimate destination.[46] This model can be refined by the addition of (4) the educational level of the population at points of migrant origin. A properly constructed model of this kind needs to be tested under different conditions and cultural settings (Drewe's work pertains only to Europe) and will need to be adapted to varying kinds of migrant behavior. The studies we examined dealt chiefly with three areas of concern: distinction among the several types of migrant behavior; selectivity of migrants from their home populations; and mechanisms of individual adjustment to and social integration into the dominant core region societies.

(1) *Types of migrant behavior.* Data from Colombia and Chile suggest that most migration occurs as a *permanent* move. This may be preceded, as evidence from Ecuador, Bolivia, and Guayana show, by temporary residence in an urban milieu, the result of military service, schooling, or prior work experience.[47] Although corroboration from other studies is lacking, some exposure to urban life would seem to increase the probability of a subsequent permanent change of residence. It should be noted, however, that much of the migration in Latin America seems to be inter-urban rather than rural to urban in character. This pattern may help to explain the absence of significant *return* migration. Inasmuch as Latin American countries are also more urbanized$_1$ than those of Africa and South Asia, one would expect a larger proportion of internal migrants to come from cities, and especially from the more stagnant, peripheral urban communities.[48] In Africa, by contrast, migrants are typically from agricultural villages.[49] The same appears to be true for South Asia.

In southern and central Africa, a good deal of migration has been described as a constant moving back and forth between home community and core.[50] Migrants involved in this *shuttle pattern* are only temporary city residents. The studies do not examine the extent to which frequent shuttle migration increases the probability of a permanent move. The continued growth of African cities suggests, however, that urban areas are experiencing a continuing net inflow of migrants.

Seasonal migration is also common. During the slack period in agriculture, adult males from the periphery frequently come to work in the cities. This type of migration may be one of the important sources of capital accumulation in the agricultural sector. Whether or not it is also a mechanism for the diffusion of urban innovations into the countryside is a question that has yet to be answered. This pattern of seasonal work has been reported for Costa Rica, Bolivia, and Ecuador as well as for parts of Africa and India. Elkan refers to it as *circular* migration.[51]

(2) *Migrant selectivity.* Evidence on the selectivity of migrant populations is contradictory, though certain common themes seem to emerge. Scholars in Latin America, Africa, and South Asia are generally agreed that migrants are positively selected from among their home populations for such characteristics as education, skill levels, and socio-economic status. Thus it is the most urbanized$_2$ groups among the population, rather than the rural poor, that appear to be most prone to migrate.[52]

The positive selection of migrants is reflected in their employment experience once once they reach the urban area. Herrick, for instance, found that migrants to Santiago are about as successful as the native population of that city in finding and keeping employment.[53] Data from Mexico City, Monterrey, Guatemala City, and San Salvador suggest a similar conclusion.[54] Comparable data are reported by Sovani for four of five Indian cities studied.[55] In Africa, on the other hand, migrant status may be less important for finding employment than is membership in tribal groups.[56]

Browning and Feindt, in their study of Monterrey, Mexico, suggest that migrants, when compared to the population in their home communities, are indeed positively selected for education and skill levels. But when these migrants are contrasted with earlier migrant groups, it appears that selectivity has been declining over time. The authors believe that, with the progressive development of the Mexican economy, pools of potential migrants came to be tapped in the traditionally more backward regions of the country. The early migrants were urban pioneers. Those who followed them came to the city with fewer adaptive skills. Similar patterns are reported for Buenos Aires and Mexico City.[57] The question of how selectivity changes over time has not yet been studied for any African or South Asian country.

(3) *Individual adjustment and social integration.* There appears to be agreement that migrants do not suffer traumatic upsets when they move to new environments.[58] Very few migrants arrive in the city as complete strangers. Most have already been introduced to many of the values, artefacts, and other elements of urban culture. They have frequently been to the city, listened to the radio, and come into contact with government officials and other urban institutions. Most rural migrants, therefore, begin the psychological transition to

core region cultures before they undertake the geographical move.[59] As suggested above, it is the most urbanized$_2$ population that is likely to leave their villages behind and opt for life in the cities.

All of the empirical evidence reviewed leads to a rejection of the Wirthian hypothesis that cities break down the primary group and substitute for it secondary associations that leave the individual isolated, rootless, and anomic. In fact, the migrant to the large city rarely leaves the security of a supporting social environment. His contacts with his home community and with fellow migrants from his home community generally remain strong throughout his stay in the city.[60]

Once he has accomplished his move, the migrant may live with family or friends who offer counsel and financial help until he finds employment.[61] In this connection, it is interesting to speculate on the prevalence of the extended family. In his study of Tepoztecan migrants to Mexico City, Oscar Lewis found the joint family more prevalent in the city than in town or village.[62] Mukherjee and Lambert suggest that in India, at least, the joint family depends more on socio-economic status than on residence. Thus, higher-status families are more likely to be members of joint families than are members of the lower castes.[63] But, in any event, the presence of extended families in cities is more pronounced than is commonly assumed.

In Africa, it is the tribal organization that transcends spatial boundaries. Through extensive kinship and ethnic ties, the tribe maintains the solidarity of its members over wide gaps of wealth and status, as well as space.[64] Migrants to African cities frequently reside in ethnic enclaves but work in the trans-ethnic environment of the central city.[65] With time, ethnic identification may weaken as new identifications crystallize around status as determined by jobs and income. Class divisions, especially for the elite groups, tend to replace the older systems of stratification based on tribal or ethnic membership. Clear evidence for this process comes from Barbados, Brazil, and West African cities.[66] Very likely, this shift to trans-ethnic groupings reflects the changes in the multiplication of established role patterns that occur as the migrant participates ever more intensively in the social system of the core region.

An important integrating mechanism, especially for African migrants, is the urban voluntary associations. Kenneth Little has interpreted these associations as socializing agencies whose importance lies in the modern social patterns they seek to substitute for the older structures of West African society.[67] According to Little, voluntary associations fall into three general categories: traditional, Christian, and modern secular. There are differences in the three types, but their organizational structure is essentially the same. The traditional associations are divided along tribal lines, the Christian associations are social groupings with more or less a religious orientation, and the modern secular organizations, although fewer in number, set a pattern of cooperation across ethnic and religious lines. All associations share the objective of fraternity and social interaction. Perhaps more important, they provide a degree of social security based on mutual benefit and aid. Assistance is provided to members when there is a death in the family or when the breadwinner is ill. Associations may also help when one of their members is temporarily unemployed or has trouble with the authorities.

Peter Gutkind, in his study of social networks in Kampala, Uganda, concludes that associations serve primarily the needs of those migrants whose native homes are not easily accessible and whose migration conforms to the circular or permanent pattern. Migrants caught up in frequent back and forth movements between city and village (the shuttle pattern) tend to rely more on ties of kinship for assistance and security.[68]

Although most of the data on fraternal organizations come from Africa, the pattern is not unknown in parts of Latin America and has been studied by William Mangin for Peru.[69]

All these studies seem to suggest that the movement from periphery to core is less stressful than is frequently assumed.[70] A variety of social mechanisms exists to ease the migrant's passage into the life of the core region. Some of these, such as temporary, circular, or shuttle migration, as well as close kinship, tribal, or ethnic bonds, are material in helping the individual migrant adjust to city life. Other mechanisms, such as voluntary associations and the extremely important "bazaar" economy with its built-in system of mutual aid, assist still more in the social integration of the migrant.[71] There is mounting evidence, however, that despite the existence of these social aids to migration, accelerated urbanization$_1$ tends to heighten the sense of insecurity, cultural deprivation, and economic exclusion.

This phenomenon has been studied intensively in Latin America under the heading of "marginality."[72] The term is applied not only to urban migrants, but also to the disenfranchised, poverty-stricken rural population. Marginality conveys a multi-dimensional meaning, covering economic, political, and cultural forms of exclusion. By making "marginality" more visible, accelerated urbanization$_1$ tends to transform it into a salient political issue. It is increasingly doubtful, however, whether the conditions of "marginality" can be eliminated through the normal processes of individual adjustment and social integration. The research suggests that successful "integration" will ultimately depend on the ability of the dominant social system to undergo a radical transformation of its own. Social aspirations, stimulated by the urbanization$_{1+2}$ experience, may outpace the ability to satisfy them. When this happens, political tensions are set in motion whose resolution may call for far-reaching structural changes in the social system itself.[73]

E. Spatial Organization: A Theoretical Formulation

Each of the urbanization processes discussed in the preceding section leads to spatial configurations and structural patterns (see Figure 1). Thus:

Process	leads to	Pattern
innovation diffusion		socio-cultural spatial pattern
migration		settlement pattern
investment		economic activity location pattern
decision making		territorial organization of power

These patterns, representing varying degrees of spatial integration, have just begun to be studied.[74] Most of the empirical research has been done on African countries, with perhaps the most impressive study being Soja's investigation in Kenya.[75] This new African nation exhibits the relatively simple organization of its social space that is typical for countries at a similar stage in their economic development. The capital city, Nairobi, is identified as the core area of the country in terms of settlement, culture, economic activity, and political power (though it is itself dependent on a higher-order, external core region in England). Those areas which most closely interact with Nairobi have the best-developed transport and communication routes, as well as the highest concentrations of income and other indicators of social achievement. Surrounding them is a belt of "transitional" culture which remains incompletely integrated with the core. And beyond this region lie the unincorporated areas that have so far escaped effective national control and are chiefly inhabited by pastoral tribes.

This spatial pattern of urbanization$_{1+2}$ still reflects the zones of original European influence. During the colonial period, European settlements played the key role in the diffusion of innovations to the largely African periphery. The lack of a well-integrated hierarchy of urban centers slowed down the flow of innovations which barely extended beyond the major modes and corridors. The result was a highly uneven geographical spread of urbanization$_2$, further complicated by the political barriers to national unification which were erected by the colonial administration.

Other research on the subject of spatial organization supports these empirical findings. The evidence allows us to arrive at conclusions that bring a unified theory of urbanization$_{1+2}$ into sight. This section will be devoted to a summary of the principal propositions which may be drawn from a comparative analysis of the existing data.

The most important discovery is that, for a given country, at a given point in time, the four spatial patterns of urbanization$_{1+2}$ appear to be isomorphic. In other words, the spatial organization of modernization (SC + TOP) tends to have the same configuration as the spatial organization of the economy (EA + S).

The reason for this would seem to be that we are dealing here with a system in dynamic equilibrium. Each of the urbanization$_{1+2}$ processes provides negative, correcting feedback which tends to crystallize the entire spatial system into an isomorphic structure. However, the continuous stream of innovations spinning out of core-regional development tends, at the same time, to nudge the system into different spatial configurations. This perpetual diffusion of innovations across the cultural landscape produces structural tensions which necessitate adjustments in the other processes until a new level of (partial) equilibrium is reached.

The spatial integration of a growing national system therefore involves a three-phased cycle of systemic change: negative feedback, disequilibrating innovation diffusion (positive feedback), and internal adjustment to the new structural conditions (negative feedback).[76]

The empirical evidence on levels of spatial integration further leads us to hypothesize the following tendencies for countries undergoing industrialization:[77]

(1) *Socio-cultural spatial pattern.* The evolution of spatial structure proceeds from relatively isolated "islands of innovation" via major communication corridors to a continuously urbanized$_2$ surface.[78] However, this surface will be uneven, reflecting earlier differences in the structure of modernization as well as newly emerging (e.g., post-industrial) patterns.

(2) *Economic activity location pattern.* The evolution of spatial structure proceeds from a highly concentrated pattern in the location of modern economic activities to one that is spatially more deconcentrated.[79] Agglomeration economies, however, will limit the extent to which deconcentration can occur. Of equal importance is a process of industrial dispersion into the urban fields of metropolitan areas.[80]

(3) *Settlement pattern.* The evolution of spatial structure proceeds from primacy in urban settlements to a log-normal pattern in the hierarchy of cities.[81] The vectors of this pattern, however, are influenced by the general structure of accessibility in the spatial system.[82]

(4) *Territorial organization of power.* The evolution of spatial structure proceeds from a highly centralized to a polycentric system of decision-making characterized by polyarchical relations among a set of interdependent authorities.[83]

Although the isomorphism of these four urbanization patterns is evident from a reading of the foregoing hypotheses, the extent to which changes in them are related through time has not been shown. As we suggested earlier, it is probable that urbanization$_{1+2}$ processes are out of phase with one another. If this is true, we may further hypothesize that changes in the territorial organization of power will tend to precede changes in the location of modern industry, while the latter will frequently come before any quantitatively important changes in socio-cultural patterns. This sequence is based on the notion that the decentralization of power acts itself as an important location factor, and that changes in values, beliefs, and attitudes tend to follow upon changes in experience and behavior.[84] Other sequences, however, may be possible. The rough isomorphism of patterns permits us to integrate the spatial organization of modernization (SOM) and economy (SOE) into an urbanization$_{1+2}$ surface representing the integration of the entire social space at a given point in time. The height of the surface would be calibrated for different measures of urbanization$_{1+2}$. Its structure would be tent-like, urbanization$_{1+2}$ falling off at various gradients with increasing distance from major peaks in the system. If we now imagine the transformation of this surface over time, we will, first of all, note the interdependency of all points on the surface and, secondly, the continuity of changes in configuration. The spatial development of a national society may, therefore, be said to be an orderly (systemic, predictable) and continuous (evolutionary) process.

An explanation of this behavior would enable us to move towards a predictive model of urbanization$_{1+2}$. Although we shall not attempt to construct such a model here, three critically important behavioral characteristics of urbanization$_{1+2}$ may be identified:

> First, prior urbanization experience determines the subsequent form of the urbanization$_{1+2}$ surface. This accounts for the stability of spatial patterns that has been frequently observed.
>
> Second, potential information exchange (a measure of relative accessbility) partly determines the rates and spatial incidence of urbanization$_{1+2}$.
>
> Third, the territorial organization of power exerts a strong influence on the rates and spatial incidence of urbanization$_{1+2}$.

As the references to these hypotheses suggest, there has been a considerable amount of supporting empirical study. For Latin America, the urbanization$_{1+2}$ surface has been investigated by Peterson and Stöhr; for Africa, studies have been done by Edward Soja (Kenya), Peter Gould (Tanzania) and Michael McNulty (Ghana), among others; and for South Asia, important contributions have been made by Brian Berry. If we add to this the mounting evidence for the United States (Brian Berry, J. H. Thompson et al.), we must conclude that we are faced here with a major cross-cultural generalization that reinforces our basic belief in the possibilities of a unified theory of urbanization$_{1+2}$.

NOTES

1. Leonard Reissman, *The Urban Process.* New York: The Free Press, 1964.

2. Richard M. Morse, "Planning, History, Politics: Reflections on John Friedmann's 'The Role of Cities in National Development,' " in John Miller and Ralph A. Gakenheimer, eds., *Latin American Urban Policies and the Social Sciences.* Beverly Hills, Calif.: Sage Publications, Inc., 1971, pp. 189-208.

3. The notion of cross-cultural generalization was introduced to social science analysis by the anthropologist Julian Steward in his *Theory of Culture Change.* Urbana, Ill.: University of Illinois Press, 1955, chapters 1 and 2.

4. The most comprehensive treatment to date will be found in Gerald Breese, *Urbanization in Newly Developing Countries.* Englewood Cliffs, N.J.: Prentice-Hall, 1966, and by the same author, *The City in Newly Developing Countries: Readings on Urbanism and Urbanization.* Englewood Cliffs, N.J.: Prentice-Hall, 1969.

5. Among these countries, the following are particularly worth studying: Soviet Russia, Turkey, Italy, Poland, England, France, Germany, Israel, Kenya, Brazil, Chile, Venezuela, and China.

6. John Friedmann, "The Strategy of Deliberate Urbanization," Journal of the American Institute of Planners, Vol. XXXIV, No. 6 (November 1968), pp. 364-373; William Alonso, "Urban and Regional Imbalances in Economic Development," Economic Development and Cutural Change, Vol. 17, No. 1 (October 1968), pp. 1-13, as well as a series of unpublished papers available from the Center for Planning and Development Research, University of California at Berkeley; Lloyd Rodwin, *Nations and Cities.* A Comparison of Strategies for Urban Growth. Boston: Houghton Mifflin Co., 1970; and several writings, both published and unpublished, by Brian Berry, Department of Geography, University of Chicago.

7. For an elaboration of these and related concepts, see Chapter 3.

8. Ibid.

9. This formulation is consistent with the integration models of Bela Balassa, *The Theory of Economic Integration.* Homewood, Illinois: Richard D. Irwin, 1961, and Jacob and Toscano, eds., *The Integration of Political Communities.* Philadelphia: J. B. Lippincott, 1964. Where Balassa is chiefly concerned with the spatial organization of the economy (and consequently with factor movements), Jacob and Toscano stress transactive flows in the political sector (because of their preoccupation with metropolitan areas, innovation diffusion is not included in their analysis).

10. See bibliography (Appendix A at back of book) for a list of the works consulted.

11. The more notable efforts include: Gerald Breese, ed., *The City in Newly Developing Countries. Readings on Urbanism and Urbanization.* Englewood Cliffs, N.J.: Prentice-Hall, 1969; Gerald Breese, *Modernization and Urbanization: Existing and Potential Relationships in The Third World.* Exchange Bibliography No. 70. Monticello, Illinois: Council of Planning Librarians, March 1969; Gerald Breese, *Urbanization in Newly Developing Countries.* Englewood Cliffs, N.J.: Prentice-Hall, Inc., 1966; Robert T. Daland, ed., *Comparative Urban Research: The Administration and Politics of Cities.* Beverly Hills, Calif.: Sage Publications, Inc., 1969; Elizabeth Eddy, ed., *Urban Anthropology: Research Perspectives and Strategies.* Athens, Georgia: Southern Anthropology Society, 1968; Sylvia Fleis Fava, ed., *Urbanism in World Perspectives: A Reader.* New York: Thomas Y. Crowell Co., 1968; Scott Greer, Dennis L. McElrath, David W. Minar, and Peter Orleans, eds., *The New Urbanization.* New York: St. Martin's Press, 1968, Part I; Philip M. Hauser and Leo F. Schnore, eds., *The Study of Urbanization.* New York: John Wiley and Sons, 1965, Part II; Joan Nelson, "The Urban Poor: Disruption or Political Integration in Third World Countries," World Politics, Vol. 22, No. 3 (April 1970), pp. 393-414; Gideon Sjoberg, *The Preindustrial City, Past and Present.* New York: The Free Press, 1960; Paul Meadows and Ephraim H. Mizruchi, eds., *Urbanism, Urbanization, and Change: Comparative Perspectives.* Reading, Mass.: Addison-Wesley, 1969.

12. The principal studies are reviewed in Chapter 9.

13. For one of the most sophisticated forecasting studies, see Lowdon Wingo, "Recent Patterns of Urbanization Among Latin American Countries," Urban Affairs Quarterly, Vol. II, No. 3 (March 1967), pp. 81-110. See also United Nations, *Growth of the World's Urban and Rural Population, 1920-2000.* Population Studies No. 44. New York, 1969.

14. One of the best statements will be found in Philip M. Hauser, ed., *Urbanization in Latin America.* New York: International Documents Service, 1961. More recent studies, however, tend to deemphasize the importance of urban social adjustment as a problem of public policy.

15. John Friedmann, *The Spatial Structure of Economic Development in the Tennessee Valley: A Study in Regional Planning.* University of Chicago, Department of Geography, Research Paper No. 39, March 1955, especially Part II.

16. Lawrence A. Brown, *Diffusion Processes and Location. A Conceptual Framework and Bibliography.* Philadelphia: Regional Science Research Institute, 1968.

17. Brian J. L. Berry, "City Size and Economic Development: Conceptual Synthesis and Policy Problems, with Special Reference to South and South-East Asia," unpublished MS, 1970.

18. Michael Safier, "The Urban System in Africa: Patterns of Disjunction and Integration," African Urban Notes, Vol. II, No. 3 (May 1967), pp. 4-11.

19. William John Hanna and Judith Lynne Hanna, "Polyethnicity and Political Integration in Umuhia and Mbale," in Robert T. Daland, ed., op. cit., and James F. Guyot, "Creeping Urbanism and Political Development in Malaysia," in the same volume, pp. 124-161.

20. Ralph L. Beals, "Urbanization and Acculturation," American Anthropologist, Vol. LV, No. 1 (January-March 1951), pp. 1-10; and Julio Cotler, "The Mechanics of Internal Domination and Social Change in Peru," in Irving Louis Horowitz, ed., *Masses in Latin America.* New York: Oxford University Press, 1970.

21. Poul Ove Pedersen, "Innovation Diffusion Within and Between National Urban Systems," Geographic Analysis, Vol. II (1970), pp. 203-254.

22. Wilbur Schramm, *Mass Media and National Development: The Role of Information in the Developing Countries.* Stanford, Calif.: Stanford University Press, 1964; and Torsten Hagerstrand, *Innovation Diffusion as a Spatial Process.* Chicago: University of Chicago PRess, 1967.

23. James F. Guyot, op. cit.

24. John Friedmann, *Urban and Regional Development in Chile: A Case Study of Innovative Planning.* Santiago, Chile: The Ford Foundation, June 1, 1969. Multilith. See also his chapter, "Urban-Regional Policies for National Development in Chile," in Francine F. Rabinovitz and Felicity M. Trueblood, eds., *Latin American Urban Research.* Vol. I. Beverly Hills, Cal.: Sage Publications, Inc., 1971.

25. Aníbal Quijano, "Dependencia, Cambio Social y Urbanización en Latinoamérica," Cuadernos de Desarollo Urbano Regional, Universidad Católica de Chile, CIDU, No. 6 (March 1968), pp. 3-48.

26. Chandler Morse, "Potentials and Hazards of Direct International Investment in Raw Materials," in Marion Clawson, ed., *Natural Resources and International Development.* Baltimore: The Johns Hopkins Press, 1964.

27. E. J. Taaffe, R. L. Morrill, and P. R. Gould, "Transport Expansion in Underdeveloped Countries," Geographical Review, Vol. LIII, No. 4, (October 1963), pp. 503-529.

28. The best discussion of the phenomenon of urban primacy is by Brian J. L. Berry, op. cit. But see also Kingsley Davis, "Las causas y efectos del fenomeno de primacia con referencia especial a América Latina," Coyoacan, Mexico: Instituto de Investigaciónes Sociales, 1962. (Offprint); and Arnold S. Linsky, "Some Generalizations Concerning Primate Cities," in Gerald W. Breese, ed., op. cit., pp. 288-294.

29. William Alonso, "Industrial Location and Regional Policy in Economic Development," Working Paper No. 74, Center for Planning and Development Research, University of California, Berkeley, February 1968.

30. André Gunder Frank, *Capitalism and Underdevelopment in Latin America: Historical Studies of Chile and Brazil.* New York and London: Monthly Review Press, 1967.

31. Dramatic instances are afforded by recent conflict in Nigeria, the independence struggle of Algeria against (metropolitan) France, and the continuing internal struggles in Indonesia and the Sudan.

32. John Friedmann, *Venezuela: From Doctrine to Dialogue.* Syracuse: Syracuse

University Press, 1965; Fred G. Burke, *Public Administration in Africa: The Legacy of Inherited Colonial Institutions.* Syracuse University Program of Eastern African Studies, Occasional Paper No. 36, 1967.

33. Akin L. Mabogunje, "Urbanization in Nigeria: A Constraint on Economic Development," Economic Development and Cultural Change, Vol. 13, No. 4, Part I (July 1965), pp. 413-438.

34. Markos J. Mamalakis, "The Theory of Sectoral Clashes," and Gilbert W. Merkx, "Sectoral Clashes and Political Change: The Argentine Experience," Latin American Research Review, Vol. IV, No. 3 (1969), pp. 9-46 and 89-116.

35. Jeffrey Williamson, "Regional Inequality and the Process of National Development," Economic Development and Cultural Change, Vol. 13, No. 4, Pt. II (July 1965).

36. Alan G. Green, "Regional Inequality, Structural Change, and Economic Growth in Canada–1890-1956," Economic Development and Cultural Change, Vol. 17, No. 4 (July 1969), pp. 567-583.

37. Douglas H. Graham, "Divergent and Convergent Regional Economic Growth and Internal Migration in Brazil–1940-1960," Economic Development and Cultural Change, Vol. 18, No. 3 (April 1970), pp. 362-382.

38. The most complete studies of interregional income stability has been done by J. R. Lasuén, Dean of the Faculty of Economic Sciences, University of Madrid, in a series of unpublished papers. For reference to some of these studies, see his article, "On Growth Poles," Urban Studies, Vol. 6, No. 2 (June 1969), pp. 137-161.

39. Wilfred Own, *Strategy for Mobility.* Washington, D.C.: The Brookings Institution, 1964; and Gary Fromm, ed., *Transport Investment and Economic Development.* Washington, D.C.: The Brookings Institution, 1965.

40. This principle was applied, for instance, to the development of Appalachia in the United States and the Mezzogiorno in Italy. Results, however, failed to bear out the early optimism among transport planners.

41. Howard L. Gauthier, "Transportation and the Growth of the São Paulo Economy," *Journal of Regional Science, Vol. 8, No. 1 (1968), pp. 77-94.*

42. Michael L. McNulty, "Urban Structure and Development: The Urban System of Ghana," The Journal of Developing Areas, Vol. III (January 1969), pp. 159-176; and Hamzah Sendut, "Patterns of Urbanization in Malaysia," Journal of Tropical Geography, Vol. 16 (October 1962), pp. 114-130.

43. Peter Gould, "Tanzania 1920-63: The Spatial Impress of the Modernization Process," World Politics, Vol. 22, No. 1 (January 1970), pp. 149-170.

44. Poul Ove Pedersen and Walter Stöhr, "Economic Integration and the Spatial Development of South America," American Behavioral Scientist, Vol. XII, No. 5 (May-June 1969), p. 8.

45. Jorge Balán, "Migrant-Native Socioecnomic Differences in Latin American Cities: A Structural Analysis," Latin American Research Review, Vol. IV, No. 1 (1969), pp. 3-29.

46. P. Drewe, "Elementary Features of Geographical Labour Mobility," Rotterdam: Netherlands Economics Institute, 1970. Mimeographed.

47. (a) William L. Flinn, *Rural to Urban Migration,* A Columbian Case, Reprint No. 19. Madison: University of Wisconsin Land Tenure Center, July 1966; Bruce R. Herrick, *Urban Migration and Economic Development in Chile.* Cambridge, Mass.: The M.I.T. Press, 1965, p. 54. (b) David A. Preston, "Rural Emigration in Andean America," Human Organization, Vol. 28, No. 4 (Winter 1969), pp. 279-286; Beatrice D. and John S. MacDonald, "Motives and Objectives of Migration: Selective Migration and Preferences Towards Rural and Urban Life in Guayana," Ekistics, Vol. 28, No. 128 (November 1969), pp. 321-327.

48. This hypothesis is suggested by Jorge Balán, op. cit.

49. A. L. Epstein, "Urbanization and Social Change in Africa," in Gerald Breese, ed., op. cit., pp. 246-263.

50. Philip Mayer, "Migrancy and the Study of Africans in Towns," American Anthropologist, Vol. 64, No. 3, Part I (June 1962), pp. 576-592.

51. Walter Elkan, "Circular Migration and the Growth of Towns in East Africa," Ekistics, Vol. 25 (April 1968), pp. 581-589. See also, Michael D. Olien, "Levels of Urban Relationships in a Complex Society: A Costa Rican Case," in Elizabeth Eddy, ed., op. cit., pp. 83-92; David A. Preston, op. cit.; and Richard D. Lambert, "The Impact of Urban Society Upon Village Life," in Roy Turner, ed., *India's Urban Future.* Berkeley, Calif.: University of California Press, 1961, pp. 117-140.

52. Bruce R. Herrick, op. cit., pp. 77-80; Harley L. Browning and Waltrout Feindt, "Selectivity of Migrants to a Metropolis in a Developing Country: A Mexican Case Study," Demography, Vol. 6, No. 4 (November 1969), pp. 347-357; Jorge Balán, op. cit.; Akin L. Mabongunje, op. cit.; Walter Elkan, op. cit.; K. C. Zachariah, "Bombay Migration Study: A Pilot Analysis of Migration to an Asian Metropolis," in Gerald Breese, ed., op. cit., pp. 360-375; N. V. Sovani, "Some Notes on In-Migrants in Indian Cities," in Urbanization and Urban India. New York: Asia Publishing House, 1966, pp. 142-155.

53. Bruce R. Herrick, op. cit., pp. 80-85.

54. Harley L. Browning, "Urbanization and Modernization in Latin America: The Demographic Perspective," in Glenn H. Beyer, ed., *The Urban Explosion in Latin America.* Ithaca, New York: Cornell University Press, 1967, pp. 71-92; Browning and Feindt, op. cit.; Jorge Balán, op. cit.

55. N. V. Sovani, op. cit.

56. Peter Gutkind, "African Urbanism, Mobility and Social Network," in Gerald Breese, ed., op. cit., pp. 389-400.

57. Browning and Feindt, op. cit.; Jorge Balán, op. cit.; Oscar Lewis, "Urbanization Without Breakdown: A Case Study," in Dwight B. Heath and Richard N. Adams, eds., *Contemporary Cultures and Societies of Latin America.* New York: Random House, 1965, pp. 424-437 (reprinted from The Scientific Monthly, Vol. LXXV (1952), pp. 31-41).

58. Oscar Lewis, op. cit.; William P. Mangin, "Mental Health and Migration to Cities: A Peruvian Case," in Heath and Adams, eds., op. cit., pp. 546-551; Brian M. duToit, "Cultural Continuity and African Urbanization," in Elizabeth M. Eddy, ed., op. cit., pp. 58-74; Peter Gutkind, op. cit.; Edwin Eames, "Urban Migration and Joint Family in a North Indian Village," The Journal of Developing Areas, Vol. I, No. 2 (January 1967), pp. 163-178; Richard D. Lambert, op. cit.

59. Bruce R. Herrick, op. cit., pp. 89-91; William P. Mangin, op. cit.

60. Oscar Lewis, op. cit.; Ramkrishna Mukherjee, "Urbanization and Social Transformation in India," International Journal of Comparative Sociology, Vol. 4, No. 2 (September 1963), pp. 178-210; Bernard Gallin, "Political Attitudes and Political Participation of Rural-to-Urban Migrants in Taipei," Paper presented at the Conference on Urban Society and Political Development in China. New York: Social Science Research Council, 1969.

61. William L. Flinn, op. cit.; Leatrice and John MacDonald, op. cit.; Oscar Lewis, op. cit.

62. Oscar Lewis, op. cit.

63. R. Mukherjee, op. cit.; Richard Lambert, op. cit.

64. Edwin Eames, op. cit.

65. William and Judith Hanna, op. cit.

66. Raymond W. Mack, "Race, Power, and Class in an Urban Plantocracy: Barbados," in Scott Greer et al., op. cit., pp. 72-88; Emilio Willems, "Urban Classes and Acculturation in Latin America," in Elizabeth Eddy, ed., op. cit., pp. 75-82; A. L. Epstein, op. cit.; Kenneth Little, "Organization of Voluntary Organizations in West Africa," Civilisations, Vol. IX, No. 3 (1959), pp. 283-297.

67. Kenneth Little, op. cit.; "Some Contemporary Trends in African Urbanization," Lecture delivered at Northwestern University on April 20, 1965, and published by Northwestern University Press.

68. Peter Gutkind, op. cit.

69. William P. Mangin, "The Role of Regional Associations in the Adaptation of Rural Migrants to Cities in Peru," in Dwight B. Heath and Richard N. Adams, eds., op. cit., pp. 311-323.

70. Most of the relevant literature on this point is brilliantly summarized by Joan Nelson, "The Urban Poor: Disruption or Political Integration in Third World Countries," op. cit.

71. Lack of space prevents us from elaborating on the extremely interesting set of institutions which Clifford Geertz has called the "bazaar" economy. See his basic study, *Peddlers and Princes.* Chicago: University of Chicago Press, 1963, as well as W. R. Armstrong and T. G. McGee, "Revolutionary Change and the Third World City: A Theory of Involution," Civilisations, Vol. 18, No. 3 (1968), pp. 353-378. Corroborating evidence for Latin America comes from Lisa Peattie's excellent study of barrio life in Guayana (Venezuela), *The View From the Barrio.* Ann Arbor, Michigan: University of Michigan Press, 1968.

72. For a summary of current thinking, see the special issue of *Revista Latinoamericana de Sociologia,* Vol. 69, No. 2 (July 1969), edited by José Nun.

73. See Chapter 5 for an illustration from recent Chilean history.

74. Some of the leading studies include: for Latin America, John Friedmann, *Regional Development Policy: A Case Study of Venezuela.* Cambridge, Mass.: The M.I.T. Press, 1966, and Poul Ove Pedersen and Walter Stöhr, op. cit.; for Africa, Edward Soja, *The Geography of Modernization in Kenya.* Syracuse, New York: Syracuse University Press, 1968, and Peter Gould, op. cit.; for South Asia, Brian J. L. Berry, Essays on Commodity Flows and the Spatial Structure of the Indian Economy. Research Paper No. 111. Chicago: University of Chicago Department of Geography 1966, and Brian J. L. Berry, op. cit.; and for the United States, John H. Thompson, Sidney C. Sufrin, Peter R. Gould, and Marion A. Buck, "Toward a Geography of Economic Health: The Case of New York State," in John Friedmann and William Alonso, ed., *Regional Development and Planning: A Reader.* Cambridge, Mass.: The M.I.T. Press, 1964, among others. Brian Berry is the leading contributor to this new genre of structural studies.

75. Edward Soja, op. cit.

76. On the subject of feedback dynamics in open systems, see Jay W. Forrester, *Principles of Systems.* Cambridge, Mass.: Wright-Allen Press, Room 516, 238 Main Street, 1968, chapter 2.

77. See also Kenneth Boulding, "Toward a General Theory of Growth," in Joseph Spengler and Otis Dudley Duncan, eds., *Population Theory and Policy. Selected Readings.* New York: The Free Press, 1956, pp. 109-124.

78. Peter Gould, op. cit.: Pedersen and Stöhr, op. cit.; Edward Soja, op. cit.; Michael L. McNulty, op. cit.; L. P. Green and T.J.D. Fair, *Development in Africa: A Study in Regional Analysis With Special Reference to Southern Africa.* Johannesburg: Witwatersrand

University Press, 1969; C.F.J. Whebell, "Corridors: A Theory of Urban Systems," Annals of the Association of American Geographers, Vol. 59, No. 1 (March 1969), pp. 1-26.

79. José Lasuén, op. cit.

80. John Friedmann and John Miller, "The Urban Field," Journal of the American Institute of Planners, Vol. XXXI, No. 4 (November 1965), pp. 312-320; Brian J. L. Berry and Elaine Neils, "Location, Size, and Shape of Cities as Influenced by Environmental Factors: The Urban Environment Write Large," in Harvey S. Perloff, ed., *The Quality of Urban Environment.* Published by Resources for the Future, Inc., and distributed by the Johns Hopkins Press, Baltimore, 1969.

81. Brian J. L. Berry, "City Size and Economic Development," op. cit.

82. Brian J. L. Berry, *Geography of Market Centers and Retail Distribution.* Englewood Cliffs, New Jersey: Prentice-Hall, Inc., 1967, p. 33.

83. For an extensive description of polyarchy in control processes, see Robert A. Dahl and Charles E. Lindblom, *Politics, Economics, and Welfare.* New York: Harpers, 1953, chapter 10 and 11.

84. William Alonso, op. cit.; Leon Festinger, *Conflict, Decision, and Dissonance.* Stanford, Calif.: Stanford University Press, 1964.

Chapter 5

HYPERURBANIZATION AND NATIONAL DEVELOPMENT IN CHILE

"... the problem for developing countries is the simultaneous creation of the *polis* and the integration of almost everyone in it."

—K. H. Silvert

THE CONCEPT OF HYPERURBANIZATION

Hyperurbanization signifies a prolonged condition of superheated urban growth. It is not an absolute measure. For the sake of establishing a base line, let us say that hyperurbanization implies a doubling rate in urban population of at least twenty years.[1] Historically, many countries, including the United States, have experienced considerably shorter doubling rates over extended periods of time.[2] In a comprehensive statistical study of contemporary urbanization, Kingsley Davis records the doubling rates for major world regions since 1950 (see Table 1).[3]

At least since 1950, according to Davis, the greater part of the world's regions have been passing through a period of hyperurbanization. The principal exceptions include North America, Temperate South America (Argentina, Uruguay, Chile), Southcentral Asia and Europe. Of these, Northern and Western Europe, Temperate South America, and North America, with current levels of urbanization of 70% or more, were, for the most part, subject to hyperurbanization

TABLE 1
DOUBLING RATES OF URBAN POPULATION 1950-1970, BY MAJOR WORLD REGIONS

	1950-1960 (years)	*1960-1970 (years)*
North Africa	16	17
Western Africa	10	11
Eastern Africa	13	13
Middle and Southern Africa	9	15
South Africa	18	20
North America	26	32
Middle America	14	14
Caribbean	23	20
Tropical South America	13	15
Temperate South America	25	30
East Asia	12	15
Japan	11	19
Southeast Asia	15	15
Southcentral Asia	26	23
Northern Europe	88	70
Western Europe	44	39
Eastern Europe	32	37
Southern Europe	35	35
Australia-New Zealand	21	25
Oceania	14	14
USSR	20	20

during *earlier* periods in their history; Japan, which has been rapidly urbanizing since the end of World War II, is about to join these "older" urban regions. Southcentral Asia, on the other hand, presents a rather mixed picture in the detail of its country statistics. The urbanization rates for India and Ceylon, for instance, have been accelerating during the past 20 years, so that these two important countries will probably join the "club" of hyperurbanizing nations during the present decade. Pakistan, with a current doubling rate of 16 years, is already there, while Afghanistan, remote from outside modernizing influences, is expanding its urban population at the relatively low doubling rate of only 30 years.

A precise reading of these rates is not important, however; their significance derives from a presumption that the accelerated growth of cities, in every case the result of migration and almost always involving a massive relocation of people from the countryside, is generating severe stresses and tensions within the society that must be successfully resolved if national development is to continue. A plausible hypothesis, to be further investigated with respect to Chile, would therefore be that the successful resolution of developmental crises engendered by hyperurbanization accelerates the processes by which a country becomes "modernized," or acquires the capacities for continuously generating and absorbing changes that, in the long run, will transform the structural bases of the society.[4] Countries failing in this effort will either stagnate, economically and socially, or become absorbed as client states by more powerful, dynamic neighbors.

If hyperurbanization is regarded as a tension-generating process that may have positive consequences for a country's development, it is necessary sharply to distinguish this concept from the well-known, if much debated, idea of overurbanization.[5] Hyperurbanization is not meant to imply excessive urbanization in relation to employment growth as does the concept of overurbanization; it is intended rather to convey the social dynamics released by the accelerated and massive transfer of rural people to cities.

Some countries, of course, may fail to undergo the experience of hyperurbanization. This group includes certain Eastern European countries (as well as Cuba) which have instituted policies aimed at avoiding massive rural-urban population transfers in the course of rapid industrialization;[6] a number of Catholic countries of Southern Europe which, together with Austria and Ireland, are still to a large extent in the sway of conservative peasant traditions; and finally a handful of non-industrial societies which have yet to enter the period of modernization (see Table 2).

TABLE 2

PERCENTAGE OF URBAN POPULATION IN 1970 AND DOUBLING RATES OF URBAN POPULATION, 1960-1970, FOR SELECTED COUNTRIES

	Percentage of Urban Population (1970)	*Doubling Rate of Urban Population, 1960-70 (years)*
A. Czechoslovakia	52.0	41
Hungary	42.7	39
Poland	55.8	27
Rumania	38.9	27
Cuba	58.4	27
B. Portugal	36.5	41
Spain	58.8	41
Greece	62.6	44
Italy	51.5	47
Austria	51.0	88
Ireland	50.7	50
C. Afghanistan	6.4	30
Ethiopia	6.6	29

SOURCE: Kingsley Davis, op. cit., Tables C and D.

Both the origins and consequences of hyperurbanization will vary according to the specific historical circumstances of each national society. The origins may be found in demographic dislocations caused by war (e.g., Vietnam), in national independence accompanied by heavy investments in building the nation's capital (as in many of the new African nations), or in rapid industrialization (as in most South American nations). The consequences of hyperurbanization will, among other things, vary with the achieved levels of *prior* urbanization and gross national product *per capita.*

CHILE: A HYPERURBANIZED SOCIETY

Chile, with an estimated 73.7% of its population living in cities, is among the most urbanized countries in South America. During the past two decades, Chile has had an overall urbanization rate that would lead to a doubling of urban population within 18 years. By our criterion, therefore, Chile may be considered a hyperurbanizing country.

Hyperurbanization in Chile has resulted in a high and steadily increasing concentration of urban activities and population. This may be established in several ways. Measuring the degree of urban primacy according to population, the relation of the largest to the second largest urban complex (Greater Santiago vs. Valparaíso-Viña del Mar) was 2.33 in 1920 but increased to 5.18 in the course of the succeeding four decades (see Table 3). For comparison, a calculation of the same index based on data for the mid-1950's, gave values of 7.49 for France, 7.15 for Mexico, 4.08 for the United Kingdom, 2.35 for the United States, and 1.25 for Brazil.[7] Taking the relation of the largest urban complex to the next three largest agglomerations, the primacy index for Chile in 1960 was 2.77 as opposed to 1.42 in 1920. Comparative data show indices of 2.68 (France), 2.87 (Mexico), 1.48 (United Kingdom), 0.96 (United States), and 0.85 (Brazil).

TABLE 3

CHILE: INDEX OF PRIMACY, 1865-1960

	1865	*1875*	*1885*	*1895*	*1907*	*1920*	*1930*	*1940*	*1952*	*1960*
2-city index	1.64	1.31	1.72	1.92	1.76	2.33	2.94	3.45	4.44	5.18
3-city index	–	1.08	1.36	1.40	1.28	1.67	2.11	2.40	2.82	3.18
4-city index	–	–	1.16	1.18	1.11	1.42	1.82	2.13	2.50	2.77

SOURCE: Calculated from the Censo de la Publación de Chile corresponding to each of the years shown.

Manufacturing employment was even more concentrated than population. The 2-city index for Chile in 1960 was 6.9 and the 4-city index 3.6. Greater Santiago had, accordingly, seven times the number of industrial workers than the country's second largest urbanized area and nearly four times the number in the next three largest agglomerations combined.

Yet another measure of primacy is in relation to the rank-size rule which describes a standardized distribution of a country's population among its cities.

In the case of Chile, the concentration of population in Greater Santiago was 52% larger in 1960 than one would have expected from an application of the rule. On the other hand, the sizes of the eight cities between 50,000 and 100,000 population were far below the anticipated values.[8] Their relatively small size is probably related to the almost exclusive dependence of these provincial capitals on agriculture or mining, activities which are not generally city-forming.

Despite its long history of rapid urbanization, Chile has managed to attain only a middling level of per capita income which, for 1965, was estimated at about $575.[9] Increases in per capita national income have been relatively sluggish and subject to violent short-term fluctuations. Since the mid-1800's, the secular growth rate of income has been calculated at 1.0 to 1.5% a year.[10]

Judging from comparative data, Chile would appear to have reached a level of urbanization that is far in excess of its level of per capita income. Indeed, few countries in the world exhibit levels of urbanization greater than 70%, and two-thirds of these have per capita incomes of $1,200, or twice the level for Chile (see Table 4). The question may, therefore, be raised whether hyperurbanization in Chile (and, by implication, in other low-income countries) stands in

TABLE 4

PERCENTAGE OF URBAN POPULATION, 1970, AND PER CAPITA GROSS DOMESTIC PRODUCT, 1965

	Percentage of Urban Population (1970)	*GDP/Capita (dollars)*
Chile	73.7	575
Uruguay	84.2	612
Argentina	70.5	670
Japan	83.2	870
Venezuela	76.4	971
East Germany	84.2	1,270
Israel	81.3	1,407
Netherlands	72.2	1,537
United Kingdom	79.1	1,790
Federal Republic of Germany	82.2	1,977
Australia	88.5	2,035
Denmark	80.1	2,109
Iceland	71.7	2,221
Canada	74.7	2,505
U.S.A.	75.2	3,536
Kuwait	80.4	4,786

SOURCES: Calculated from Kingsley Davis, op. cit., Table C and Everett E. Hagen and Oli Hawrylyshyn, op. cit., passim.

any *causal* relationship to the level of economic development achieved. In exploring this question, I shall borrow Manfred Halpern's analytical scheme for the study of modernization.[11]

Halpern distinguishes three broad classes of society: traditional, transitional, and modernizing. *Traditional* societies are not faced continuously by system-transforming demands, but when so confronted, "are capable of responding only by disintegrating or by creating a new closed system."[12] *Transitional* societies, on the other hand, are overtaken by structural changes and demands set loose by the uncontrolled forces of modernization but are either incapable or lack the will to deal effectively with these new forces. *Modernizing* societies, finally, have achieved this will and capacity on part of the political authorities and are managing continuously to integrate structural transformations into an ongoing system of interdependencies.

Chile, it will be shown, has successfully completed one major social revolution in this century: it has irrevocably passed from a traditional to a transitional society in Halpern's terms. The critical date here is 1920, when Arturo Alessandri ascended to the Presidency, though the revolutionary process has continued to have repercussions down to the present day. Chile must now accomplish its second revolution. It is still too early to say whether its efforts in this direction will be successful.

The attempt to carry out a revolution of modernization may be dated back to the election of President Eduardo Frei in 1964; it is still in progress under Salvador Allende's Popular Unity government elected in 1970. The major thesis of the analysis which follows is that hyperurbanization was, to a large extent, responsible not only for the relatively early transformation of Chile into a transitional society but also for the subsequent political and economic changes which led to a major sociopolitical crisis in the 1950's and brought the country to the limit of incipient modernization in the mid-1960's. The emphasis will be on the role of urbanization in the growing discrepancy between the political and economic development of Chile and on the resulting "crisis of inclusion" whose resolution is now being faced in a style altogether new in Chile's historical experience.

For lack of space, the study will necessarily be brief and synoptic. The conclusions will be stated in the form of a series of interrelated hypotheses which should encourage more detailed investigations not only of Chile's historical evolution during the last several decades but also, on a comparative basis, of the evolution of other countries in relation to their urbanization processes.

REASONS FOR HYPERURBANIZATION IN CHILE

The first question which needs to be answered is why hyperurbanization occurred in Chile; the second pertains to the consequences of this process for national development. The two questions are not unrelated but, for purposes of exposition, they will each be dealt with separately.

The principal facts of Chile's ultra-rapid urbanization are documented in Tables 5 and 6 which present a series of static comparisons of Chile's urban structure since 1865. The following conclusions may be drawn from these data:

TABLE 5

DISTRIBUTION OF CHILE'S URBAN CENTERS BY MAJOR SIZE GROUPS, 1865-1960

City Size Classes (in thousands)	*Number of Cities*									
	1865	*1875*	*1885*	*1895*	*1907*	*1920*	*1930*	*1940*	*1952*	*1960*
20-50	–	1	3	3	4	6	9	14	15	22
50-100	1	1	–	1	1	2	2	1	4	8
100-250	1	1	2	1	1	1	1	1	1	1
250-500	–	–	–	1	1	–	–	1	1	1
500-1,000	–	–	–	–	–	1	1	1	–	–
1,000 and over	–	–	–	–	–	–	–	–	1	1
Total	2	3	5	6	7	10	13	18	22	33

SOURCE: Calculated from Censo de la Población corresponding to each of the years shown.

(1) Since 1865, Chile's urban population has been doubling about every twenty years.

(2) Since 1895, the largest urban complex has consistently accounted for about one-half the total urban population, while cities of between 20,000 and 50,000 have contributed about one-fifth of the total.

(3) Since 1920, the number of cities with over 20,000 inhabitants has tripled. Practically all of these "new" cities had less than 100,000 population in 1960.

(4) Since 1940, population has increased more rapidly in cities with 20,000 to 100,000 inhabitants than in any other size group. Nevertheless, the smaller cities already accounted for 29.3% in 1907. This compares to 33.2% a half-century later.

(5) Since about 1940, cities in the intermediate size groups (100,000 to 500,000) have experienced a relative decline.

(6) As late as 1960, Chile had only three urban complexes with more than 100,000 population (Greater Santiago, Valparaíso-Viña del Mar, Concepción-Talcahuano).

(7) Despite increasing complexity, the relative structure of urban size distribution has remained remarkably stable throughout the entire period. Furthermore, the average size of cities has stayed nearly constant, rising by only 9% in over a half-century, or from 105,000 in 1907 to 114,000 in 1960.

We must look for the origins of this impressive flow of population to cities in the latter part of the nineteenth century. The economy's heavy dependence on the production and export of minerals, especially nitrates and copper, gave a strong initial push to urbanization during that period. The transfer of funds from mining operations via the public treasury resulted in heavy expenditures for public works, with particular emphasis on Santiago and the principal port cities of the country. At the same time, the urban upper class grown wealthy in mining, banking, and commerce adopted a sumptuous style of living based chiefly on imports from overseas. This style set the pattern not only for those families whose wealth was primarily in agriculture but also for those who came to prefer the commodiousness of living in the nation's capital; it was eventually

TABLE 6
DISTRIBUTION OF CHILE'S URBAN POPULATION BY MAJOR SIZE GROUPS, 1865-1960

	City Size Classes (in thousands)							
	20-50	*50-100*	*100-250*	*250-500*	*500-1,000*	*1,000 and Over*	*Total*[a]	*Total (thousands)*
Percentage of Total Urban Population								
1865	–	37.9	62.1	–	–	–	100.0	185.8
1875	8.3	39.7	52.0	–	–	–	100.0	249.9
1885	19.8	–	80.2	–	–	–	100.0	373.6
1895	18.5	9.3	24.7	47.5	–	–	100.0	539.7
1907	19.7	9.6	25.6	45.1	–	–	100.0	737.3
1920	17.2	13.4	20.8	–	48.5	–	100.0	1,045.1
1930	21.0	10.6	17.4	–	51.0	–	100.0	1,398.4
1940	23.5	2.8	6.6	15.1	52.0	–	100.0	1,829.5
1952	19.2	8.7	6.9	12.0	–	53.2	100.0	2,538.8
1960	19.5	13.7	6.2	9.8	–	50.8	100.0	3,752.4

SOURCE: Calculated from Censo de la Publación corresponding to each of the years shown.
a. Totals may not add to 100.00% because of rounding.

adopted by the rising urban middle class as well. The city offered an agreeable setting for graceful living. So long as mining flourished, imports continued unrestricted, and the city's poor made no effective claim for a just share in the amenities of urban living, there was little need for concern with the state of domestic production.[13]

But in the end, this claim was to be made. In the course of the present century, the accelerated movement to cities was facilitated by the completion of the North-South railway system in 1914 and by the improved national road network a decade or two later. Once the city became accessible, there was little to hold back the swelling mass of rural population, especially after 1940, when increasing mechanization in agriculture made large numbers of farm workers redundant. Nearly one-half the urban expansion between 1940 and 1952 was accounted for by migration alone.

Two conditions contributed strongly to the ease with which rural people abandoned the countryside. First, Chile had never known a peasantry rooted to land by sentiment and cultural tradition. Chile's feudalistic society depended principally on landless labor for production. As late as 1960, 71% of all rural families were either landless workers or small producers (usually operating under some share-cropping arrangement) having an average annual family income of only 636 escudos (approximately $53 a month).[14] Many of these families, of course, maintained personal ties to the master of a large estate, but the ties were those of dependence and servitude and had been made secure chiefly by the absence of more attractive alternatives. Now the city, which had become accessible, offered just that. To be sure, the risks of casting oneself upon the mercies of the city might be great, but the promise of a genuine improvement in life chances weighed heavily in the balance.

The second contributing factor is to be found in the exceedingly low living conditions of the rural workers. In all the relevant indices–income, diet, housing, schooling, health–the rural poor rank at the bottom of the social scale.[15] And their situation is getting worse. For over a quarter of a century, at least, real wages in the rural areas of Chile have been declining both in absolute and relative terms. In 1964, for instance, the official minimum wage for rural workers had only 76% of the purchasing power eleven years earlier.[16] This systematic impoverishment of the rural masses was due, first, to a total lack of concern with the problem on the part of the landholding oligarchy and later, to the unwillingness of the urban middle sectors to challenge the interests of the oligarchy in any fundamental way.

The rural proletariat, then, were a forgotten people. It is scarcely surprising that many of them decided to try their luck in the cities; for even though the possibilities of finding a decent job there would be small, there were none at all in the physically lovely, but socially brutalizing environment of the countryside. In the city, the government might be induced to provide housing and schools; sooner or later, some work would appear; one's daughter might be taken in by a rich family as a servant; the boys might be fortunate and find a secure post as messenger or clerk in one of the ministries with all the guaranteed privileges attending this position. In short, in the lottery of life, the odds for a modest bit of success were clearly greater in the city. This view is supported by the fact that return migration to rural areas has apparently been negligible; the vast majority of the migrants to the city stay there.[17] And once established, they are in a position to help their relatives and friends to follow: migration is a cumulative process.[18]

Concluding this discussion of the reasons for hyperurbanization, mention should be made of Chile's expanding pioneering areas in the extreme south of the country. The province of Magallanes, for instance, may be classified as a resource frontier region whose development, as in other parts of the world, is chiefly based on cities.[19] During the past forty years, Chile's frontier regions have provided relatively little outlet for rural populations.

CONSEQUENCES OF HYPERURBANIZATION

We are now ready to approach the second question. What were the results of Chile's urbanization process for national development? The survey will have to be a hurried one.

1. The middle and organized working classes gained early access to political power. Until 1920, Chile was safely governed by an elite of large landowners and allied banking interests. (Mining was largely in the hands of foreign companies.) Although everything seemed relatively calm on the surface, the traditional structure was, in fact, gradually being weakened in its foundations. Under the impact of rapid urbanization, a new class of professionals, bureaucrats, businessmen, and clerks was forming and had begun to push toward the top, demanding influence and power. In 1920, taking advantage of the economic crisis brought on by World War I, the new middle sectors moved to gain control of the government. The election of Arturo Alessandri Palma to the Presidency, amidst glowing promises of a new era,

> "had all the aspects of a social revolution, albeit a bloodless one. The bitter middle class made common cause with laborers in mine and factory to unseat the aristocracy. When the counting of the ballots failed to give a majority to Alessandri, the Conservative senate prudently decided in favor of this man, who was the clear popular choice."[20]

Chile had at last begun to move into a transitional society.

The new middle sectors were still numerically weak, however. In 1920, they probably accounted for little more than 10% of the total population. Thirty years later, they had doubled their relative strength,[21] and by 1960, they had risen to one-quarter of the total population. Somewhat less than half their number was concentrated in Santiago.[22]

These figures are significant in two respects. First, they suggest that although the middle sectors were small relative to population, they were vastly superior in number to the landed aristocracy whom they replaced in the political management of the nation. Chile's traditional upper class has been estimated at less than 1% of population.[23] On the other hand, the middle sectors had an insufficient base for political maneuver. If they wished to prevail, middle-class politicians would have to follow a shifting pattern of alliances. And one of their early allies was the organized working class.

The unionization of Chilean labor had proceeded slowly. As late as 1965, only 12% of the approximately three million men and women in the "active" population were members of trade unions.[24] The political representation of the working class, however, was far more powerful than these figures would lead one to believe. Already by 1938—scarcely eighteen years after the first election of President Alessandri—the parties of the Left had grown sufficiently strong to join with the Radicals in forming a Popular Front government. For a variety of reasons, this coalition did not last very long. Nevertheless, in the Presidential election of 1964, the combined Marxist parties (FRAP) obtained 39% of the national vote. The urban component of this vote made up 70% of their total strength.[25] Rapid urbanization had speeded the pace of effective working-class participation in the political life of the nation.[26] In 1970, a coalition of Marxist parties finally achieved control of the government.

The arrival of new social strata on the political scene led to a series of important national reforms. Chief among these was the new Constitution of 1925 which established a strong presidential system in Chile and accomplished the effective separation of church and state. Other reforms followed. In the main, they secured economic privileges and rights to those urban sectors which were capable of expressing and defending their interests with vigor. The resulting social gains were impressive; but from 60 to 80% of the country's population did not share in them to a significant extent.

For nearly half a century, beginning with the first period of Alessandri's administration, the politics of the middle and working classes served primarily as an instrument to advance the partial interests of their component groups. This tendency to elevate class interests over the interests of the nation as a whole, and the failure to evolve a persuasive class-transcending *national* ideology and a corresponding political system, made it exceedingly difficult to steer a steady course toward modernization. The fundamental struggle was for a bigger share of the total wealth, for individual security, and for social status.

The pattern of political action established with Alessandri's election set the model for decades to come. Reduced to elemental terms, it was this. Party coalitions would help to elect a President with personal charisma and strong popular appeal; once he had assumed office, however, the parties would withdraw their support, the coalitions would be unmade, and the President would be left to promote the national interest as best he could, guided by little more than intuition and political wisdom. Although the parties might on occasion vote in favor of popular legislation, they were careful not to impair the interests of the groups they represented. The Radicals, for instance, might align themselves with the Communists to vote for a high schedule of public works, but they would staunchly vote with the Conservatives and Liberals in opposing higher taxes.

It was in this way that Chile evolved towards a "pluralistic bartering system" in which, according to Manfred Halpern, "the dominant decision-making groups . . . share a broad agreement regarding system maintenance and modification, and a general opposition to system transformation."[27] However, Halpern continues,

> "The consensus built on the aggregation of existing dominant interests under conditions of great scarcity and rising frustrations is likely to grant inadequate place to newly emerging social classes, to the uprooted. Interest bargaining among those already possessed of power and influence tends to concentrate on the distribution of scarce resources and the reinforcement of existing relationships, rather than on the sacrifice of vested interests and accustomed beliefs, values, and behavior which is the unavoidable price of productive transformation at this stage of development."[28]

In Chile, this situation continued until the election of 1964.

2. Economic development was retarded. Hyperurbanization acted directly on economic growth, by lowering the propensity to save, by discouraging agricultural production, and by diverting scarce resources from investment in highly productive activities to those with low ratios of output to capital invested. The net effect was a retardation of the rate of economic growth.

During the past six decades, Chile's rate of increase in gross national product per capita has remained extremely low. From 1908 to 1927, it progressed at an annual rate of 1.45%; from 1927 to 1940, at 0.94%; and from 1940 to 1957, at 0.96%.[29] During the subsequent seven years, the rate declined to 0.8%, and in two of these years, production per capita actually declined.[30]

It would be rash to place the entire blame for this discouraging performance on hyperurbanization alone. Nevertheless, the peculiar characteristics of urbanization in Chile must share in the total responsibility.

Much of the appeal of middle- and working-class politics was in its aim to obtain immediate improvements in the levels of consumption and, even more, in securing social prestige for their varied clienteles. The Chilean middle sectors, it will be recalled, had their origin neither in a pre-industrial artisan tradition nor in manufacturing, but in the burgeoning white-collar occupations of a hyperurbanized society. And the chief interest of these new groups was in conquering the symbols of social status which the landed aristocracy held up to them; conspicuous consumption was among the most attainable of these.[31] Contrary to European experience, the early emergence of the middle sectors in politics, and of the organized urban workers a little later, did not lead to a growth in savings and investment. Nothing could have been farther from their mind than to submit to a postponement of immediate gratifications in order to make increased production possible. And when, in 1939, an ambitious industrialization program was finally launched by the slim margin of one vote, the strategy to promote national industry as a substitute for imported consumer goods was adopted almost as a matter of course. The national production of machinery and equipment, though a feasible alternative, was not made the main thrust.[32]

The extraordinarily low propensity to save was apparently justified by an inflation that reached an annual average rate of 18% in the 1940's and rose to a spectacular 36% in the succeeding decade.[33] In point of fact, the actual rate of capital formation during this period fluctuated between 8 and 12% which is little more than half of this investment which went for the replacement of worn-out plant and equipment.

The discouragement of production in agriculture was a second economic effect of hyperurbanization in Chile. Stagnation set in about 1930 and has continued ever since. The long-term trend in agriculture was less than population growth during this period and led to a steep increase in the volume of agricultural imports.[35] Between 1949 and 1963 alone, such imports grew by three times over a base of $64 million.[36] Though some investment occurred, the rate of improvement in the efficiency of the agricultural sector after 1930 was only one-fifth to one-tenth the equivalent rate in the United States.[37]

The reasons for this deficient performance were various. The middle sectors, needing to rely from time to time on Conservative support, were disinclined to make a frontal assault on the structural problems affecting Chilean agriculture by advancing a forthright reform of the agrarian establishment. At the same time, they reduced the profitability of agricultural and especially of livestock production through price and subsidy policies which systematically favored the urban consumer.[38] The lower-than-average returns to capital in agriculture, in turn, encouraged a flight of capital from this sector to more profitable ventures.[39]

Finally, hyperurbanization deflected large amounts of capital from investments of relatively high productivity to the expansion of social infrastructure in urban facilities. The massive inflow of migrants to the city obliged the government to make large-scale investments having extremely low ratios of output to capital. In addition, the location of the government establishement in Santiago made the monumental beautification of the urban landscape a not unattractive alternative for the use of capital. Assuming that urban social investments account for 30% of all investments and that Chile's proportion of urban population has been about double the "normal" level in relation to per capita income, approximately 15% of the available capital resources might have been applied more productively. The cumulative effect of these savings on the national growth rate would have been significant.

3. The politics of "readjustment" helped to maintain hyperinflation. Inflation has been endemic to Chile for nearly a century. But hyperinflation with annual rates of up to 84% set in only in 1940, shortly after the Popular Front government had raised all wages by a flat 20% and had intensified Chile's industrialization program under the stimulus of World War II.

Few subjects in Chile's economic history have been more hotly disputed than the high rate of persisting inflation.[40] There are nearly as many theories attempting to explain this puzzling phenomenon as there are authors. Nevertheless, it is significant that hyperinflation started when Chile was already suffering from the tensions produced by a serious gap between the levels of urbanization and per capita income. Hyperurbanization provided the setting for hyperinflation, and many causal links relate the two phenomena.

At one level of analysis—though by no means the only one that is relevant—inflation is induced by a critical imbalance between supply and demand. As shown in the preceding pages, the conditions affecting both these variables had an intimate connection with the process of hyperurbanization. On the *supply* side, mention must be made of the low index of growth of total production, of the steady decline in agriculture relative to population since about 1930, and the planned industrialization program after 1939, which emphasizing consumer over production goods and aiming chiefly at substituting for overseas imports, resulted in steep price increases due to protectionist policies, production inefficiencies, and the monopoly position of many of the new firms.

This low overall output of the economy was balanced, on the *demand* side, by a relatively high and persisting demand for consumer goods. As I have tried to suggest, the high propensity to consume which had been the traditional pattern for a small landed and financial elite was aggravated by hyperurbanization as new classes gained in importance and began to imitate the behavior of the wealthy. The resulting disinclination to save was also reflected in the reluctance of the governing groups to raise the level of taxation and, in particular, to change its structure from a regressive to a progressive incidence of the tax burden. Unwillingness to raise the proportion of taxes in total revenues, together with changing fortunes in the world market for Chile's principal export item, copper, made it necessary to finance a good deal of the government's investment program through successive expansions of the money supply in excess of production. And, as we have seen, an important element of national investment was public works to supply the needs of an urban population growing at a steady 4% a year.

Periodic attempts to halt the onrush of inflation were made, but they were singularly unsuccessful except for short-lived victories in a few isolated years. One of the principal engines keeping inflation going and, at times, accelerating its headlong pace, was the emergence of a *politics of readjustment.* The inability of the government to bring inflation to a halt was to a large extent due to the unwillingness of the organized sectors of the population to forego the opportunity for periodically attempting, through various political stratagems, to improve their income position *relative to that of other groups.*[41]

The mechanism worked somewhat as follows. Every year, each organized group would make claims for a wage or salary readjustment more or less commensurate with the increase in the consumer price index during the preceding twelve months. Some groups were obviously more successful in this game than others. In 1941, a law was passed requiring annual revisions of the minimum wage. In 1952, public functionaries received the right to an automatic readjustment of their salaries, though this right was later revoked. By 1965, employees of the private sector were similarly covered by legislation. Strikes and artificial shortages of consumer goods were the main weapons in the struggle to obtain differential treatment in the annual round of readjustments. Now this, now the other group was ahead, and a victory secured this year might become the basis for next year's negotiations. A general forward movement in prices was thus maintained, interrupted every so often by a brief spell of exhaustion and recovery. The game was being played for high stakes, and no group was willing to beg out so long as it saw any chance for getting ahead. And without general public support, no Chilean government was strong enough to push through a program of restricting consumption and increasing investment. Hyperinflation in Chile had become a cherished national institution.

But not on everybody's part. For one of the main results of the politics of readjustment has been an income transfer from the poor to the more fortunate, from the marginal groups of society of those better able to defend their interests in the political arena. The regressive effect of inflation on the agricultural labor force has already been mentioned. A similar effect, though difficult to document, must have impinged upon the urban proletariat as well.

4. A large marginal sector of the population appeared. What somewhat antiseptically is called the marginal sector refers, in fact, to the disinherited, the poor, who live alone in their misery, defenseless, and are squeezed from above.

They do not share in the society that runs the country with an eye primarily to their own profit: economically, they lack the means to live beyond a level of bare subsistence, politically they are excluded from most of the decisions which bear on their welfare, culturally a vast gap separates them from the "children of light." In a phrase which is often heard in Chile, the poor are the objects of history and not its subjects.[42]

There have always been the poor, of course. But recognition of the phenomenon of marginality is something new. It was triggered by the massive importation of poverty from the countryside to the city via migration. For in the city, not only did the poor become *visible* to the ruling sectors, they also began to claim their right to participate in society as full-fledged members. This historical claim was eventually supported by a minority among the middle sectors who nourished a profoundly democratic and generous ideology. And once marginality was perceived in the cities, it was not long before it was recognized to exist in the rural areas as well. Paradoxically, the very movement of rural people to the city may be interpreted as an effort on the part of a pre-marginal population to acquire marginal status and, from this precarious base outside the prevailing social system, to gain a foothold within it. However, as a matter of historical record, their upward mobility has been exceedingly slow.

One important reason for this has been the inability of the urban economy to absorb all of the newly recruited labor force in productive employment. The situation is well known throughout Latin America; it has been especially severe in a country such as Chile, where the gap between urbanization and economic development has been steadily increasing for decades. According to a study of the United Nations Economic Commission for Latin America,

> "Since employment opportunities in high productivity sectors were scarce, the labor force was obliged to shift to ill-paid, unproductive employment. . . . This process varied significantly during the fifties between one Latin American country and another. Thus, the most dynamic sectors absorbed only 18 percent of the manpower increment in Argentina, Chile, and Uruguay, 33 and 31 percent in Brazil and Mexico repsectively, and 29 percent in Venezuela."[43]

Many of the new arrivals in the city, therefore, had no other choice but to join the ranks of the unemployed or the various forms of shifting marginal employment.

The periodic surveys of employment in Greater Santiago conducted by the Institute of Economics of the University of Chile permit some insight into this situation. The results for June 1966 are summarized in Table 7.[44] Nearly one-fifth of the potential labor force at that time was either wholly or partially unemployed.

Marginality, however, is a more complex social phenomenon than these figures suggest. There are degrees of marginality and multiple aspects which cannot easily be reduced to a common denominator without distortion. Gino

TABLE 7

FUNCTIONAL UNEMPLOYMENT IN GREATER SANTIAGO, JUNE 1966

	n	%
1. Unemployed	51,500	5.4
2. Partially employed (less than 20 hours a week)	34,400	3.6
3. Inactive population desiring work (more than 20 hours a week)	88,300	9.3
4. Total functional unemployment (1 + 2 + 3)	174,200	18.3
5. Active population or labor force	862,400	–
6. Potential labor force (3 + 5)	950,700	100.0

SOURCE: See note 44.

Germani calculates that 79% of Chile's population in 1950 had low occupational status.[45] This figure is essentially the same as the proportion of farmers, fishermen, miners, drivers, artisans, operatives, manual, personal service, and "other" workers in the 1960 labor force.[46] But not all the population in these categories need be considered marginal in the strict sense. In his excellent study on marginality, Armand Mattelard establishes five classes of social integration (or disintegration). Applying this scheme to Greater Santiago and using data from the 1960 Census, he finds 60% of the population, aggregated into fifteen municipalities, experiencing some degree of marginality, while 24% lived in the six municipalities with the lowest indexes.[47]

Whichever figures are used, it is clear that marginality is extensively present in Chile and in some measure probably affects a majority of the population. We also know that it was the marginal population that experienced a sustained decline in real income as the result of hyperinflation and the regressive politics of readjustment.

If we now combine the three fundamental conditions of a hyperurbanized society—a slow-growing economy, increasing misery, and rising expectations within the context of a democratic ideology that affirms the right of the poor to an effective stake in the society—we must conclude that sooner or later a major socio-political crisis will ensue. This crisis, which I shall call the *crisis of inclusion,* confronts the society with the radical challenge to modernize itself. In Chile, the crisis set in during the late 1940's; it has reached a decisive stage now.

THE CRISIS OF INCLUSION

To put it very simply, the crisis was produced by an insistent demand on the part of the urban poor to be seated at the table of the rich. But their demand could not be met within the existing arrangements. There was insufficient food and wine to be passed around if the size of the feast were to be doubled. Slowly people came to realize that the politics of readjustment had reached its inevitable limits in a stagnant economy. The lesson was perfectly clear: production would have to be stepped up, but this would mean, at the same time, putting an end to hyperinflation which was both a cause and effect of readjustment politics and tended to reinforce the traditional social preference for consumption over investment.

But learning a lesson and acting in accord with it calls for an emotional commitment. And the will to change was not yet strong enough. It was clearly impossible to demand a major sacrifice from each of the competing groups unless the sacrifices were properly adjusted to each other within a comprehensive frame of social justice and an inclusive ideology of nationalism and development. Since the late 1940's, one government after another had tried to control inflation and failed.[48]

Hyperurbanization had enhanced class consciousness to the point of producing increasingly embittered classes among peasants, workers, the several sectors of the middle class, and the upper socioeconomic groups. The number of street demonstrations and strikes increased; intellectual attitudes shifted to the left; and Marxist parties made striking gains in popular support.[49]

The crisis of inclusion not only created general unrest and spread a climate of uncertainty throughout the country, it also led to an accelerated disintegration of the traditional authority structure which up to then had been an important integrative force in the national culture.

During the preceding decades, hyperurbanization had prepared the ground by undermining the values of what K. H. Silvert has called the "classical Mediterranean tradition," above all the widely shared belief that "loyalties must be vertically ordered in accord with authority, and that religious and primary loyalties supersede secondary ones in the determination of social actions."[50] As a consequence, many middle- and upper-class families had come to modify

> "both action and ideology to accommodate themselves to a certain kind of industrialization.... These 'modern' traditionalists think in syndicalist or corporativist terms, essentially but a complication of the hierarchical order of mediaeval society.... The good society pictured by the syndicalist would have the individual firmly rooted in his newly complicated institutional place. His representation in government would not be a result of his individuality or of his mere nationality, but rather a function of his place in the institutional order of events. Public decisions would then result from the interplay of the institutional oligarchs, and not from the deliberations of groups and men elected at large from a citizenry escaping its occupational and class bonds in an act of political selection and decision formally and somewhat substantially indicative of equality.... The major social purpose of the syndicalist approach is to find a way of subsuming the new class complications of modernism to hierarchy, preserving a kind of Latin *Führerprinzip,* leaving inviolate the privileges and powers of the traditional, and thus escaping the secularization and, to their eyes, immorality of the nation-State."[51]

Not unsurprising in view of these beliefs, the "decent" people of Chile shrank in fear from the consequences of threatening disintegration; they were certain that the breakup of the consecrated principle of authority would only plunge the country into chaos. The alternative interpretation that this might represent the beginning of a radical social transformation leading to individual self-determination, fundamental democratization, and a new sense of dignity was not widely shared.

Nowhere is this brought out more clearly than in an editorial, written in 1966, of Santiago's leading newspaper, *El Mercurio,* a daily of generally moderate views. The editorial adopted an unusually severe tone on this occasion, perhaps because basic values were being threatened.[52]

> "Every collectivity, from the family to the individual firm and on the university and the country, must divide itself into a majority that obeys and a minority that orders and disposes. . . . From the moment that an individual lives in society with others, he limits his freedom and takes it for granted that everyone's actions must be subordinated to the common good. This also means that he renounces part of his will and transfers it to those who retain the power to make decisions for him and to oblige him to carry them out. . . . Democratic liberty to deliberate, and to elect and judge the actions of the authorities is one thing; quite another thing is the disorder, the lack of discipline, and the anarchy that make of everyone his own authority and, by the same token, eliminate all other forms. . . . In a country such as Chile . . . the sense of discipline and respect for authority are irreplaceable ingredients of national preservation and progress. . . . The principle of authority is an expression of the structure of society in which those for whom it is proper to do so assume the responsibility to command and thus to achieve the general good. Subversion and lack of discipline frustrate these achievements and, under the pretext of granting unrestricted rights, bring about an environment of anarchy in which even the most elemental rights will shipwreck."

This editorial was printed on the occasion of the first student strike in the history of the Catholic University of Chile.

The reaction in defense of established authority was a desperate one; but it pointed in the wrong direction. What was needed by the times was a reintegration of society around a set of values that would be more in keeping with the processes of *continuous transformation* that are typical of modern national societies. And among these values would necessarily be those of nationalism. Their significance is described by Silvert and Bonilla:[53]

> "Common to all already developed countries is the implicit assumption that the state is the ultimate institutional repository of worldly loyalty and thus also the last resort for the settlement of human dispute. This premise of conduct is further based upon the acceptance of the value that certain areas of human endeavor can be rationally assessed best in terms of total social interest than of only class interest. Nationalism . . . is the key value of social cohesion related to industrial development as it has historically occurred."

And a little later they add:[54]

> "The presence or absence of this value may be determined operationally in the testing of attitudes toward social conflict, in the ability of individuals to project themselves into situations of others at a relatively far remove socially, and in the acceptance of an area susceptible to empirical and rational human decision through the agency of a political mechanism viewed as at least partially neutral and objective. This mutual loyalty also must evidence itself, to be truly effective, in the recognition of minimum rights considered as common to all in the extended society."

In the situation of mounting inter-group conflict that marked the 1950's and early 1960's, the politics of Chile veered sharply to the left. Communists and Socialists joined in a loose coalition (FRAP) and pooled 30% of the national vote for President in 1958 and an astounding 39% in 1964; with this record, it emerged as the second strongest political alignment in the country. In the same elections, however, the Christian Democrats (PDC) jumped from third place in 1958 with 21% of the vote to first place with a clear majority of 56%.[55]

The feat was repeated in the Congressional elections the following March, when the PDC obtained an overwhelming majority of 82 seats in the Chamber of Deputies. According to Orville Cope, this was the first time in over 120 years that a single political party had received such a clear mandate.[56] And he continues:

> "The emergence of the Christian Democratic Party, replete with mass organization, is one of the most significant steps toward political modernization in Chile. . . . Senator Frei relied upon a well-organized political party that had descended the socio-economic scale of Chilean society and had sought out the vote in all geographic areas."[57]

The PDC was an urban-based party. But despite the fact that its leadership derived mainly from the middle-class intelligentsia, it represented a wide cross-section of major social groups in all parts of the country. National in its composition, it was committed to bringing about a series of basic structural reforms that were meant to solve the crisis of inclusion.

In spirit and intention, Chile's *Revolución en Libertad* was reminiscent of Franklin D. Roosevelt's New Deal. In its economic aspects, the government's program stressed the necessity to shift from consumption to production and declared its determination to bring hyperinflation under control. Politically, priority was given to the so-called Chilenization of the copper industry, agrarian reform, income redistribution, regional planning, and housing and education policies that would favor the urban poor. Doctrinally, the government advanced the ideas of *promoción popular* and *el pueblo organizado* with the intention of encouraging a wider and more effective participation of the mass of the people in national affairs.[58]

The following six years were a period of great accomplishments.[59] But the transition to a modernizing society in Halpern's sense was not accomplished. Popular demands outran the capacity of the government to fulfill them. Political polarization occurred as the conservative forces in the country regrouped; the

Christian Democratic Party divided into left-wing, center, and right-wing factions; and the parties of the Left (primarily Communist and Socialist) united, as they had done before, behind their perennial candidate, Dr. Salvador Allende. Allende carried the Presidential election of September 1970 by a slim margin and subsequently initiated a period in which econmic and social policies were oriented along socialist principles, including the large-scale nationalization of manufacturing and mining, and a vigorous policy of income redistribution. It is still too early to tell whether Chile is at last on the verge of achieving a breakthrough to a syncretic modernizing system marked by a "willingness of the traditional elite to accept, at any time, as much modernization as seems required to maintain its place in the evolving elite."[60] But whatever the ultimate outcome, hyperurbanization may in large measure be credited for bringing Chile to the threshold of modernity within a democratic framework. The developmental crises it produced have been overcome successfully so far with acts of political statesmanship uncommon among Latin American nations. The present Popular Unity government is yet another phase in Chile's evolving history of national development.

SUMMARY OF SOME HYPOTHESES

The basic thesis of this chapter may be summed up in a series of related propositions. Suggested by Chile's historical experience, there is no reason to suppose that they have general validity. Nevertheless, they may be useful in directing students of comparative urbanization to investigate certain relationships between hyperurbanization and national development that might otherwise remain invisible and so contribute to a theory of urbanization.

(1) Hyperurbanization in Chile has tended to accelerate the process of political development in the sense of an early replacement of the traditional rural landowning elites by urban-based middle- and working-class groups. It has thus assisted in shifting society from a traditional to a transitional system in terms of the systems's ability to cope with continuing forces of change.

(2) In the early period of transition, hyperurbanization in Chile tended to encourage a pluralist interest-bartering system characterized by popular social reforms and by opportunistic political alliances which, although subordinating national to group interests, nevertheless attempted to preserve most of the existing social and economic arrangements.

(3) Hyperurbanization in Chile has tended to retard the rate of economic growth by encouraging high consumption, discoraging agricultural production, and diverting capital investments from economically productive to less productive social sectors, especially in urban public works.

(4) Hyperurbanization in Chile has tended to reinforce inflationary pressures in the economy by encouraging a politics of readjustment that favored the better organized sectors of the population.

(5) Hyperurbanization in Chile eventually led to the formation of a large,

unorganized urban proletariat that, until recently, was effectively excluded from participation in affairs affecting its livelihood and future—economically, politically, and culturally—and that was becoming increasingly conscious of its "marginal" role in the society.

(6) Acting through the processes of a politics of readjustment, hyperurbanization in Chile has resulted in regressive income transfers from the marginal sectors of the population to the participant and ruling groups of the society.

(7) Eventually, hyperurbanization produced a "crisis of inclusion" characterized by general unrest, political agitation, and a disintegration of traditional elitist values. This crisis was the result, specifically, of growing class consciousness, the urban demonstration effect on the poor, a spreading ideology of participation, and the growing impotence of the traditional politics of readjustment under conditions of prolonged economic stagnation.

(8) The crisis of inclusion may be faced in a number of ways. One would be to push for a breakthrough into modernization via a left-of-center national reform party operating within the framework of a democratic tradition (a "syncretic modernizing system" in Halpern's terminology). Alternative options include: (a) political chaos (a system of "structured instability"); (b) a totalitarian dictatorship of the remaining traditionalist elements ("extremist system"); or (c) a system of "radical authoritarian modernization."[61] In recent Chilean history, the "syncretic" alternative was attempted under President Frei but did not succeed. The present government of Dr. Allende seems to be evolving towards alternative (c), although the future may still bring any of the remaining alternatives into play.

(9) If the crisis of inclusion is successfully overcome and economic breakdown avoided, the tensions produced by hyperurbanization in the society will eventually be reduced, as the limits of relative population concentration in cities are reached (absolute maximum of 100%), and further income gains will bring urban-economic relations back into the range corresponding to industrial and post-industrial societies.

NOTES

1. This is equivalent to an annual growth rate of 3.5%. For a suggested method of calculation, see Richard W. Chadwick and Karl W. Deutsch, "Doubling Time and Half Life: Two Suggested Conventions for Describing Rates of Change in Social Science Data," American Behavioral Scientist, Vol. II, No. 4 (March-April 1968), pp. NS-9 to NS-11. The twenty-year doubling rate is established here as an arbitrary threshold. It could well be that doubling rates of between 20 and 30 years would confront the country with somewhat similar problems as those discussed for Chile in this chapter.

2. Adna Ferrin Weber, *The Growth of Cities in the Nineteenth Century: A Study in Statistics.* Ithaca, New York: Cornell University Press, 1963. (Originally published in 1899).

3. Calculated from Kingsley Davis, *World Urbanization 1950-1970.* Vol. I: Basic Data for Cities, Countries, and Regions. Berkeley, California: University of California, Institute of International Studies, 1969, Table D.

4. This definition of "modernization" is taken from Manfred Halpern, "The Revolution of Modernization in National and International Society," in Carl J. Friedrich, ed., *Revolution.* Nomos, Vol. VIII. *Yearbook of the American Society for Political and Legal Philosophy.* New York: Atherton Press, 1966, pp. 178-216.

5. N. V. Sovani, "The Analysis of 'Over-Urbanization,' " Economic Development and Cultural Change, Vol. XII, No. 2 (January 1964), pp. 113-122.

6. For a discussion of these and other measures at controlling internal migrations, see Chapter 12.

7. Jack P. Gibbs and Hanley L. Browning, "Systems of Cities," in Jack P. Gibbs, ed., *Urban Research Methods.* Princeton: Van Nostrand, 1961, pp. 436-461.

8. For a description of the method of calculating expected size distributions on the basis of the rank-size rule, see Gibbs and Browning, op. cit.

9. Everett E. Hagen and Oli Hawrylyshyn, "Analysis of World Income and Growth, 1955-1965," Economic Development and Cultural Change, Vol. 18, No. 1, Part II (October 1969), Table 5B.

10. Marto A. Ballesteros and Tom E. Davis, "El Crecimiento de la Producción y el Empleo en Sectores Básicos de la Economía Chilena," Cuadernos de Economía, Vol. II, No. 7 (September-December, 1965), pp. 5-29.

11. Manfred Halpern, op. cit.

12. Ibid., p. 180.

13. Anibal Pinto Santa Cruz, *Chile: Un Caso de Desarrollo Frustrado.* Santiago de Chile: Editorial Universitaria, 1958, pp. 79 ff.

14. Comité Interamericano de Desarrollo Agrícola, *Chile: Tenencia de la Tierra y Desarrollo Socio-Económico del Sector Agrícola.* Santiago: n.p., 1966, Table XII-1. In subsequent footnotes this publication will be referred to as the CIDA Report.

15. Ibid., Chapter XIII.

16. Marcos Mamalakis, "Public Policy and Sectoral Development: A Case Study of Chile 1940-1958," in Marcos Mamalakis and Clark Winton Reynolds, *Essays on the Chilean Economy.* Homewood, Illinois: Richard D. Irwin, Inc., 1965, Part I, Table 31; also CIDA Report, op. cit., p. 193.

17. Bruce Herrick, *Urban Migration and Economic Development in Chile.* Cambridge, Mass.: The M.I.T. Press, 1965, pp. 54-55.

18. Ibid., p. 91.

19. John Friedmann, *Regional Development Policy: A Case Study of Venezuela.* Cambridge,: Mass. The M.I.T. Press, 1966, pp. 76-87.

20. Hubert Herring, *A History of Latin America.* New York: Alfred A. Knopf, 1955, p. 559. For a political history of the period, see also John J. Johnson, *Political Change in Latin America: The Emergence of the Middle Sectors.* Stanford: Stanford University Press, 1958, Chapter 5.

21. Gino Germani, "The Strategy of Fostering Social Mobility," in Egbert de Vries and José Medina Echevarría, eds., *Social Aspects of Economic Development in Latin America* Vol. I. UNESCO, 1963, p. 224.

22. CIDA Report, op. cit., p. 30.

23. Germani, loc. cit.

24. Osvaldo Sunkel, "Change and Frustration in Chile," in Claudio Veliz, ed., *Obstacles to Change in Latin America.* London: Oxford University Press, 1965, p. 127.

25. Orville G. Cope, "The 1964 Presidential Election in Chile: the Politics of Change and Access," Inter-American Economic Affairs, XIX, No. 4 (Spring 1966), 3-30.

26. James Petras, *The Politics of Conflict and Stagnation: Emergence of Working Class Politics in Chile.* Master's thesis, Department of Political Science, Berkeley: University of California, December, 1963. See also Ernst Halperin *Nationalism and Communism in Chile.* Cambridge, Mass.: The M.I.T. Press, 1965.

27. Halpern, op. cit., p. 189.

28. Ibid., p. 190.

29. Ballesteros and Davis, op. cit., p. 24.

30. Banco Central de Chile, *Boletín Central No. 462* (August, 1966), p. 1201.

31. Sunkel, op. cit., pp. 130-31.

32. Mamalakis, op. cit., Chapter 4

33. Albert O. Hirschman, *Journeys Toward Progress: Studies of Economic Decision-Making in Latin America.* New York: The Twentieth Century Fund, 1963, p. 160.

34. Mamalakis, op. cit., pp. 69-72.

35. Marto Ballesteros, "Desarrollo Agrícola Chileno, 1910-1955," Cuadernos de Economia, II, No. 5 (January-April, 1965), 13-15.

36. CIDA Report, op. cit., p. 23.

37. Ballesteros, op. cit. p. 24.

38. Mamalakis, op. cit., Chapter 3.

39. Ibid., p. 140.

40. Major writings in English include: Hirschman, op. cit., Chapter 3; Joseph Grunwald, "The 'Structuralist' School on Price Stability and Development: The Chilean Case," in Albert O. Hirschman, ed., *Latin American Issues: Essays and Comments.* New York: The Twentieth Century Fund, 1961, pp. 95-124; Mamalakis, op. cit.; and, most recently, Tom E. Davis, "Inflation and Growth in Latin America: Theory, Performance, and Policy," Economic Development and Cultural Change, XIV, No. 4 (July, 1966), 506-11.

41. Hirschman, *Journeys Towards Progress,* op. cit., pp. 185-99.

42. The best single study on the phenomenon of marginality in Chile is by Armand Mattelard and Manuel A. Garretón, *Integración Nacional y Marginalidad: Ensayo de Regionalización Social de Chile.* Santiago de Chile: Editorial del Pacífico, 1965.

43. "Structural Changes in Employment Within the Context of Latin America's Economic Development," Economic Bulletin for Latin America X, No. 2 (October, 1965), 165-68.

44. Banco Central de Chile, *Boletín Mensual,* No. 461 (July, 1966), pp. 1021-1024.

45. Germani, op. cit. Table 2.

46. Herrick, op. cit., p. 62.

47. Mattelard and Garretón, op. cit., p. 158.

48. For a history of these efforts, see Albert O. Hirschman, *Journeys Towards Progress,* op. cit., Chapter 3.

49. The average number of strikes per year in Chile during 1947-1950 was 121, affecting 45,000 workers; in the course of the following decade, the annual average increased to 201 strikes, involving 90,500 workers. Since many of these strikes affected key industries and services, each one of which was capable of paralyzing major sectors of the country's economy, the overall effect of this trend in the frequency of strikes was exceedingly disruptive. See Petras, op. cit., pp. 126 and 168.

50. K. H. Silvert, "National Values, Development and Leaders and Followers," International Social Science Journal, XV, No. 4 (1963), 562.

51. Ibid., pp. 562-63.

52. "El Principio de Autoridad," editorial, El Mercurio (Santiago de Chile), May 18, 1966, p. 3.

53. K. H. Silvert and Frank Bonilla, *Education and the Social Meaning of Development: A Preliminary Statement.* New York: American University Field Staff, 1961, p. 15.

54. Ibid., p. 17.

55. Petras, op. cit., p. 139.

56. Cope, op. cit., p. 5.

57. Ibid., p. 10.

58. For the official statement of the government's program, see *Segundo Mensaje del Presidente de la República de Chile.* May 21, 1966. Santiago: Departamento de Publicaciones de la Presidencia de la República, 1966.

59. James Petras, *(Politics and Social Forces in Chilean Development.* Berkeley: University of California Press, 1969) provides much useful background to the Chilean social situation and provides some benchmarks for an evaluation of at least the first half of the Frei administration.

60. Manfred Halpern, op. cit., p. 192.

61. For a precise definition of these terms, see Halpern, op. cit., pp. 188-193.

Chapter 6

SOCIAL DIMENSIONS OF URBAN DEVELOPMENT

According to a widely accepted theory, the economic growth of a depressed or newly settled region is initially brought about by expanding its production for export to other regions.[1] The income generated by this activity within the region, as well as the stimulus given to local enterprise by supplying the endogenous export sector with certain commodities and services or by enabling it to purchase the output of the export sector at a lower cost, will tend to increase the rate of regional production. As a result, production for intraregional demand will gain relative to production for export so that the economy becomes increasingly self-sustaining and less dependent on extraregional markets.

This theory may be challenged on two grounds. First, the regional multiplier effects of the initial export activities may be very low. This typically happens when the export sector employs a small amount of highly skilled labor, as in modern mining operation, or where it uses larger amounts of relatively unskilled labor at low wage rates, as in plantation economies.[2] The regional multiplier effect will be further reduced where the export investment is in the hands of extraregional interests that will channel most of the profits earned to places outside the region. Finally, "backward" and "forward" linkages of the initial export activities may be established with industries that are located, not within the region itself, but in one or more of the existing cores (for example, processing facilities for raw material obtained from the periphery).

The second challenge skirts these particular difficulties but argues that the regional multiplier depends chiefly on the capacity of a region to organize itself for economic growth.

The ways leading to the economic growth of a region are limited. They include:[3]

(1) the discovery and capture of new markets for old products;
(2) the introduction of new ways to produce old products (that is, changes in the production function giving more output per unit of labor and capital input);
(3) the production of new or improved products and services;
(4) the establishment of new types of organizations for production (for example, cooperatives, corporations, associations or firms);
(5) the buildup of local physical infrastructure directly relevant for production activities (for example, roads, supply of electric energy, port facilities);
(6) the creation of local savings and investment institutions;
(7) the development of human resources: labor supply and its skills;
(8) the development of local natural resources and improvement of locational advantage; and
(9) the development of institutions and services to provide more or better information and knowledge useful in planning and production (for example, universities, research stations, telex, market research services, planning offices, mass media).

Some of these ways require the existence and ingenious use of individual entrepreneurial capacity in the sense of Joseph Schumpeter's classical model of risk-taking entrepreneurship. But their majority calls for cooperative, organized activity. Cooperation is needed, for example, to pull outside capital into the region for investments that will strategically advance the long-term interest in economic growth (for example, a new university, a tourist hotel, a shipyard, a research center). Organized action is also needed to mobilize local resources for special purposes or put existing capital to more efficient use (for example, a regional planning office, development-conscious regional and local authorities, improved management of natural resources, local institutions for investment credits to small- and medium-sized industries). Where concerted action fails to appear in sufficient volume, where it is misdirected, or where participation is restricted to only a small and self-serving elite of families, self-generated, sustained regional economic growth will fail to occur.

On the basis of these amendments to regional growth theory, it becomes possible to distinguish between "active" and "passive" peripheries (core regions are, by definition, active). For present purposes, passive peripheries may be treated as residual areas after the active and potentially dynamic regions have been identified.

I shall refer to active peripheries as *social development poles* or areas of concentrated settlement (usually cities of some minimum size) having a high potential capacity for organizing themselves to achieve sustained economic growth. I propose to call this their capacity for social development. This concept of social development is meant to include all processes endogenous to a region that contribute and lead up to increases in production and productivity in the use of regional resources. Within a given area, this may include a growing ability to perceive and seize economic opportunities given in the environment, to mobilize resources for production, to diagnose common problems accurately, to organize successfully and carry through programs looking towards an "improvement" of the physical environment and, more generally, to experiment with new ideas, new approaches and new forms of organization in dealing with the tasks at hand. Among sociologists today, the capacity of an area to generate and absorb continuous change is accepted as a distinctive sign of successful modernization. Social development, then, refers most generally to continuing increases in the capacity of a territorially based subsystem of society to deal with problems of innovative change. In most cases, this will involve responses on the part of broad segments of the population.[4]

If we accept this point of view, it is clear that economic growth may be indirectly achieved by promoting and strengthening certain facilitating social processes internal to a community. The alternative to this would be to implant artificially into an area productive facilities which can be sustained with only minimal involvement of the local community. This might occur, for instance, with the construction in some remote region–say, in the Chilean province of Magallanes–of an oil refinery whose capital is entirely in the hands of interests external to the region and which can be operated with only minimal adjustments as a "foreign" enclave in the region. More generally, it is the case of all "neopatrimonial" cities whose growth is the result of initiatives originating in the national capital with little, if any, meaningful participation of local populations.

Social development is conceived here as occurring primarily in cities or micro-regions. To the extent that active peripheries become linked to each other as well as to core regions, the economic growth of an entire system of cities and regions would occur.

THE PATTERN VARIABLES: PSYCHOSOCIAL INFRASTRUCTURE OF A REGION

We know very little concerning the configuration of conditions that makes some regional "communities" more capable of organizing for economic growth than others. The following complex variables, each composed of subvariables forming a pattern, are suggested as a basis for working hypotheses in the empirical study of social development:

(1) *Regional consciousness,* or the desire and will of populations to promote the economic growth of their area; a group-focused image of economic growth.

(2) *Information transfer,* or the "openness" of leading sectors of the population to information external to the region and their capacity to reinterpret and use this information in ways that will make it relevant to local economic growth.

(3) *Autonomy,* or the ability of local populations to make effective decisions relating significantly to the economic growth of the area.

(4) *Innovative propensity of the social system,* or the affinity of local leadership for patterns of social organization conducive to sustained innovative activity.
(5) *Organizational capacity,* or the ability of local populations to organize themselves effectively for the solution of common tasks related to economic growth.

Following Janusz Ziólkowski, we may call these pattern variables of social development the "psychosocial infrastructure" of a city.[5] To clarify the relations among them, the following two-by-two matrix may be set up.

Pattern Variables	*Individual*	*Collective*
Motivational	Regional consciousness Innovative propensity of local leadership	Regional consciousness
Facilitating	Information transfer Organizational capacity (enterprise)	Information transfer Organizational capacity (community) Autonomy Innovative propensity of social organization

These pattern variables refer exclusively to ecological subsystems of society (micro-regions, cities). They are therefore measures of a region's internal capacity for social development. The matrix of the larger societal system (nation) in which they are embedded establishes the basic norms from which regional subsystems will deviate. If the political organization of the system is highly centralized, for instance, nearly all regions will, to a large degree, be externally dependent and capable of organizing primarily to petition the central authorities for assistance in local projects. The number of significant deviant cases, as well as the degree of deviation, will be relatively small. More open and decentralized systems, however, will give the regions greater initiative and power to act on their own behalf. System parameters, therefore, must be viewed as establishing the limits to social development at subsystem levels. If these limits are both rigid and narrow, structural change in the parameters is the only path that is likely to permit more rapid economic growth in the periphery. But if they are relatively wide and flexible, the number of deviant cases—and the degree of deviance from system norms—will increase, and social development in the active periphery will be more likely to lead to the results desired.

Each pattern variable at the regional level is, in turn, composed of a constellation of subvariables or major components that are identified and discussed below.

A. Motivational Variables

1. INDIVIDUAL

a. Regional consciousness

(1) Political participation: in democratic polities, local interests find expression through the political process and specifically through the election of politicians to office. Active participation in local politics, as measured for instance by voting behavior, may be taken as a rough indicator of concern with public affairs; a nonparticipant population will tend to be alienated from the larger community and difficult to enlist for social development.

(2) Regional commitment (a group-focused image of change); some pride in and identification with a local area would seem to be prerequisite for most kinds of socially oriented action. Regional commitment, particularly on the part of local leadership groups, has its national counterpart in the ideology of nationalism. A dependent internal proletariat, as well as noncommitted residents whose principal orientation is towards centers and groups outside the area, will not be likely to devote much energy to promoting local economic growth.

b. Innovative propensity of local leadership

(1) Leadership attitudes (dominant): every social system may be divided into a dominant and dependent sector. Attitudes of the dominant groups towards change will be important for economic change. Also relevant are elite attitudes towards the dependent sectors of population, the legitimate aspirations of these groups and expectations regarding their proper relationship to the dominant sectors. Elite understanding of society as a fluid, open, participant system, characterized by free collaboration rather than by authority and compliance will have positive meaning for the social development of an area.

(2) Leadership attitudes (dependent): attitudes similar to those for the dominant leadership are important here. Of special relevance will be the perceived—as well as desired—degree of autonomy of dependent groups, their sense of social power, and their desire to collaborate freely in the process of regional economic growth.

2. COLLECTIVE

a. Regional consciousness

(1) Institutional orientations: institutions such as local centers of higher education, local political party organizations, and business and manufacturing concerns may be oriented predominantly towards advancing their own interests, the interests of the region or extraregional interests; they may also be more or less inclined to accept changes in the traditional social order. Both are important motivational variables in the process of social development.

(2) Mass media orientations: questions similar to those immediately above

may be raised also with respect to the performance of the principal mass media (newspapers, radio, television) which serve the local population.

B. Facilitating Variables

1. INDIVIDUAL

a. Information transfer

(1) Educational level: the educational level of local populations will be one of the determining factors in the process of innovation, facilitating or hindering the transfer of information from sources outside the immediate area. In general, an educated population will tend to be better prepared for generating and absorbing innovations than one that is less educated. Its perceptions of problems and opportunities for action will also tend to be more realistic.

(2) Number of professionals in the population: professional people and technicians nearly always have a university education. As such, they play a special role in the transfer of information, being in more direct contact with the outside world and generally better equipped than others to lead the process of community change. Their specialized skills are also needed for many types of social development.

b. Organizational capacity (enterprise)

(1) Available "managerial" skills: the distribution of management abilities in the population—whether in private enterprise, the cooperative movement, civic organizations, labor unions, housing projects and the like—serves as a significant index of the "ability to organize" for economic growth. Communities deprived of significant organizing skills will tend to have a low potential for dealing effectively with forces of external change.

2. COLLECTIVE

a. Information transfer

(1) Centers of higher education: universities, technical schools, and similar institutions will generally function as "transmission belts" for scientific, technical, and cultural information from outside the local area. They will also be capable of rendering specialized services to the community. Their presence, size, vocation, and efficiency in carrying out their local mediating role constitute significant measures of the social development potential of a region.

(2) Mass media: the number and types of mass media available to a community and the size of their audience are all relevant indicators of social development potential.

b. Organizational capacity (community)

(1) Community organizations: the number of self-generated organizations in

an area—neighborhood unit councils, clubs, civic associations, labor unions, housing cooperatives, and so forth—constitutes, together with their membership, a direct measure of local organizing ability and of experience with collaborative action. Important additional information should be obtained concerning their programs, stability, intensity of activity, and ability to satisfy the specific needs of their members.

c. Autonomy

(1) Extent of autonomous decision-making powers: if social development is conceived as a process occurring in distinct localities, it may be reasonable to insist that a significant proportion of the vital decisions which affect economic growth must be made by local authorities and other organized groups in the region. This ability is partly a function of the institutional framework for decision-making and the financial resources and skills available to the region for specific undertakings.

d. Innovative propensity of social organization

(1) The structure of social organization of a city will be a major factor in its development. Two basic models of organization may be distinguished: one emphasizes hierarchy, authority, permanence, and limited access to information (the elitist, bureaucratic model); the other stresses horizonal relations, shifting patterns of collaboration, free association and radical openness to information (the democratic, nonbureaucratic model). The second of these models tends to reveal a higher propensity for innovative change than the first. To the extent that its social organization shows conformity with the second model, a city will tend to have a high degree of social development potential.

The foregoing account of pattern variables may be summarized by saying that subsystems of society in which some or all of these structural conditions are present will be expected to demonstrate a significantly higher capacity for social development than subsystems where only a few or none of them exist, or where their intensity is weak.

As a corollary, the presence of the structural components identified would be reflected, first, in the character of local economic development (self-directed, responsive to local needs and aspirations) and, second, in the rate of local economic development (a function partly of the ability to mobilize resources). These two sets of consequences, however, need not coincide. It is entirely conceivable that local economic development is intense without being responsive to local interests (where investments, though locally promoted, are exogenous and controlled by core regions); it is also possible to imagine a situation where local development, though responsive to local needs, will be of relatively low intensity.

The economic growth of an ecological subsystem of the society may therefore be either *exogenous* (imported and outer-directed) or *endogenous* (locally generated and self-directed). In both cases it may appear as a result of social development, but only in the second instance would this always be true. The study of the conditions for social development and the processes by which social development occurs is thus of fundamental importance for industrializing societies that, in their entirety, may lie within the peripheries of international core regions. It is of special importance for the formulation of appropriate policies directed at coping with underdevelopment in national peripheries. For if peripheral regions are to break out of their condition of colonial dependency, they will have to choose the path of social development to economic growth.

The nature of the data required is such that they do not readily lend themselves to factor analysis. What is needed, therefore, is the construction of community "profiles" composed of the 14 variables listed and an interpretive judgment on the meaning of each city profile for social development. Cumulative evidence should make it possible to assign proper weights to each component and reduce the number of relevant indices of an area's "psychosocial infrastructure" to those that are truly critical for its ability to organize itself for sustained economic growth.

SOME RESULTS OF EMPIRICAL RESEARCH

Exploratory studies on the conditions affecting the capacity for social development of two cities in southern Chile were carried out under the guidance of the author and Ernesto Cohen.[6] Valdivia and Puerto Montt were selected on the basis of several considerations. Both were settled initially by German populations; their sizes—73,000 and 66,000, respectively, in 1960—were roughly comparable; they are located within three hours' driving distance from each other; both are coastal cities serving a larger region; and both are distant from Santiago (855 and 1090 kilometers, respectively). Despite these external similarities, however, the two cities showed marked internal differences in their score on a number of components of the pattern variables described. Of 28 cities examined, Valdivia ranked 8 and Puerto Montt 27 with respect to the sum of 11 surrogate measures used.[7] There was thus a presumption that Valdivia would exhibit a fairly greater capacity to organize for economic growth than Puerto Montt. The overall quantitative index of social development was not taken too seriously, however. The research attempted rather to shed light on the validity and meaning of the pattern variables for social development.

Four individual studies were carried out in the two cities. Clinton Bourdon's interest was in the performance of the local economic elite groups; Diva Dinovitzer studied organizations of the so-called *sector popular,* that is, the dependent, low-income population; Marguerite Diggs did a study of local responses to a devastating earthquake which struck the southern part of Chile in 1960, in the hope of discovering significant differences in the responses of Valdivia and Puerto Montt to a disaster that left no sector of the population unaffected; and Nancy Zimmerman concentrated on the region-serving role of the Universidad Austral, a medium-sized private university located in Valdivia.

Before proceeding, it is necessary to describe some salient features of the national system which, it was subsequently discovered, exerted a decisive influence on certain aspects of social development in the two regions. Economically, Chile occupies a middle range between the already industrialized and newly industrializing nations. Its per capita income is estimated at about $575, placing the country near the top of a list of Latin American nations, immediately following Argentina, Venezuela, and Uruguay. Politically, Chile is considered a unitary state. The only government mediating between the citizen and the national government is the municipality which has an elected city council; provinces and regions are simply units serving administrative convenience. Municipalities, however, are exceedingly weak and are able to spend an average of only $10 per capita per year, most of which goes for the salary payments of local officials. Nearly all effective power is therefore wielded from locations in Santiago, where it is divided among the Chief Executive, the National Congress, national political parties, the national administration, and a number of nationally organized sectoral interests. Local interests are not clearly represented in the Congress, whose legislative acts are steered principally by partisan and ideological considerations that have no clear relationship to local needs. A good deal of the time, however, the system operates on the basis of reciprocal favors through extremely complex networks of personal contacts which help to meliorate what otherwise would be a rather inflexible pattern of administrative behavior.

The structure of Chilean society has traditionally been held together by the so-called principle of authority, according to which it is the right of the few to govern and the duty of the many to obey. It is a system of ascriptive status, cast in the mold of a bureaucratic model of organization. Although the system is now being challenged from many sides, it still retains considerable strength and vitality.

Our studies in Valdivia and Puerto Montt revealed the immense weight and influence of this national system on local social organization. The major conclusions with respect to the social development model presented earlier are best summarized by looking separately at the responses of the dependent and dominant sectors of the population.

The dependent sector accounts for between 65 and 95% of the population in both cities; a more precise estimate is not possible because an unambiguous criterion of "dependency" is lacking. The dependent sector might also be called the passive or excluded population. In Chile it is common to refer to this sector indiscriminately as *población marginal* or *sector popular.* Armand Mattelart, a Belgian demographer, uses employment characteristics of the active male population as part of a broader definition (Table 1).

According to Mattelart, approximately 75% of the male labor force in both cities can be classified as blue-collar. Family income statistics, unfortunately, are not available. If they were, a possibly smaller percentage than Mattelart's might reasonably be called "dependent." In any event, the phenomenon of depend-

TABLE 1

OCCUPATIONAL STRATIFICATION OF ACTIVE MALE URBAN POPULATION, 1960 (in percentages)

	Valdivia	*Puerto Montt*
Urban managerial groups, medium-sized and large enterprises	0.8	1.3
Upper administrative class	4.5	3.9
Independent professionals	0.5	0.5
Self-employed managers *(empresarios)*	1.2	1.6
Lower-middle administrative class	8.2	8.2
Autonomous lower-middle class	8.4	9.5
Subtotal, urban middle class	23.6	25.0
Dependent working class	67.9	62.8
Autonomous working class	8.5	12.2
Subtotal, urban working class	76.4	75.0

SOURCE: Armand Mattelart and Manuel A. Garretón, *Integración Nacional y Marginalidad* (Santiago: Pacifico, 1965), Table 18.

ency is a matter of degree rather than of absolute judgment. What can be asserted with confidence is that it refers predominantly to low-income groups whose social participation is limited to membership in such organizations as soccer clubs, mothers' centers, syndicates, neighborhood juntas, cooperatives and mutual aid associations. These organizations are typically small, local, and often unaffiliated. In 1966 about one-half of the sector popular in the two cities studied were not members of any organization and consequently were excluded from any meaningful participation in the affairs of their community. Between them and the all-powerful central government, there were no mediating institutions of any sort. But the participation of the remaining 50% was also exceedingly restricted. For certain types of organization, only small numbers of the population from which recruits might potentially be drawn in fact held formal membership in them (Table 2). Less than 15% of the adult women belonged to mothers' clubs; between 40 and 60% of the men did not participate in athletic organizations; less than one-fifth of the workers were organized into unions; less than 40% of the families in the dependent population were represented in neighborhood councils; less than 10% belonged to mutual aid associations. The resultant defenselessness, if not anomie, of the vast majority of the low-income population can be readily imagined.

The data in Table 2 also show that the "dependent" populations in the two cities showed no significant differences in the overall pattern of their organization. Furthermore, and important from the standpoint of our concept of social development, popular organizations of the type found in Valdivia and Puerto Montt had very little to do with economic growth as such or, for that matter, with any of the truly vital concerns of the dependent population. Sport clubs organize to play soccer; mothers' centers receive cooking and sewing

TABLE 2

DATA ON COMMUNITY ORGANIZATIONS FOR THE DEPENDENT SECTOR, 1966

	Number of Organizations		*Average Number of Members*		*% of Organizations Affiliated*		*Total Membership*		*% Members of Relevant Population Base*	
	V	*PM*	*V*	*PM*	*V*	*PM*	*V*	*PM*	*V*	*PM*
Mothers' centers	82	68	35	28	76	66	2,902	1,958	15	11
Soccer clubs	82	72	129	104	63	49	11,432	7,551	58	41
Syndicates	55	23	69	99	46	39	3,803	2,286	19	12
Neighborhood juntas	38	41	112	102	50	44	4,287	4,190	35	39
Mutual aid associations	16	20	88	100	50	75	1,411	2,008	4	6
Others[a]	26	27	73	106	58	56	1,910	2,886	5	8

SOURCE: Census of *Organizaciones de Base,* Consejería Nacional de Promoción Popular, 1966. Percentages in the final column were estimated on the basis of the relevant population group from which membership might conceivably be drawn and represent rough orders of magnitude rather than precise measurements.

a. Includes cooperatives and cultural, youth, and welfare organizations.

classes; neighborhood organizations form to present their needs—the paving of a street, the construction of a playground, the provision of emergency housing—to the authorities; labor unions organize according to craft on a shop basis to press limited demands against employers or to plead their case directly with the authorities. There is no conceivable way by which these modest collective activities can be joined to matters that relate to life in the larger "community"—the city as a whole. The street, the neighborhood, perhaps the barrio command some loyalty and are capable of being conceptualized, but not the city and its future. Only one case was discovered where syndicates and neighborhood juntas collaborated in any sense with the dominant elites on a matter pertaining to the economic growth of the area. This was a special committee set up in Valdivia to pressure the government into locating a shipyard in the area. This collaboration was clearly an instance of complementary interests. A shipyard would create jobs and income for workers who therefore acquired a stake in the project.

More notable differences in the capacity to organize for economic growth were found between the dominant elite groups of the two cities. In neither city did the elites exploit existing possibilities for concerted action, but Valdivia clearly showed more ability in this regard than Puerto Montt. Valdivia's superior showing could be ascribed, in part, to a greater identification of the local elites with the region, a more accurate perception of problems, and a more adequate understanding of what would constitute effectively instrumental action, a greater capacity for concerting interests, a large managerial tradition, and the presence in Valdivia of education and research institutions that could be relied upon to give technical advice when needed.

Even so, the difference is only one of degree. The elite groups of both cities tend to organize, if at all, primarily to achieve specific economic ends through petition and pressure on the central authorities. In spite of their greater organizational skill, for instance, the elites of Valdivia have so far failed to assure themselves of a long-term supply of high-quality forests for their wood product industries, nor have they taken any steps that would mobilize local capital resources for industrial expansion in the region. And although the University of the South came into being as the result of local initiative by a few wealthy Valdivian families, this center of higher education has traditionally served only individual clients by making its specialized expertise in agriculture and forestry available to those able and willing to purchase it; only recently has it begun to play a more active leadership role in the region.

In both cities, the dominant elite groups appear to find the conceptualization of city- or region-wide interests exceedingly difficult. Under the pressure of disaster, as immediately following the devastating earthquake in 1960, the atomistic character of the local social system becomes strikingly clear. Individual families and groups of related families suddenly revealed themselves as the basic and, indeed, the only active agents in the locality. Oblivious of wider concerns, they advanced their own self-interest quite without shame by exploiting as best they could the network of reciprocal favors that operates outside, but parallel with, the formal system of the state bureaucracy. The rhetoric they used was communitarian, but the benefits they sought accrued only to themselves.

Thus, although Valdivia appears to have the edge on Puerto Montt in the city's potential for social development (in the highly specific meaning of the term used in this chapter), the difference is nowhere as pronounced as we had initially supposed on the basis of purely formal criteria. We had given insufficient weight to the influence of the national system as setting up the limits for effective local organization. In Chile these are unusually narrow and inflexible. We had also overlooked the importance for social development of the internal social and political structure of the areas we studied. This structure was evidently not neutral with respect to the ability to organize effective actions for economic growth. The mass-elite structure, common to both Valdivia and Puerto Montt, effectively excluded the great majority of the population from any meaningful participation in the larger affairs of their communities and left them with a small number of highly personal concerns of relatively marginal significance to their own sense of well-being. More pluralistic or communitarian structures of local society—American and Yugoslav models come to mind as prototypes—might lead to a more active involvement and to reorientation of basic societal values. In Chile, where economic power is to a large extent in the hands of the State, effective action is restricted to a small group of elite families that has direct access to the nerve centers of the State management apparatus. The internally dependent masses do not have significant economic power, are consequently at the mercy of the State and of the elites linked into its machinery for compliance and are limited to sporadic interventions through direct action, such as strikes and demonstrations.

Where economic power is concentrated in the hands of the State, self-generated local development will be limited essentially to an ability for organizing petitions to channel public investments to the region. To the extent that economic power is concentrated in the hands of a small group of elite families, self-generated local development will depend on certain characteristics of the local elite groups, namely, their education level, their cosmopolitan outlook, their connections with national elite groups, the extent to which their behavior is self- or group-oriented, the degree of their class or community consciousness, the extent of their identification with the future of the local area and their capability for rational appraisal of local opportunities and limiting conditions. Where elite groups define their interests narrowly, do not identify with the local area and are more class- than community-conscious, the local economy is likely to stagnate. Where elite groups are poorly connected with national elite groups, little social development can be expected. Where elite groups are unrealistic about what can be achieved, their efforts are likely to be misdirected.

FURTHER THOUGHTS ON SOCIAL DEVELOPMENT

The Chilean study reported above might seem to invalidate the original hypothesis that local communities differ in their abilities to respond to opportunities in their environment in ways that will promote their economic growth, and that these differences vary systematically with certain characteristics of these communities. An alternate explanation of the results, however, would stress the importance of the national environment for community response as a significant intermediate variable.

In the Chilean case, this environment was characterized by a pattern of decision-making so centralized that local differences in potential response ability were completely overshadowed by the economic and political dependency of both communities on national power centers located in Santiago. If this finding is accepted, as I believe it should be, the original hypothesis is neither invalidated nor confirmed; it is simply expanded. The critical tests remain to be performed.

In any event, superficial evidence would seem to support the hypothesis. Within the same country, some communities deploy a great deal of energy and ingenuity in promoting their own development while other communities remain relatively inactive. Juanusz Ziólkowski formulated his concept of "psycho-social infrastructure" from Poland's experience with accelerated industrialization in the decade of the fifties. He writes:[9]

> "The transformations taking place in Poland are often described as the results of industrialization. The latter is regarded an an independent variable, while various

phenomena and processes—economic, demographic, social and cultural—are regarded as dependent variables. An intensive industrialization of the country, embarked upon about 1949, was informed by the principle of an even distribution of the forces of production, aimed at raising the economic level of the most backward regions. By having set in motion the mechanism of spatial and social mobility and by moving large masses of rural population to the towns, where they were introduced to the system of industrial work, it has revealed—often in a dramatic manner—the deep differences existing between individual regions in relation not only to the material and technical but also to the psycho-social infrastructure. Among the distinguishing characteristics of this psycho-social infrastructure are the ability of organized cooperation, tradition, culture and discipline of work, sense of responsibility, and general standards. It is precisely this infrastructure, so frequently and lightly dismissed in investment policy, that largely determines the economic effects of this policy."

Similar differences have been observed elsewhere, witness the difficulties experienced by certain cities, such as Buffalo, New York, and Pittsburgh, in making adaptive responses to the changing character of national economic development in the incipient post-industrial era after World War II. While Pittsburgh managed its transition from ninteenth-century industrialism successfully, Buffalo remains to this day an economically and physically depressed area.

Traditional theories of regional economic development have assumed that regional multiplier effects based on initial export activities happen automatically. I have tried to suggest that the size of this multiplier depends, among other things, upon a community's capacity for social development and, further, that this capacity will vary not only with characteristics of its social structure but also with the degree of autonomy enjoyed by the community for development decisions.

If this hypothesis is correct, it follows that, other things being equal, urban-regional development will flourish under conditions of local autonomy and that the economic development of the nation will be favored by a system which combines, in optimal fashion, both centralized and decentralized decision-making processes. It also suggests that local autonomy will not produce similar responses on the part of all communities, and that it should therefore be granted, as a matter of public policy, only to those communities which appear to have structural characteristics conducive to social development. It remains to be seen whether public policy is able, in addition, to improve the conditions of a community's psycho-social infrastructure through appropriate action.

NOTES

1. Douglas C. North, "Location Theory and Regional Economic Growth," in John Friedmann and William Alonso, eds., *Regional Development and Planning. A Reader.* Cambrige, Mass.: The M.I.T. Press, 1964, chapter 12.

2. R. E. Baldwin, "Patterns of Development in Newly Settled Regions," in Friedmann and Alonso, eds., op. cit., chapter 14.

3. This list was suggested to me by Clinton Bourdon to whom I am also indebted for many new insights into the process of regional economic growth.

4. For a full statement of the implications of this model, see N. S. Eisenstadt (El Estudio del 'Fenómeno de Urbanización en Venezuela' y los Problemas de Desarrollo de una Sociedad Moderna," Ensayos y Exposiciónes. Third Series, No. 4. Caracas: CENDES, 1967). Eisenstadt's model, however, suffers from being excessively elitist in its orientation and from not giving sufficient emphasis to the creative role of social conflict.

5. In his illuminating paper, Janusz Ziólkowski ("Sociological Problems of Regional Development," *Proceedings of the First Scandinavian-Policy Regional Science Seminar.* Warsaw: PWN-Polish Scientific Publisher, 1967, pp. 16-32) describes contrasting forms of recent industrialization efforts and related urbanization in Poland–one which was imposed on a town in neopatrimonial fashion by central authorities and another which grew in an "organic" fadhion. Ziólkowski's article is especially worth reading not only for the support it lends to the model presented in this chapter but also for salient points of difference.

6. These studies were supported by the Centro Interdisciplinario de Desarrollo Urbano y Regional (CIDU) of the Catholic University of Chile.

7. The measures included road distance from Santiago; population increase between 1940 and 1960; average annual municipal income per capita, 1960 to 1965; number of students registered at local universities, 1967; percentage of population with secondary school, technical and university education in 1960; percentage of eligible population voting in the Congressional elections of 1965; and percentage of the labor force in mining and manufacturing industries, 1960. Despite the inadequacy of these measures, they were thought capable of discriminating between high and low social development capacity.

8. A. Mattelard and M. A. Garretón, *Integración Nacional y Marginalidad.* Santiago: Pacifico, 1965.

9. J. Ziólkowski, op. cit., pp. 22-23.

Part Two

POLICY

Chapter 7

NATIONAL URBANIZATION POLICY: SCOPE AND OBJECTIVES

THE SCOPE OF URBANIZATION POLICY

The field of urbanization studies is so new, that a common terminology has yet to emerge. This chapter, introductory to Part II, might have been given other titles, such as National *Urban* Policy, National Policy for *Urban and Regional* Development, National Policy for *Spatial* Development, National Policy for *Subnational Area* Development, and National Policy for *Regional* Development. These designations, though each is different, refer to much the same set of ideas. It would be helpful, however, and certainly an aid to clarity, if basic conceptual distinctions could be drawn.

Some specialists in policy analysis insist on differentiating between urban and regional development. This distinction often has its counterpart in institutional arrangements. In Chile, for instance, responsibility for urban policy is vested in the Ministry of Housing and Urban Planning, while regional policy is centered in the National Planning Office. To a considerable extent, the Ministry is staffed by architects who make a professional claim to urban design and the physical layout of cities. The National Planning Office, on the other hand, is staffed by economists who view regional planning as a branch of applied economics. Conceptually as well as practically, however, there is little basis for distinguishing between the two types of policy planning, except for the scale of the phenomena with which each group of planners is concerned. Urban planners, working with the internal spatial arrangements of cities, must take into account the economic base of city growth, while regional planners must include in their studies not only the economics of urban expansion, relating this to regional resources, but also the spatial organization of economic activities within regions. Both types of study are, in turn, rooted in the common ground of economic location theory.[1]

Because professional education in architecture does not prepare one for work with the tools of locational analysis, architects wishing to devote their energies to urban development policy must, in effect, cease to think as architects and study the economics relevant for guiding development within a spatial framework that includes cities as parts of regions and similar territorial units, up to the level of the nation as the largest system appropriate for policy analysis.[2]

The institutionalized separation of urban and regional policy planning, as in Chile, based as it is on professional traditions instead of theory and logic, makes a meaningful coordination of policies and programs immensely difficult, particularly since the Ministry of Housing commands vast financial resources, whereas the National Planning Office has none and, in order to be effective, must rely primarily on the persuasiveness of its ideas.

Unfortunately, the separation of urban from regional planning is being perpetuated in academic circles. A recent three-volume set of collected readings sets out to distinguish among urban, regional, and spatial economics, but fails to make any theoretical case for this division.[3] A look at the selections included suggests that urban and regional economics may at best be considered as somewhat arbitrarily defined subfields of spatial economics which, in turn, provides an encompassing framework for both.

At the level of public decision-making, spatial economics may be translated into spatial (or urban-regional) development policy. This designation manages to transcend the discrete and awkward dichotomy of city (small area) and region (large area) by substituting for both the continuous variable of spatial relationships.

Spatial policy is concerned with the organization of activities in territorial space, and is often decisive in solving broader questions of national development.[4] Its two-dimensional character, however, has some drawbacks by excluding from its theoretical framework such issues as housing policy, political participation, administrative organization, and social integration which, we have seen, are central issues in national planning.[5]

These additional policy concerns may be brought into the analysis by raising spatial development to the yet more inclusive level of urbanization policy. As I have noted in Chapter 1, an adequate theory of urbanization does not as yet exist; this makes urbanization policy a fuzzy set of related questions, although the paradigm for the study of urbanization, presented in Chapter 3, does indicate the principal variables which must be considered for inclusion:

innovation, innovation diffusion, socio-cultural differences in population characteristics, decision-making and control, the territorial organization of power, migration, and settlement patterns, capital flows, and economic location. More than this however, urbanization policy focuses on processes rather than structures. This shift in emphasis is crucial for policy, since to produce a change in structure, policy must influence the appropriate processes; it can never act directly upon the structures themselves.

In the three chapters which follow, the terminology is not as neatly drawn as it is here. Urbanization, urban-regional, spatial, and subnational development policies are, to some extent, used interchangeably. In general, however, I have striven to include the most comprehensive set of considerations for public policy in subnational development. The failure of terminology to correspond precisely to the notions set forth above is owing to the general lack of an established consensus on the meanings of these terms and the fact that the chapters were written at different times for different occasions.

OBJECTIVES OF URBANIZATION POLICY

Policy prescriptions for urbanization will vary with a number of contextual variables, including: (a) the distribution of power in society, (b) the political commitment to particular value propositions, (c) the information available, and (d) the country's level of development.[6]

The first will determine the instruments available for implementation. It is still widely believed that the formulation of public policies is primarily a question of stating the proper objectives for governmental action. This view is incorrect, however. To begin with, the distribution of power in the society will determine whether a command, policy, corporate or participant style of planning will be adopted.[7] In the first, direct are substituted for the indirect controls of policy planning, while corporate planning is largely a process of bargaining among powerful organized sectors of the population and depends chiefly on the moral pressures participants can bring to bear upon each other to abide by the decisions negotiated; participant planning further decentralizes the process of decision-making to the smallest viable units in the society. In both corporate and participant styles, the objectives of urbanization policy are not centrally determined but result from the interactions of the participants in the planning process. But even in command and policy planning, objectives are not established prior to a knowledge of the possible instruments for effectuating policy and the consequences of their use.

This is so fundamental a point, and one so frequently misunderstood, that an example will be helpful. It would be relatively easy to determine some national pattern of settlement that is, for whatever reason, considered "optimal" for the country's development. For example, urbanization policies frequently attempt to lay down a rule that small- to medium-sized cities should grow more rapidly than the supposedly oversized national capital region. But an objective of this sort makes little sense unless the ways of achieving it are first considered. Only when the means whereby the location decisions of business establishments and families may be influenced in the desired direction have been evaluated, is it possible to resolve the question of whether the objective in view is also realistic and attainable.

The political commitment to particular value propositions is a second condition for policy. Without this constraint, the prescription of policies is trivial. The announcement of a policy for urban decentralization, for instance, either implies a public commitment to change the settlement pattern in the desired direction, or it is meaningless. Whether a commitment of this sort can be sustained over a sufficiently long period of time will depend on the value integration of the society and the role and predispositions of the existing political elites. No policy can ever hope to be more successful than the supporting political consensus.

The third contextual variable is information or knowledge. A simple rule covers this situation: *do not prescribe what should be done in more detail than a knowledge of the consequences permits you to do.* Actual behavior is usually quite the reverse of this: the less they know, the more do planners try to prescribe for everything in sight. Soviet planners, for instance, were unable selectively to control the Russian economy so long as a theory of socialist economics did not exist.[8] They proceeded instead by a crude method of material balances to determine physical production targets for every nut and bolt in a set of projected final demands. In spite of its tremendous human and administrative costs, this policy resulted in an economy that lurched forward from crisis to crisis, as shortages and surpluses followed each other in rapid and disorderly succession. On the other hand, the more the consequences of potential interventions are known, the more elegant, parsimonious, and precise can be the form of public intervention. The problem lies in the detection of a small number of controlling variables to which the system as a whole responds in ways that are at least to some extent predictable. Lacking such knowledge, planners should be constrained from recommending any policy whatever, for the only justification for giving a decisive voice to planners is their superior understanding of the relevant relationships. Their claims to a superior understanding of essential relationships is thus the sole criterion for distinguishing planners' recommendations from those of non-professionals.

This understanding has two facets. The first is theoretical (hence the importance attached in Part I to an adequate theory of urbanization); the second is factual and relates to the informational basis for planning which may be either good or bad in relation to the problems to be solved. Information in many of the newly industrializing countries permits only the grossest kind of policy objectives to be formed. Few planners, unfortunately, are willing to abdicate their presumed right to prescribe what is good for others, even where they lack even the slightest basis in knowledge for doing so.

The fourth condition—level of national development—is the most complex of all the contextual variables considered. Nevertheless, it is decisive for formulating the right kind of policy objective. In the broadest possible terms, the following generalizations may be ventured. Countries in the early phases of industrialization must make *spatial integration* the overriding policy objective; they are still striving to achieve a minimum measure of national unity. At a later stage, *regional inequities* become predominant and will have to be balanced against the requirement of *efficiency* in location. Still later, well into the *post-industrial* period of development, policy objectives for urbanization will be primarily concerned with questions of *environmental quality* and the structure and organization of major metropolitan regions.[9]

In the chapters which follow, I shall address myself primarily to the problems of countries in the *early and intermediate stages of industrialization.* They are assumed to be countries sufficiently large in size to permit their treatment as a system of cities in which the distribution of power favors a relatively small political elite committed to the values of national development and modernization. A minimum of relevant knowledge is also assumed to be available.

Chapter 8 sets forth a framework for policy analysis, identifying the crucial questions with which urbanization policy must be concerned. It is probably the first time that the scope and range of a national urbanization policy has been so laid out.

Chapter 9 presents a hypothetical discussion between the Head of State of a rapidly urbanizing country and his Minister of National Planning, touching on four major questions of urbanization policy: the rate of rural to urban migration, the spatial pattern of investment and migratory flows, the rate of innovation diffusion to non-urban areas, and the territorial organization of power best suited to the implementation of a national urbanization policy. The various arguments for (and against) alternative policies are presented, but the reader, like the Head of State, is left to make his own choices.

Chapter 10 presents a case study of policy for the spatial development of the Capital Region of Chile. This study, originally commissioned by the National Planning Office of Chile, was prepared in the course of a seminar at the Center for Urban Studies (CIDU) at the Catholic University of Chile. It not only helps to illustrate some of the more abstract notions discussed in the chapters preceding it, but also demonstrates how policy analysis for urban development might be carried out. Because this chapter forms part of a larger study, it does not include many types of information which were covered by my colleagues in the Seminar.[10] Still, even in this limited form, it may be of value to those who are wondering what sort of analysis would be required for determining a policy of spatial development.[11]

NOTES

1. Peter Haggett, *Locational Analysis in Human Geography.* London: Edward Arnold Ltd., 1965; and William Alonso, *Location and Land Use. Toward a Theory of Land Rent.* Cambridge, Mass.: Harvard Univeristy Press, 1964.

2. In rare instances, the largest system may be a multi-national grouping of cities and regions.

3. Robert Dean, et al., *Spatial Economic Theory.* New York: The Free Press, 1970; David McKee et al., *Regional Economics.* New York: The Free Press, 1970; and William H. Leahy, et al., *Urban Economics.* New York: The Free Press, 1970.

4. John Friedmann, "Regional Planning as a Field of Study," Journal of the American Institute of Planners, Vol. XXIX, No. 3 (August 1963), pp. 168-174; and Edwin von Böventer, "Spatial Organization Theory as a Basis for Regional Planning," Journal of the American Institute of Planners, Vol. XXX, No. 2 (May 1964), pp. 90-99.

5. See chapter 5.

6. The social context will determine the styles of planning. For a discussion of this point, see John Friedmann, *Retracking America. A Theory of Transactive Planning.* New York: Doubleday, 1973, chapters 3 and 4.

7. Ibid., chapter 3.

8. See, for example, Oscar Lange, "On the Economic Theory of Socialism," in B. E. Lippincott, ed., *On the Economic Theory of Socialism.* Minneapolis: University of Minnesota Press, 1938. See also, by the same author, *Essays on Economic Planning.* London: Asia Publishing House, 1960; and L. N. Kantorovich, *The Best Use of Economic Resources,* Cambridge, Mass.: Harvard University Press, 1965.

9. A fifth contextual variable may, in certain extreme cases, be the size of country measured according to the distribution of the population. Very small countries offer quite different circumstances for urbanization planning than do normal-sized countries.

10. Some of the papers which emerged from this seminar, though by no means all, were published in a special number of *Revista Latinoamericana de Estudios Urbano Regionales,* Vol. I, No. 1 (October 1970). This new journal is published at the Catholic University of Chile under the auspices of the Latin American Social Science Research Council (CLACSO).

11. The following works bearing on urbanization policy have come to my attention too late for inclusion in writing these chapters: E.A.J. Johnson, *The Organization of Space in Developing Countries.* Cambridge, Mass.: Harvard University Press, 1970; Jos G. M. Hilhorst, *Regional Planning.* Rotterdam: Rotterdam University Press, 1971; Tormod Hermansen, *Spatial Organization and Economic Development.* The Scope and Task of Spatial Planning. Mysore: University of Mysore, Institute of Development Studies, January 1971; Walter Stöhr, "Regional Planning as a Necessary Tool for the Comprehensive Development of a Country," Paper presented at the United Nations Interregional Symposium on Training of Planners for Comprehensive Regional Development, Warsaw, June 14-28, 1971. Antoni Kuklinski, ed., *Growth Poles and Growth Centres in Regional Planning.* The Hague: Mouton, 1972; and A. Kuklinski and R. Petrella, eds., *Growth Poles and Regional Policies: A Seminar.* The Hague: Mouton, 1972.

Chapter 8

THE URBAN-REGIONAL FRAME FOR NATIONAL DEVELOPMENT

According to conventional practice, national planning is divided into two principal parts, global and sectoral. In the former, the express concern is with establishing overall balances among variables such as consumption and investment, exports and imports, and the supply and demand of employment opportunities. In addition, global planning establishes national priorities for resource allocations to the broad functional sectors of the economy, such as agriculture, manufacturing, transportation, health, and education. Sectoral planning, on the other hand, working within this framework of target figures and budget constraints, is chiefly concerned with drawing up specific projects and combining them into more comprehensive programs. To a significant extent, the major sectors of the economy correspond to the administrative structure of national government, with its ministries and independent agencies, each of which is devoted to functionally specialized tasks. They also coincide with the structure of national income accounts. This correspondence of functional sectors with administrative and accounting structures facilitates the formal integration of sectoral with global planning.

Sectoral planning remains incomplete, however, so long as it lacks a spatial dimension. Each sectoral project has a definite geographic location and varying impacts on developments and levels of welfare in different parts of the nation. These differential effects of sectoral investments and related action programs must be taken into consideration in program planning in order to arrive at a full social accounting of costs and benefits. But the spatial, or urban-regional frame, as I shall call it, contributes to planning in a number of other ways as well.

First, it focuses the attention of planners on projects and programs to an even greater extent than in sectoral planning, thus contributing to the implementation of national objectives. Second, it emphasizes coordinative relationships among projects and programs in particular localities by drawing the attention of planners to linkages, multipliers, and efficiency considerations in public investments. Third, by working within the urban-regional frame, planners may be led to the discovery of new combinations of resources which, if properly exploited, will result in an expansion of the total productive capacity of the economy. Fourth, the urban-regional frame reveals a set of entirely new issues for policy planning, such as settlement patterns, the territorial division of powers, and problems of community participation.

Where urban and regional planning have been introduced at the national level, they are frequently seen as efforts parallel to sectoral planning, with global planning providing the overall integration. This view is incorrect, however. The urban-regional frame, in fact, allows for a second level of synthesis or integration. One may think of these relationships taking the following form:

global planning . . . . Synthesis I

↓ ↑

sectoral planning

↓ ↑

urban-regional planning . . . . Synthesis II

As a lower level of synthesis, urban-regional planning helps to bring the abstract thinking of global and sectoral planners to the ground. Because it represents a new and different level of synthesis from global planning, it requires the formulation of policy objectives appropriate to its own level of synthesis. The urban-regional frame provides for more than just a translation of global objectives into the language of concrete experience. Constituting a normative realm in its own right, it must somehow be incorporated into the structure of objectives as these appear at global and sectoral levels of national planning.

Urban-regional planning deals with matters that are closer to the immediate concerns of people than either of the other two approaches. For this reason, it also moves closer to the political pulse beat of the nation. Its concern is with the points of impact of investment decisions, and it is precisely here that the bitterest conflicts are engendered. Because of this intimate connection with problems that move people to action, a major task for urban-regional planning is to offer new perspectives and an objective frame for ordering the preferences of local communities within a context of national requirements.

Seven policy areas may be included within the urban-regional frame: (a) regional economic policy, (b) migration and settlement policy, (c) urban development policy, (d) housing policy, (e) urban land policy, (f) administrative and political development policy at subnational levels, and (g) social development policy. All of these areas cut across more than one functional sector and show a high degree of interdependence among themselves. They are consequently not covered by standard sectoral analysis, nor are they explicitly considered in global planning: the mesh of the global planner's net is far too loose to catch the regions and cities of a nation. Each of the seven policy areas is separately described in the following pages, providing a comprehensive view over the scope and substance of urban and regional planning in national development.

REGIONAL ECONOMIC POLICY

The basic question of economic policy at the subnational level is *what* investments should be located *where.* It might be asked whether there exists a theoretical basis for allocating investments to specific geographic areas of a nation. Cost-benefit analysis may be of help in the case of individual projects, but it is less useful as a guide where projects are so interrelated that their total (as distinguished from their marginal) social impact must be assessed. The difficulties are compounded where project details fail to be provided as so often occurs in developing countries where a shortage of capital may be less of a constraint than a shortage of specific projects capable of being funded. In the case of programs where the projects themselves are poorly identified, a thorough knowledge of regions, their structure and development potential, is more immediately useful in deciding budgetary allocations than a set of formal criteria for helping decide particular investments.

If a national perspective is to be maintained, the consequences of location decisions must be evaluated not only in regional terms but also from a standpoint of national objectives. The overriding objective in most developing countries is to strengthen the national economy through a closer integration of its several regional economies. Integration in this sense involves territorial specialization, increased interregional trade, the diminution of regional differences in per capita income, and the transfer of labor from low-productivity to high-productivity regions.

I shall refrain from a discussion of the technical issues involved in making analyses of this kind, except to highlight the importance of a concept of core-region development for regional analysis. Core regions may be thought of as equivalent, in spatial terms, to the leading production sectors of an economy. The main idea underlying a core-region strategy of regional development is that investments should be concentrated in a small number of areas identified as having a high potential for economic growth. These areas are usually centered on cities of metropolitan size. The strategy for these areas envisions a planned program of coordinated investments in infrastructure and more directly productive activities which will be carried out over a period of time sufficient to permit the core economy to be pushed across the threshold of self-sustaining growth. Core-region development is a way of encouraging city-ward migration and of engendering structural changes in the countryside, as traditional agricultural practices are subject to modernizing influences in response to increased urban demands for the products of agricultural labor and rising labor costs.

However desirable such a strategy may be, considerable difficulties are encountered in implementing its precepts. Because the development of core regions requires an intensified mobilization of resources, it suggests a corporate administrative structure. For areas other than core regions, a less dramatic approach may be more appropriate. Its success will hinge primarily on the ability of planners to control the regional allocation of the national capital budget.

An important aspect of regional development policy is the coordination of individual investment projects. It is not enough to achieve formal coordination at the budgetary level; coordination in the field is equally necessary. For this reason, it is well to think of a network of planning offices at the regional level, linked to the national planning mechanism, which can prepare detailed regional studies, offer advice to regional administrative offices, defend and promote regional interests, and serve as coordinating centers.

MIGRATION AND SETTLEMENT POLICY

It might be thought that the pattern of migration and settlement will be determined exclusively by pursuing regional development policies as described above. Identifying migration as a special policy issue, however, serves to move into focus a different dimension of the problem which cannot be dealt with solely on the basis of considering alternative investment budgets by regions.

Migration policy is concerned with guiding both the volume and direction of internal population movements. Regional planners are lacking sharply honed criteria for determining a policy that will ultimately result in a changed spatial distribution of settlements ordered according to function and population size. This fuzziness of decision rules for determining an optimum settlement pattern for the nation should not, however, deter planners from seeking greater precision in their analyses. It is evident by now, that the spatial distribution of settlements is not simply a passive variable, indifferent to the processes of development. We are familiar with the extreme conditions where there is either an insufficient concentration of the forces of production for agglomeration economies to occur or an excessive concentration, so that the costs of further growth in the same area fail to produce commensurate social benefits. The excessive concentration of economic activities in a country, usually in the national capital or principal port, is capable of slowing down the development of the nation as a whole, by draining the periphery of its productive resources and reducing it to a quasi-colonial status. Under these conditions, only those resources of the country's periphery are likely to be developed which are needed for the further expansion of core-region economies. This will lead to a less than optimal development for the total set of regions in the nation.

By drawing attention to the pathological conditions of insufficient or excessive concentration, regional planners implicitly support policies for strengthening the economic base of intermediate cities in the country's periphery. But that is only one aspect of the problem. Major core regions will themselves have to be laid out in space, so that its diversity of individual communities creates a livable and efficient metropolitan environment. And finally, at the lower end of the urban hierarchy, a relatively large number of small towns and cities needs to be strengthened to carry out the function of serving rural areas as marketing, distribution, and service centers. A migration and settlement policy must encompass the full spectrum of central places in the nation.

The role of small- and medium-sized centers is especially important for what I should like to call the *urbanization of the countryside.* Agricultural development means to bring traditional peasant cultures into closer contact with urban values and implies the commercialization, mechanization, and rationalization of agricultural production, the organization of farming into cooperatives, increased literacy, improved health care services, and increased opportunities for the young. These objectives are best achieved by concentrating appropriate programs on rural towns and by drawing scattered farming populations into villages. Settlement policy is based on the fact that the national economy is organized through its urban centers, each of which has a set of functions and a geographic area of influence appropriate to its level in the hierarchy.

A migration and settlement policy may also play a significant role in opening new frontier regions to colonization and in moving populations from unpromising areas to zones of greater economic opportunity. It may be carried out through information, subsidies, and action programs of various kinds. A good deal of experience has been built up in this regard; I refer specifically to resettlement schemes in Israel, Germany, and the Soviet Union where migration policy has been used, in part, as a major tool for national development.

URBAN DEVELOPMENT POLICY

This dimension of urban-regional policy is specifically concerned with the internal structure and form of urban, metropolitan, and regional communities to the extent to which these act as total environments for living. Here, social are joined to economic considerations, as the attempt is made to build, at whatever scale may be appropriate, a fitting home for man. The pattern and web of relationships is of great concern in this connection. Where are the jobs to be located with respect to the residences for individual families? Is the internal road network of the city adequate? Is the city efficiently tied into the economy of other urban centers through routes of transportation? Do existing community facilities, such as schools, hospitals, commercial areas, recreation spots, serve the needs of an expanding population? Is the social mixture of the city in conformity with desired norms, such as inter-occupational mobility, interclass harmony, or acculturation? In which physical direction should the city grow? How can conflicts in land use with adjoining agricultural areas be minimized? Are public services and utilities sufficient to take care of the expected growth of cities?

Because staff time is always limited, and nowhere more so than in developing countries, a nation that undertakes to formulate an urban development policy will have to concentrate at least initially on core regions identified at earlier stages in the analysis. Urban development policy thus becomes a means for linking metropolitan development to national policies and programs.

The relevance is often immediate and direct. Many countries are considering rapid transit systems for their capital cities. The investment required for these may account for an appreciable proportion of the total national investment budget and an even greater proportion of the foreign exchange budget for many years. It is, therefore, important to know whether such a system should be built at all, what form it should take, in what sequence it should be constructed, whom it should serve, and when its construction should be started. Such decisions cannot be made without taking into account many related considerations having to do with the rate of growth of the city, its physical spread, and its internal structure. A decision such as this throws into sharp relief the interconnected nature of many policies and decisions.

In this brief account of urban development policy, I have purposely left out two policy issues which, because of their magnitude, are best discussed under separate headings. I refer to policies for housing and land development.

HOUSING POLICY

Housing policy is basically concerned with (1) the provision of shelter and essential public services (water, waste disposal, electricity, and transportation), and (2) the building of community environments that will facilitate constructive social processes (family formation, upbringing of children, acculturation to urban ways of living, social mobility, and social integration). These are needs whose satisfaction depends, in part, on public intervention.

Rapid urbanization makes the formulation of housing policy unavoidable. Urban growth rates of 4-7% are common throughout the underdeveloped world, and the growth rate of shantytowns is even greater, rising above 15% in certain areas. It is evident, therefore, that countries undergoing accelerated urbanization must come to grips with the problem of how to absorb the rapid influx of population from rural areas. Since the capacity to pay for housing services is very low among migrant populations, massive public investments are generally called for. At the same time, the urban housing problem has its counterpart in rural districts where housing investments may serve two related purposes: first, to slow down, if possible, the rate of city-ward migration and, second, to assist in the urbanization of the countryside. An urban-regional frame for national development provides new ways of viewing these complex issues.

One set of questions deals with housing policy as a *tool for development.* Housing is normally regarded as a form of social investment and is therefore treated as a residual category remaining after urgent economic priorities have been taken care of. The distinction between social and economic investments, however, is an essentially arbitrary one. For housing investments can also be used to promote a number of important economic objectives. Four of these are mentioned below.

a. Mobilization of Unused Resources

Housing policy can be used to promote savings and to draw unused labor reserves into production. It is a well-demonstrated fact that people are generally willing to save for a home more readily and with greater persistence than for nearly any other purpose. And once in possession of shelter adequate to their needs, they will also save for appliances such as furniture, stoves, refrigeration, and television sets. Housing programs may therefore be devised to serve as an incentive to save, an important considerations in countries where the rate of savings is low. The use of unemployed labor, on the other hand, refers to the more efficient use of the education, skills, and energy stored up in human beings. The introduction of self-help measures and of wide community participation during the planning, construction, and management phases of housing schemes is one of the chief means for valorizing existing human resources.

b. Channeling of Migration

Housing investments may be used to guide migration to points of maximum development potential and possible also to reduce the flow of migrants from the countryside. The relationship between housing and migration has not been proved statistically as yet, and its existence may be accepted as only a reasonable hypothesis. The hypothesis states that one of the major attractions of migration to the city is the possibility of gaining access to superior housing services. The judicious location of these services would therefore tend to attract migrants in larger numbers. To furnish equivalent services in rural areas might, by the same token, hold a larger proportion of the population in villages and rural service centers, easing the pressure of population on the cities. The assumption here is that the costs of rural housing investments will be significantly lower per family than their equivalent in cities.

c. Increasing Labor Productivity

This relationship also needs more thorough documentation. It is widely believed that the provision of some minimum of housing services is necessary to obtain the maximum labor contribution from a given work force. Such services would tend to reduce absenteeism, provide direct incentives to work, improve health conditions, and in general tend to make for more responsible attitudes towards work. This is undoubtedly the reason why priorities are often assigned to worker housing in key industries and why employers often express a direct interest in the housing conditions for their workers.

d. Developing the Construction Industry

The construction industry is one of the most important sectors in an economy, and housing accounts for between 65 to 75 percent of its total output. The industry is also highly related to manufacturing (building materials, appliances, furniture); it has a high value added; it uses a high proportion of labor relative to capital equipment; and it has very low import requirements. Given these characteristics, housing policy may aim at stabilizing the construction sector which normally displays quite an erratic growth pattern. This must be done at a sufficiently high level of overall activity not only to meet the minimum needs of the population for housing but also to use the expanding capacity of related industries and to serve the interests of a general policy for reducing unemployment.

A second set of questions in housing policy relates to the building of healthy communities. Here, emphasis is given to the integrative role of community facilities, such as schools, health clinics, social centers, sport fields, places of worship, shopping areas, and administrative buildings (police, fire, civil register, etc.). The absence of these and other public services turns low-income housing areas into the dismal, anti-social slums they are. Housing-development schemes would do well to begin with providing the central facilities for the community and to relate the building of shelter space to them. Related issues are the mixture of social groups to be housed within the same community; home ownership versus rental housing; and location of housing schemes in relation to potential job locations.

The urbanization of the countryside is related to these questions. It envisions the provision of urban-type community facilities and public services to rural populations. Since it is difficult to serve dispersed settlements efficiently, a rural housing program implies a strengthening of villages and rural service centers. The objectives would be to raise the educational and health levels of the rural population, to introduce them to commercial practices, to raise their aspirations, to facilitate their eventual absorption into urban society, to facilitate their organization into farmers' unions and cooperatives, and more generally, to increase the participation of peasant populations in national life.

URBAN LAND POLICY

Housing and urban land policy must be closely coordinated. Through control over land use, the growth pattern of cities may be guided, especially the pattern of urban densities. The objectives would be to control speculation and to manage land in the interest of the whole community.

A special problem occurs here in connection with new migrants to the city. Traditionally, they settle on the fringes of the built-up area of the city, establishing squatters' rights in vast and insalubrious shantytowns. The land onto which they move, however, is not their own, and being illegal occupants of the land, they are rarely provided with even the most fundamental social services. In some cases, squatters' districts may simply be left out of official city maps; as far as the government is concerned, they do not exist.

Urban land may be so managed, however, that new arrivals to the city are brought into reception areas where basic services are provided and where the migrant is assisted in establishing his first foothold in the city. He may be assigned a parcel of land in areas to be newly urbanized; he may be organized into self-help groups and rudimentary cooperatives; he may be assisted in finding a job; he may be given elementary medical attention; and his children may be sent to school. Obviously, this requires more than just the proper management of land, but such management is an important element in the total program.

Acquiring legal title to a piece of land, either in outright ownership or on long-term lease, is a basic condition for accommodating the recent migrant to urban society and for preventing the formation of a sub-proletariat excluded from the mainstream of social life. This is really what squatting is all about. To come into inexpropriable possession of a plot of land, measuring a few square meters, is among the most difficult tasks for recent arrivals to the city. Devising a method for allocating land to families and financing its acquisition is consequently a crucial aspect of urban land policy in developing countries.

ADMINISTRATIVE AND POLITICAL DEVELOPMENT POLICY

With some notable exceptions, the typical situation in developing countries is an extreme degree of centralization of political and administrative authority. Consequently, provincial and municipal governments are weak and ineffective.

The conflicts produced by this pattern must be resolved, if only because after a certain point in the complexity of economic relations is reached, further centralization tends to slow down rather than stimulate further economic and social progress.

In countries where cultural regionalism is strong, pressures for decentralization are especially severe and may pose a direct threat to national unity unless demands for greater autonomy are met. The forces engendered by economic development strike at root ideas of nationhood and will engender, if development is continuously successful, a new pattern of political relations. This may be characterized by reduced central dominance, a wider sharing in political power by governmental units, greater diversity of political opinion and attitudes, greater responsiveness of government to local interests, and greater flexibility of action.

What are the key issues in administrative and political development policy at subnational levels? One concerns the type of administrative arrangement appropriate for regional development. By what means is core-region development to be achieved? How are major resource development programs to be organized? How much power should be devolved to regional and local offices for the coordination of centrally financed development? What should be the nature of these offices?

A second issue is closely related: what geographic units are optimal for the administration of development and possibly also for the organization of political power below the level of central government? This question may be studied with respect to metropolitan districts, special development districts, and groups of provinces.

A third issue concerns the degree of participation by provincial and municipal governments (or by newly constituted local authorities) in development decisions. At question is the extent to which central authorities should exercise control over development processes in local areas.

Of all the policy areas, this is the least tractible from a planning standpoint. Historical circumstances differ so greatly from country to country, that generalizations are of little value. A certain amount of expert knowledge can be brought to bear on each of the questions posed (measures of regional interdependency, administrative organization of regional development corporations), but the overriding considerations remain political.

SOCIAL DEVELOPMENT POLICY

This is the last of the policy areas dealing with subnational development. Social development cannot be represented in a national product account. It is a process that, although rooted in economic relationships, clearly transcends them. The principal objective of social development policy is the elimination of social marginality and the active involvement of all social sectors in affairs that are vital to their interests.

Who is included among the socially marginal sectors? In the extreme case, it is the Indian in Peru who is excluded from Spanish and mestizo society, who is purely an object of history and not yet a self-motivated actor on the national stage. The North American Negro is in an analogous position, especially in the southern states, but he has started to overcome his passivity and has begun to act on his own behalf. The poor, illiterate masses anywhere in the world are, for the most part "marginal" to the dominant society. They are not organized, they are not informed, they lack material means to defend and even to express their interests. They are acted upon but not acting. Where significant numbers of the population are thus divorced from the central processes of decision-making and control, it is possible to speak of nationhood in only a relative sense. There is the effective nation which includes a small class of aristocrats, professionals, businessmen, and perhaps clerics, and there are the inarticulate and vaguely structured masses who somehow belong to the nation in a legal sense—they have to pay taxes, they must serve in the armed forces—but who are excluded from the actual management of public affairs though they may be entitled to a vote in national elections.

Social development acquires significance within the urban-regional frame because the phenomenon of marginality has an important geographic dimension. Rural migrant populations living in the shantytowns of the big cities are, for the most part, marginal to the society at whose doorsteps they are camping. Peasants and farm workers tend to be marginal to the dominant society, especially where feudal conditions still obtain; peripheral populations are marginal with respect to the core. Degrees and types of marginality will vary and often the matter of definition will be difficult. Still, these broad relationships seem to hold in many societies. And where they do, the urban-regional frame requires the choice of a social approach to problems of economic, physical, and politico-administrative development.

This may not be immediately apparent. Let us suppose, therefore, that a scheme of irrigated farming is being considered which, combined with rudimentary industrialization, mostly involving the processing of agricultural raw materials, will raise the living levels of farm families involved in the program by fifty percent. But which families will participate in the program? The most capable, the most successful ones, to be sure. And the marginal peasants and day laborers? They will scarcely be touched, unless special programs are designed to take their needs into account and so to raise them gradually from the obscurity of poverty to a new level of dignity. Such programs must at the very least include provisions for expanded economic opportunities of the marginal population, an explicit redistribution of real income in their favor (chiefly through provision of subsidized social services—housing, education, and the like), and a system of organized representation of their interests (local councils, social centers, unions, cooperatives). The formulation of these programs and the principles on which they are based, at all appropriate levels, from locality to nation, constitute the substance of a genuine social development policy.

THE ADMINISTRATION OF URBAN-REGIONAL POLICY PLANNING

The temptation is to enclose the seven policy areas described above within a structure of comprehensive planning. Because they are obviously closely related to each other, it is easy to argue that all specific policies should be filtered through a single intelligence and coordinated from beginning to end. This attitude, widespread in planning circles, is reinforced by the attractiveness of a systems approach to development planning which emphasizes the maintenance of a series of equilibrium relationships in the course of economic growth and transformation.

Insistence on this point, however, ignores the existing institutional framework for policy planning and the limitations of administrative control. The policy areas desribed are functionally related, but each of them also has a distinct focus. Relations between social development and regional economic development, for instance, are not immediately apparent. Their objectives, underlying assumptions, and instrumentalities for implementation are quite different. The spatial framework accommodates them both, but administrative organization for planning and programming should be separate for each. Indeed, we are now in a position to see more clearly that urban and regional development is coextensive with national development; it is simply national development brought to the ground. Approaches to policy formation must therefore be adapted to the administrative structure that exists for all forms of development.

This should not be read as an argument against institutional innovation. On the contrary, the effective realization of urban and regional policy planning requires a new organization base. But it is desirable to start by regarding each major policy area as a separate (and separable) subsystem for this purpose, allowing these subsystems to articulate their relationships as the work advances and experience is gained. The need for coordination must be felt if countervailing measures are to succeed. Coordination always implies a cost, not the least of which is the cost of freedom of action. No administrative agency will therefore acquiesce to being coordinated simply because such relationships are postulated theoretically. Coordination will be sought only when the purposes of an agency are blocked by a failure to coordinate its program.

To allow a separation of functions to occur is also a means for minimizing risks. In no case can equal advance in all policy areas be guaranteed (aside from the fact that it may not always be possible to tell just what would represent an "equal" advance). By isolating effects through minimizing coordination in advance, the success of some programs may be assured even if other programs should fail or lag behind. Because each major policy area refers to a subsystem of the urban-regional frame, the margin of potentially independent action within each subsystem is quite large. As the total system matures, interrelations will tend to increase in volume, frequency, and strength, imposing greater needs for coordination. But that comes during later stages, when a considerable body of experience has accumulated, and each subsystem has evolved to the maximum extent possible for a given pattern and degree of coordination.

Information is a key to the operation of the planning system as a whole. Establishing appropriate information processes is therefore a leading task for policy planners. The difficulties of this undertaking should not be underestimated where interagency jealousies are great, administrative secrecy is habitual, and information is considered a weapon in the struggle for administrative survival. Nevertheless, it must be attempted. The information referred to may include empirical studies, statistical reports, program progress reports, and planning documents which set forth targets and priorities for different areas. All these should be made freely available to agencies which have responsibilities for different aspects of urban and regional development. Routines must be established for this purpose. As one of the principal users and generators of planning information, the national planning agency is strategically positioned to establish regular communication flows by insisting on standardized reporting procedures.

Chapter 9

TOWARDS A NATIONAL URBANIZATION POLICY*

The positive contributions of urbanization to national development were slow in being recognized. Conservative politicians and social scientists saw ramshackle, makeshift housing pile up along the fringes of the traditional city and hastily concluded that migration to cities had been excessive and ought to be rolled back. They saw the large metropolis grow ever larger, claiming a giant's bite of high-skilled jobs and income, and reacted to bigness as though it were some rare variety of megacephalism. They saw the city seething with popular discontent, and prophesied that the established orders would soon yield to the rabble in the street.[1]

The vision of the older urban critics was fixed on the individual city and the life within it in a foreshortened, partial perspective. Development, they thought, should be a process of gradual, harmonious advance. But a radical turnabout occurred when the time horizon relevant for judgment was extended further into the future, when the individual city came to be regarded as only one element in a complex network of transactions, when process was substituted for artifact as the primary object of social inquiry, and when modernization came to be viewed as a more encompassing concept than economic development.

This change in perspective produced a series of entirely new questions. If national development were to be regarded as a process of social transformation, was it, in fact, reasonable to expect that it might occur without substantial strain and conflict? What were the sources of innovation in society? How did

*This chapter was written with the assistance of Eileen McGlynn, Barbara Stuckey, and Chung-Tong Wu.

innovations spread, and what tensions might the unequal diffusion of innovations across the national territory produce in the social system? Might not these tensions be productive of further transformations in a self-propelling cycle of challenge and response, and would they not ultimately lead to a society capable of *continuously* generating and absorbing innovations.[2]

This reassessment may be dated back to 1966, when a conference on "The Role of the City in the Modernization of Latin America" was convened at Cornell University. The organizers of the conference had decided that the focus would be on what they called the positive functions performed by the city in the modernization process. Their aim, candidly stated in the preface to the conference proceedings, was "to identify the components of urbanization that could be accelerated without unduly exacerbating those problems normally (perhaps even necessarily) associated with rapid urban growth."[3]

Substantially, this is also the point of view adopted in the present volume. The accelerated growth of cities, if properly guided, may lead to greater openness in the system of power and improve the probabilities for innovation.[4] The city has not only invented the concept of unlimited opportunities, it has also made it possible and has internalized it as the basis for its own unique form of culture.[5] And the accelerated growth of cities may hasten the coming of a national mass politics based on bargaining and compromise among a growing number of competing interest groups.[6]

In this view, urbanization is no longer regarded as an unfortunate by-product of planned industrialization whose consequences governments should learn how to constrain, but as a set of powerful developmental forces in its own right. The following two case studies may help to illustrate how urbanization in both its socio-cultural and demographic meanings may exert a positive, constructive influence on developmental processes. The first concerns the transformation of a backward, semi-nomadic tribal community in Turkey into a thriving township; the second describes the progressive spatial integration of a new national society in Kenya. In both instances, urbanization processes proved to be decisive for the results.

THE DIFFUSION OF INNOVATIONS: THE TRANSFORMATIONS OF A TURKISH VILLAGE[7]

In 1940, Erdemli was an insignificant, semi-nomadic village in southern Turkey, clinging precariously to the shores of the Mediterranean Sea. Its inhabitants maintained only minimal contacts with the outside world. The community was tradition-directed and almost completely self-sufficient. Communications with the government were sporadic and consisted almost entirely of messages and instructions from the central administration. Attitudes toward strangers, including government officials, were hostile.

Two decades later, Erdemli had grown into the bustling administrative center of a prosperous agricultural region specializing in citrus and truck farming. Population had multiplied several times and had been augmented by migrants from adjoining areas who had come to Erdemli in search of economic fortune. From its meager beginnings on the margins of Turkish society, Erdemli had been pulled well into the periphery of the national system. Its relationship to the central authorities continued to be tutelary, and few local actions could be undertaken without the approval and material support of the central government. But increasingly, Erdemli was coming to share in national decisions and to take advantage of opportunities correctly perceived and interpreted. It had attained the status of a municipality and administrative center largely by its own perspicacity. Its local economy, expanding with amazing speed, was managed primarily by local people. By 1960, the town had become a political factor to be reckoned with in Ankara.

How had this transformation been achieved? How had the new demands for urban necessities such as radios, bicycles, running water, electricity, and schools been generated? How had a people who only a short while ago had herded sheep and goats acquired the skills for successfully manipulating national political institutions? The story is a complex one, but a beginning might be made by looking at the major agencies of change that has been introduced to Erdemli during the "heroic" period of its history.

Major impulses for modernization were apparently transmitted through a small number of institutions and facilities that acted as efficient channels for the diffusion of innovations from Ankara and Istanbul. Special importance must be attributed to the establishment of two credit institutes, a cooperative serving primarily the farming community and a bank which made possible the phenomenal increase in specialized business ventures; to the paving of a highway connecting Erdemli to the nearest larger city which not only reduced to a small fraction what had formerly been an arduous eight-hour journey but brought the local economy within the radius of national (metropolitan) markets; to the establishment of Erdemli as administrative center for a local district which led to the arrival in the area of an assistant governor and other bureaucratic specialists; to the creation of political party organizations in the district which provided a direct link to national centers of influence and created a new level of political sophistication among the population; to the introduction of the radio and metropolitan newspapers which served as sources of information about the outside world; and to the establishment of commercial enterprises—coffeehouses, restaurants, gas stations, and even a hotel—where information might be exchanged and opinions formed, and which set up the lowest link in a chain of economic relations that led up to the highest centers in the urban hierarchy.

The combined effect of these "implants" was to incorporate Erdemli into the economic, political, and cultural space of the nation. The people of the area responded to this new situation with great energy and imagination. They were able to translate newly felt desires into concrete projects serving their own interests. Although social traditions had changed more gradually, the physical appearance of Erdemli in 1960 presented a radically new aspect. The erstwhile village had grown to the size of a substantial town, tents had given way to houses of stone, electric lights were glimmering in the streets at night. The diffusion of core region innovations–banking, political democracy, education, mass media, automotive transportation–had affected other urbanization processes–control, migration, and investment–in a major way. Dependency relations to the core had multiplied, but authority had become more responsive to local demands. Migration to Erdemli had increased at a rate faster than the movement of local people to outside urban centers, and the flow of national capital to the region had been augmented. Although by 1965 Erdemli was still far from being incorporated into the national core itself, the earlier nomadic village had acquired a future in which both individual and collective effort might be rewarded with effective change.

THE SPATIAL INTEGRATION OF KENYA[8]

By subjugating the dominant African tribal societies at the beginning of the twentieth century, European colonization of the Kenya highlands paradoxically laid down the conditions for the emergence, a scarce two generations later, of a modern nation-state on the same soil. For under the shadow of the modernizing, impulse-generating core areas centered on the city of Nairobi and controlled essentially by Europeans, there grew up an African periphery which, though in all respects dependent on the core, was yet integrally a part of the nascent economic system. Beyond these two interlinked areas lay the traditional grazing lands, sparsely populated, and nearly unaffected by the profound changes that were proceeding apace in the rapidly expanding "islands of innovation." Even today, the primitive people inhabiting these vast savannas lie beyond the pale of national control. The history of contemporary Kenya was forged exclusively in the dialectic struggle between the modernizing core and dependent periphery of the spatial system.

That part of the predominantly African periphery which lay closest to the European core received the direct impact of the modernization processes as these spread inland along the major transport and communication corridors from the coastal city of Mombassa to interior Nairobi and beyond. This diffusion process has been succinctly postulated by Whebbell:[9]

> "Corridors comprise the historical-spatial system of innovation diffusion, the progress of which leads . . . to differentials in the mix of material, organizational, and philosophical innovations from place to place. Such differentials are postulated to have a degree of regularity termed 'culture gradient' which extends downwards along the corridor from the senior town most accessible, historically, to Europe.
>
> "Secondary diffusion takes place laterally from the corridor into the areas between and away from the corridor. Socio-economic space, in this conceptualization, consists

> of irregular cells of which the bounds are corridors of movement and diffusion. The interior of such cells shows a pattern of socio-economic alignment towards the senior corridor boundary and a dearth of large central places, and thus the emphasis in urban systems is seen to be marginal rather than central *(sensu stricto).*"

In Kenya, this pattern was modified only to the extent that the rather insalubrious lowlands intervening between Mombassa and the highlands were judged unsuitable for European settlement.

The African periphery of Kenya's emerging spatial system provided a convenient pool of cheap labor to the European settlers who governed the country from administrative seats in the principal urban centers. The majority of settlers, however, were farmers. There was little industry as yet in Kenya, and the principal functions of the towns were political and commercial. But towns were also the principal transmitters of imported innovations, and the penetration of the periphery by urban innovations–a money economy, a common (though foreign) language, a common administrative system, modern rail and motor transit, health services, postal and radio communications–was inevitable. Reinforced by spatial proximity and interaction with the core area, the periphery was penetrated by modernizing influences that created what must have been an entirely unexpected and certainly unwanted result for the dominant elites: the birth of an energetic African counter-elite capable of challenging the prevailing authorities and the order they had established. The ensuing conflict led to several unsuccessful attempts at violent repression and concluded, barely a decade ago, with the replacement of the white elites by an African counter-elite who physically moved in to take over the existing core settlements, declare the nation independent, and announce a new political agenda. The classical drama of core-periphery conflict had thus once more been reenacted.[10]

At the time of independence, the effective national territory of Kenya–its modernizing regions–constituted only about one-fourth of the national area; the remainder was "functionally not part of the Kenya nation in that the critical level of consensus that would permit centralized planning to be accepted and take effect has not yet been reached. . . . Political allegiance is unsettled and erratic. But, in addition, the level of participation is extremely low and, more important, the frequency of internal conflict is such that the impress of the central government appears weak and ineffective."[11] By comparison, the effectively incorporated portions of Kenya constituted the principal transaction space of the nation. Composed of a terraced hierarchy of regions–core, national nuclei, participant, and transitional areas–the organization of the spatial system reflected the internal logic of the basic urbanization process at work. A national periphery still exists, of course, though the dominant elites today are African, and their main objective is to accelerate the spatial integration of a still-fragmented nation.

Six decades of colonial rule had served to accentuate the gap between ethnic groups by superimposing upon them a highly uneven pattern of modernization.

These differences, translated into political action, appeared to threaten national stability. No effort therefore was spared to consolidate the participant and transitional zones, and tie them securely into the spatial system. Postal facilities were dispersed to equalize service throughout Kenya. Urban administration was reorganized by centralizing many of the functions of local government in the former White Highlands, while new authority over local affairs was granted to many of the townships in the former African periphery. A national radio and television network was created. New roads connecting major urban centers were built; others were improved at great expense. The national school system was expanded. Attempts were made to guide the location of new industries away from Nairobi to new development areas. And a vast program of agricultural resettlement was launched in the former White Highlands. It may be uncertain whether these efforts will achieve all of the results intended, but the stability and cohesion of Kenya vitally depend on them.

SOME QUESTIONS FOR NATIONAL PLANNERS

If the influence of urbanization on national development, operating through the agencies of control and investment, innovation and the diffusion of innovations, and migration are indeed as powerful as they appear to be, ought they not to be made the subject of public policy and selective intervention? Economic growth and social equity have become accepted as major objectives of governmental policy throughout the world. Spatial integration may now be elevated to the same high level of national consciousness as a goal worthy of political commitment and essential to the realization of sovereign nationhood. In this sense, national urbanization policies are likely to become as important to this generation of political leaders as industrialization policies were to the last.[12]

Accumulated experience with urbanization policies has not so far been able to provide unambiguous guidelines to policy makers, and the theoretical basis for urbanization policy is still weak. This ambiguity, if not uncertainty, about the proper course is best brought out in dialogue. Imagine, therefore, the Head of State of an imaginary country deep in the throes of industrialization, modernization, and national development confronting his Minister of National Planning with a series of penetrating questions preliminary to working out the bases for a national urbanization policy. Being a person accustomed to proceed systematically in the analysis of complex issues, the Head of State has arranged his inquiry into four broad topics: the rate of rural to urban migration, the spatial pattern of investment and migratory flows, the rate of innovation-diffusion to non-urban areas, and the territorial organization of power best suited to the implementation of national urbanization policies. How might he have formulated his questions, and how might his minister have answered him? I shall select only portions of the dialogue.

1. The Rate of Rural to Urban Migration

The Head of State speaks:

> "Should the movement of population to our major urban centers be encouraged or slowed down? Is not large-scale urbanization at a rate several times that of the increase in the per capita gross national product an extraordinarily costly process, not only in economic, but in social and political terms as well? Is not the amount of social capital required for urban housing and transportation, for example, excessive, given that our cities are currently unable productively to absorb more than a fraction of the newcomers from rural districts and small towns? Is there not an inherent danger for political stability in a vastly mounting urban subproletariat, empoverished and poorly integrated into the national society? How can urban growth proceed in an orderly fashion so long as migrants are sprawled out, chaotically, on the edges of our large cities, occupying land illegally and falling prey to social alienation and crime? And if, after all, the Government should decide to undertake a major rehousing effort, to settle tens of thousands of these squatters in permanent communities under reasonably sanitary conditions and with a minimum of public services provided, would not this growth of social amenities simply accelerate the inflow of migrants, leaving the final situation as bad, if not worse, than before? Under these conditions, would it not be altogether preferable to attempt holding urban migration to the rate at which employment oppotunities can be created in the modern sector of our economy, especially in manufacturing? Should not the Government assign priority to an energetic policy of rural development (and industrial decentralization) capable of holding farmers and their families on the land or, at least, channel as many of the migrants as possible–since some migration is probably unavoidable–to the smaller towns and villages whose economies are closely intertwined with agriculture and the rural way of life?"[13]

And the Minister of National Planning replies:

Your Excellency, may I ask you to consider an extreme and, to be sure, quite hypothetical question which, I believe, lies implicit in your own thinking. Are large cities necessary? Can we carry on with our current efforts at industrialization without continued urbanization? Can the present trends be reversed?

As Your Excellency is undoubtedly aware, urbanization is a universal phenomenon. In the Western world, it has been closely associated with the expansion of an industrial base. But we also know that some of the industrially advanced countries, such as Switzerland, have been able to preserve an essentially small-town culture. Can Switzerland's experience be replicated? The answer to this question, I believe, is that it can not. We would need a small country in area and population size that has had an ancient tradition of town dwelling, where local government has exercised important prerogatives for a long time, where the people pride themselves in being citizens of small places like Solothurn, Graubunden, Uri . . . none of which has more than 20,000 inhabitants. Power in Switzerland is highly dispersed. Much autonomy resides at municipal and Canton levels: the Federal government has been purposely kept weak. Yet Switzerland is also a modern country. Her economy is held together by one of Europe's best road and railway systems. Telecommunications tie even the smallest village into a national network. These, I believe, are a few of the reasons why Swiss cities have not grown to a large size: the different parts have grown so interdependent that, in a certain sense, the country has become a single city.

Now it is probably true that the fundamental reason for urbanization is the universal desire for a high degree of "access" to a variety of goods and services. The existing access pattern may be measured by an index of "potential information exchange."[14] Given a measure of differential access extending over the landscape, activities will sort themselves out according to their relative urgency to be near to other activities. This sorting-out process, however, will only tend to reinforce existing access differentials. Much time is needed to bring about significant changes in the spatial structure of a social system.

In order to industrialize a country without extensive rural to urban migration, all inhabited parts would have to gain nearly equal, and increasingly high, access to the goods of a developing society. To make this possible, it would not only be necessary to lay down the physical means for high access (transportation and communication), but to balance the rate structure governing the interaction process so that populations living further away from desired access points would be compensated for their relative locational disadvantage; the time factor, could not, of course, be entirely eliminated. In any event, some physical control over the location of the mobile factors of production would probably be needed in order to reduce the movement of labor to cities (except on a temporary basis) and to steer the location of economic activities in accord with the relatively fixed distribution of labor. Any competitive disadvantage arising from such a policy would have to be made up by subsidies. A similar policy would need to be followed with regard to the location of cultural facilities and services, such as schools, research institutes, and even the Government itself, all of which would have to be radically decentralized.

As Your Excellency can see, the provision of even reasonably equal access at our present stage of development might turn out to be prohibitively expensive. Primitive transport and communication means would prevent industrialization; more adequate means, however, would absorb so large a percentage of national capital resources that little would remain for increasing productivity in the economy.

Given the important role that Government would play in the construction and maintenance of such a system—assuming that it might be done—it is clear that an essential condition for the successful implementation of an industrialization program *without* urbanization would be the effective decentralization of power, since concentrated power in a nation such as ours acts by itself as a major influence on location. Aside from the question of its feasibility, such a governmental structure might devolve essential power to the more traditional social elements, such as rural landlords, and tribal chieftains who might be reluctant to encourage industrialization. Furthermore, the policy contains an internal contradiction: the development of equal access must be guided by a powerful central authority. Yet the system requires that power be locally retained. I doubt whether this contradiction can be successfully resolved.

Since with technological change (and changes in external demand) certain regions would necessarily be favored (a spatially randomized economy is impossible to conceive), the Government would have to carry out a conscious regional income redistribution policy. This would consistently penalize the more "progressive" or "successful" regions in order to benefit their poorer neighbors. If every gain were immediately to be taxed away, however, the entire system would lose its internal drive to respond to new challenges, unless a different set of motivations, more powerful than income, could be invoked. This, however, would presuppose the imposition of very tight central controls and a justifying ideology.

What I am arguing is this: labor mobility and, specifically, urbanization, is a means to economize on resources and thus permits the more rapid build-up of productive activity. It may also save on the psychic costs of development. A system such as I have described would require a totalitarian government (though this solution is paradoxically contradicted by the other requirement of decentralization).

Very small countries such as Hong Kong, Puerto Rico, Israel, Cuba, and, of course, Switzerland, might go a long way to "urbanize" the entire country all at once, treating the national territory as one continuous urban field. But larger national entities are not in this fortunate position.

The reason, then, why urbanization appears to be a universal phenomenon is that men have always preferred efficiency to inefficiency–taken in a gross sense–and mobility to local permanence. This conclusion is contrary to your thesis, Your Excellency; it is ruralization, not urbanization that is the socially expensive process.

Your Excellency also referred to the concentration of an empoverished subproletariat in the large city as posing two quite different though related issues, employment and housing.

The presence of the poor in the large city is indisputable. They are here in large numbers; they are inadequately housed and poorly served; many are unemployed; still larger numbers are only fractionally employed in low-paying jobs. But there is little evidence that their lives are in any sense chaotic, that their social patterns are disorganized, that politically their behavior is either volatile or dangerous to the security of the State. On the contrary, the radical, extremist parties find more support among the older residents who have attained some status in the city but find their further progress blocked.[15]

The urban social system of the poor is identifiably distinct from the parallel to the dominant system. For some, it can also be a bridge to the dominant system. The problem for policy is how to facilitate the passage across it, how to permit a larger number of the poor to gain a foothold in the modernizing sector of society and so accelerate expansion of this sector. An equally important problem, of course, is how to provide for increased upward mobility *through* this sector.

One of the most characteristic aspects of the urban social system of the poor is what the American anthropologist, Clifford Geertz, has called the bazaar economy.[16] The bazaar economy offers a way for sharing poverty; it has an almost unlimited capacity to absorb labor; without its cushioning effects, migration to the city would fall off to a mere trickle. In almost every way, it constitutes a system distinct and separate from the technologically advanced sector of the social economy. Witness the minute business firms, production units, and service enterprises, where the internal division of labor is carried to an extreme of specialization. In this way, risks are fractionalized, and the system can maintain a high degree of stability in spite of frequent individual bankruptcy. The bazaar economy operates on the basis of kinship and social relations. Its sliding price permits the individual entrepreneur to try his virtuosity in bargaining and the exploitation of competitive advantage. A small initial loan obtained from a kinsman or friend may be skyrocketed into handsome profits or quickly lost through ineptness. But the system of reciprocal favors keeps most of the participants in the game. The system is highly individualistic and oriented to short-term gain; it is also permeated by a strong sense of mutual assistance.

The bazaar economy is found in all of our cities, but especially in the nation's capital. It probably explains why the official unemployment figures collected by the Statistical Bureau over the past ten years have remained fairly stable. Many of our unemployed are young people venturing forth to find their first job in the modern sector; for years before, they may have been submerged and statistically invisible in the low-productivity, subsistence economy of the bazaar. We have estimated that as many as 40 percent of the metropolitan population may be incorporated into this sector. Nevertheless, every month, thousands make it across the invisible bridge, finding employment in industry, transportation, and government. But for every one who leaves, there is at least one newcomer to the bazaar. What we seem to be faced with here is a system of equilibrium unemployment: so long as the queue at the bridgehead does not exceed a tolerable waiting period, migration continues to maintain a given level of unemployment. Should the queue lengthen, migration from the countryside will tend to fall; should it decrease, because of the more rapid expansion of the modern sector, migration will tend to rise. This theory would explain why registered unemployment tends to be the greatest in the most dynamic regions of the country.[17] Where a large labor surplus exists, as it does in our rural economy, the net rate of rural to urban migration appears to be governed by an equilibrium rate of urban unemployment.

Your Excellency also referred to the housing question. Here we have learned a great deal over the past few years. The housing priorities of the poor are rather differently ordered from our own. They reflect a stronger preference for access to job information, transportation, and public utilities.[18] The bazaar economy permits people to save: when the poor feel secure on their own land, they are likely to channel substantial amounts of savings into the improvement of their own homes. They remain piled up in physical slums only for as long as they are threatened by sudden expulsion and loss of property. They occupy land illegally. But if they could obtain firm title to the land, and if the Government decided to extend basic utilities and services into these new settlements, we should witness a remarable physical transformation taking place, as solidly built-up sections of the city would spring into being in only a few years. The house so consolidated into a neighborhood, would give the family a firmer economic and social base for its own advancement and its more complete integration with the dominant society.

I have been unable to go into all facets of the argument concerning the rate of migration. It would appear, however, that the process of city formation is not only essential to rapid industrialization but not nearly as disruptive as it is generally believed to be.

2. The Spatial Pattern of Investment and Migratory Flows

The Head of State speaks:

> "Assuming that migration to urban centers continues, should not the Government make a concerted effort to guide the location of new investments into areas suitable for development in the interior of the country, away from the national capital, which is already three times the size of the next largest city and accounts for nearly two-thirds of all the manufacturing employment in our country? Before the last election, I travelled extensively throughout the provinces, and wherever I went I found the same thing to be true: the people demanded schools and roads and factories; they asked for trucks to carry their produce to markets, for electricity to light their homes, for water to irrigate their crops. As Chief Executive, I feel responsible to all the people, and I firmly believe that a sense of equity and fairness should direct the actions of my Government. Is it not, therefore, true, as the opposition claims, that the capital region has grown too large, that it is draining other parts of the country of their capital and manpower without giving anything in return? Can we continue to afford this systematic spoliation of the country's periphery? Our people are not yet fully united as a nation; they speak more than a dozen dialects and follow many different customs; economic backwardness in many areas is little short of catastrophic. Certain regions have scarcely been touched by changes which we, living in the capital, have come to take for granted. Now there are those who argue that we should not concern ourselves with these conditions, that economic growth and social progress will spread to all parts of the country in due course. But rumors of a secret separatist movement in the South have recently been brought to my attention, and it is common knowledge that the opposition is promoting dissension in the regions. Is it not imperative, therefore, that the Government act vigorously to extend the benefits of economic growth as widely as possible, as soon as possible, to make our people truly feel a part of the nation? Should not the Government promote a radical decentralization in the distribution of investments and jobs? Should we not reinvigorate the hundreds of small towns and villages of our countryside?"

And the Minister of National Planning replies:

The arguments you make, Your Excellency, turn chiefly on the question of the size of cities. Are our large cities too large? Is large size efficient for economic growth? Does large size contribute to regional conflict? Permit me to proceed in order.

The size of our national capital is such that geographers would classify it as a primate city. We share this condition of primacy with small and densely populated countries, that are low in per capita income and have export-oriented and agricultural economies, a colonial history, and rapid rates of population growth.[19] But primacy is more properly viewed in a historical context. There seems to be growing evidence from both cross-sectional and longitudinal studies that primacy is but a temporary and functionally necessary stage in the evolutionary process of city formation. Primate cities, in this view, are seen as energizers of national systems. Professor Brian Berry, who has conducted the most systematic studies, explains the rise and decline of primacy in these terms:[20]

> "From a situation of low degrees of urbanization and modernization with a high degree of entropy (many small equally sized urban places, or a lower-level steady state), primate cities proved the largest discrete reductions in entropy possible (i.e., the greatest gains in systematic order through organization) as the system is moved from one level of organization to another. . . . Once the quantum leap has been made, there can be a gradual convergence on a new steady state (lognormal if the same proportionate growth processes are extended over the entire system of cities)."

This formulation, of course, presumes the unimpeded expansion of the urban system. Whether this is the most likely outcome or whether primacy will tend to close off further system growth depends in part on how we answer your remaining questions.

Large cities are claimed to be inefficient spatial systems. But the evidence on which such claims are made is very slim indeed. All studies that have attempted to probe this correlation have looked exclusively at the cost side of urban size. Generally they have found per capita costs for such things as investment in urban infrastructure and municipal services increasing up to some limit of size and maintaining cost levels thereafter. But only direct monetary outlays were considered, and the benefit side of the equation has so far resisted most attempts to measure it.[21] And yet, only a complete assessment of inputs and outputs, costs and benefits, would give us a basis for judging the efficiency of cities at *any* size. Until this is done, the question of optimum city size must be held in abeyance.

But if we study the phenomenon of city size in the more restricted perspective of industrial and business location decisions, we seem to be driven to acknowledge that the large city in a country such as ours may indeed be the most rational location for most enterprise. This finding reinforces our earlier conclusions on primacy.

The case is argued by Professor William Alonso.[22] Private entrepreneurs, he says, will find total market access greatest in the large metropolis, not only because of the nature of the transport network, which is frequently held together by the central metropolis, but also because the highest proportion of disposable consumer income is concentrated there. Deliveries in the interior of the country are costly, time-consuming, and uncertain. Large inventories would have to be stocked. Repair parts may be difficult to obtain. Communication with the metropolis may raise administrative costs substantially over the lower levels prevailing at central locations. The social and cultural amenities of the metropolis exert a powerful attraction upon managers and technicians who may refuse to work in remote and smaller towns where, in addition to presumed inconveniences, they would be far away from the mainstream of national life and the political forces that activate it. Other things being equal, the cost of a location in the central metropolis is lower because the density of information is greater there. But the most trenchant reason supporting a rational preference for central locations is the existence of significant external economies which allow the individual entrepreneur to draw on other firms and assume only a small fraction of the cost of maintaining extensive communication and transportation facilities. In a sense, the businessman shares some of his expenses with the population at large. Because the entry of certain services into a market is determined by threshold sizes that correlate with the minimum efficient size of given firms, the large city justifies the establishment of such supporting services as specialized shippers and jobbers, financial agencies, trade publications, repair services, specialized printing, consulting services, equipment leasing, advanced educational facilities, laboratories, etc. In addition, by spreading uncertainties and risk over a large number of firms, operating in the large city is more predictable and less risky for the individual firm. Large numbers, as Professor Alonso points out, permit much greater flexibility and adaptability to changing and often unforeseen circumstances because more varied opportunities are open. In the same vein, bulk transactions made possible in the large city tend to reduce the costs of these transactions.

But perhaps most important of all is the necessity for face-to-face contacts in the conduct of business. Entrepreneurial success often depends on the cultivation of kinship, friendship networks, and relations based on reciprocity of favors. In a provincial location, isolation from these contacts would make successful operation of a business enterprise extremely difficult.

From a standpoint of individual interest, therefore, there is a strong presumption in favor of the efficiency of large cities. This argument, however, overlooks the question of equity. Do individual and social interests coincide in supporting the rationality of large cities as a principle of public policy? At first sight it would appear that this is not the case. Large size leads to regional inequities. The metropolis grows at the expense of other areas. To the extent that the Government must be concerned with questions of equity, the growth of large cities must be restrained.

I should like to propose, however, that equity is not an appropriate consideration for national urbanization policy. What matters is not so much the equality of welfare conditions among regions, but among people. Measures of welfare in urban-regional analysis must be aggregated over a given spatial unit. These measures, however, depend on the contours of a region's boundaries and its size. By the simple expedient of changing the alignment of boundary lines, one can show an increase or decrease in regional welfare conditions. It is inconceivable, however, that all areas of the nation, however they may be defined, can show approximately equal conditions. As an issue of urban policy, therefore, the criterion of regional welfare equality must be discarded. More relevant as a policy criterion is the spatial integration of the national society.

Your Excellency referred to certain rumors of separatist tendencies in the southern provinces. To the extent that these rumors have any basis in fact, we can demonstrate, I think, that they are the result of a long-standing condition of political and economic dominance by the national core region. The consequences of this condition have become aggravated in recent years and can probably be traced to such events as the completion of the all-weather highway connecting the southern parts of the country to the nation's capital, the literacy campaigns the Government has helped support, and the founding of the Provincial University. These moves have contributed to creating a stronger regional awareness among the local population, have helped to bring into being a regional intellectual elite, and have facilitated the outward flow of migrants to the central region. Yet the future of the South is still being decided by the central Government, by central banking institutions, by central industries whose branch plants have started operating there but whose headquarters are located in the metropolis. Powerful desires have been stirred up by these changes, as urban influences have begun to spread into the region, but the means for realizing the newly awakened aspirations are not yet within reach of the population.

May I add a fact well known to Your Excellency. The large mining operations in the South are controlled by foreign interests. Except for the wages paid to miners, no visible economic benefits accrue to the region from this activity. The national Government, of course, receives its share of taxes from the Company, but very little of this, perhaps not more than 15 percent, finds its way back into the region as development capital. From a regional point of view, then, one of the major resources of the South is being depleted and no permanent bases for economic progress are being put in their place.

If primacy is a temporary condition, as has been asserted, the speed at which the condition of primacy is being eroded must truly be called glacial. While modernization and economic growth are racing ahead in the national core area, this is occurring at the expense of the periphery, and the political indicators are beginning to flash appropriate warning signals. While the diffusion of innovations is proceeding apace, the noose of central control is being drawn ever more tightly around the neck of the periphery. More capital flows from the periphery into the core, and migrants follow because only here can they gain "access to good hope." Unless these imbalances can be redressed, the rumors of separatism will start to appear as front-page headlines in the nation's press. The entire program of your Government may be jeopardized.

I wish to stress again that regional equity is not the primary issue here. Rather, we must consider the spatial integration of society, an increase in the volume of transactions, a wider sharing in the national culture, and greater reciprocity in the exchange of goods, services and ideas. We should cease to worry about the magnitude of urbanization; instead we should start seriously to think about the *direction* of this process. The question is not how much urbanization to allow, but where. What Professor Berry calls a log-normal distribution of cities as the most probable outcome of development in the long run is not going to happen here, in our country, unless the Government itself undertakes to encourage regional development as a location factor.[23] The promotion of secondary core regions in the periphery has become urgent business on the agenda of the nation.

This proposal is substantially different from that suggested by Your Excellency. You wished to carry out a radical deconcentration of development that would carry economic progress directly into the villages and towns of the periphery, by-passing large-scale urbanization. But I trust I have persuaded Your Excellency that large-scale urbanization cannot be easily by-passed. At this time in our history, the optimal strategy is not to encourage the even spread of development among small, rural communities but deliberate urbanization in only a few places. By coordinating the necessary governmental programs and by channeling a larger volume of investments into the South, urban growth may be promoted there which, through linkages to the national core region, would give the periphery its own centers of gravity, reduce primacy, break down the stark division—dominant core: dependent periphery—into a more finely grained pattern, and encourage greater interdependencies among regions.

Policies to promote the development of subsidiary core regions have been the subject of heated debate in the academic journals, both here and abroad.[24] In a few countries such as Turkey, Italy, France, Venezuela, Brazil, and Chile, they have been tried on a significant scale. Both the discussion and the experiments, however, show a good deal of confusion in the use of terminology and concept, and the experiments are still too recent to yield unambiguous results. But enough evidence has accumulated, I believe, to warrant taking the calculated risks of initiating a policy of deliberate urbanization now. The process of modernization in the periphery will be speeded up as the spread effects of core-region innovations penetrate more deeply into the periphery. Rural to urban and interurban migration will be re-directed towards the new areas of rapid growth. Agriculture, at least in the vicinity of the new growth areas, will attain new levels of productivity.[25] The unity of the nation will be strengthened.

Core-region policies have generally stressed large-scale industrial complexes as the "propulsive" element in regional economic growth. These leading industrial sectors, generators of powerful linkage effects, are expected to set in motion an upward spiral of both internally generated and externally supported growth.[26] This policy is not necessarily wrong, and where the opportunities for large-scale industrialization exist, they should be exploited to the fullest possible extent. On the other hand, this strategy overlooks the long-term consequences of building up an entrepreneurial tradition that has its roots in the bazaar economy but ties directly into the technologically advanced sector of the economy. As Your Excellency is aware, most of our industries are small, having less than ten employees. Many have barely reached beyond the artisan stage of production. Nevertheless, there are good reasons for encouraging small-scale modern enterprise as part of a new core-region development policy. Such a program would strive to build up capacity for producing a wide range of goods and services appropriate to a modern society. In the words of Professor Richard L. Meier, "it would fill in the interstices between the major enterprises, producing for small to medium-size local markets which are created in the course of growth."[27]

If a program for encouraging small-scale modern industry can be successfully carried out—and the difficulties should not be underestimated—it would accomplish four things that are essential to the sustained development of core regions. First, it would permit the industrial sector of the economy to absorb a larger share of employment than is now the case. Small-scale industry is generally less capital-intensive than the largely automated plants of large-scale enterprise that vastly improve the productivity of labor but generate few jobs within the plant itself. Second, it will broaden the base for capital formation and increase the amount of productive saving in the society. Although the bankruptcy rate among small industrial ventures is generally high, the firms that emerge successfully from their first period of trial and error will have a strong incentive to reinvest their earnings. Most of the new plant and equipment will tend to be local and would thus contribute to a further expansion of the core-region economy. Third, a pattern of diversified, small-scale production would offer invaluable experience to an emerging entrepreneurial class, thereby increasing the pool of talent available for the next round of industrial expansion. The beginnings of an entrepreneurial spirit are already found in the bazaar economy. But this start can come to a quick end unless opportunities are made available for the exercise of entrepreneurial skills on a wider, more challenging scale. As Professor Bert Hoselitz has so eloquently expressed it, "the creation of a 'climate for entrepreneurship' depends . . . on the one hand on establishing social institutions which made possible the exercise of independent, individual enterprises, and on the other allowing the maturation and development of personalities whose predominant orientation is in the direction of productivity, working, and creative integration."[28]

Finally, the encouragement of small-scale modern industry would help to increase the innovative capacity of core-region populations. As the American writer, Jane Jacobs, points out, new work grows out of the old; innovations are partly the result of solving problems that occur in the process of production.[29] The more such opportunities are afforded, the greater will be the probability that innovations will, in fact, occur. And a rise in innovativeness is one of the distinguishing features of core-region development.

3. The Diffusion of Urban Innovations

The Head of State speaks:

> "Is there anything the Government can do, short of large-scale investment in economic and social infrastructure, to achieve a substantial measure of national unification and to create a wider consciousness of developmental problems among the population? Modern development is visible in our cities some of which, such as the nation's capital, have doubled in size within the last ten years. Yet only a small proportion of our population has been affected by these changes. The spread affects of urban change have been extremely weak; the city terminates abruptly at the outskirts, and modern life stands in sharp contrast to the traditional ways in the villages. On my last visit abroad I was deeply impressed by the extent to which even peasants and farmers have come to participate in the life of their nation. They listen to the radio; they vote in elections, they are active members in their own associations; they form farm labor unions; they establish producer cooperatives; they send their children to school and even to the university; they deal with banks that specialize in loans to farmers. In a sense, the city has been brought into the countryside, not the whole city, to be sure, for the people continue to live on their own farms and in the villages, but vital elements of it are present in rural areas. What should we do to accomplish essentially the same objective? Is extensive modernization indeed a feasible objective, given that so much of our population is still dependent on the soil? Is there no alternative to increasing the pace of city-ward migration if we desire to modernize the nation with all due speed? Should we build, perhaps, more highways and rural roads to improve access to markets and encourage the commercialization (and capitalization) of agriculture? Should we undertake a vast program of school construction in rural districts? Should we carry electricity into our agricultural areas on a massive scale? Should we propose legislation that would facilitate the establishment of peasant cooperatives and the syndicalization of farm labor? Should we promote new systems of rural credit? Should we develop more adequate central services in our provincial cities? What priorities should we establish as a guide in formulating a set of interrelated, coordinated policies to accelerate the diffusion of urban innovations, to 'urbanize the countryside?' "

And the Minister of National Planning replies:

Your Excellency is right: compared to the industrially more advanced nations, the process by which urban innovations have diffused throughout our national territory has been sluggish. The reasons for this are rather complicated; they have to do with the higher degree of "friction" imposed by spatial distance in a country such as ours, with its poorly developed transport and communication facilities, its low level of education, and its non-adaptive social institutions. They are also related to the poorly evolved system of urban places which, as we have commented, labors under the difficulties of urban primacy, so that innovation diffusion *down* the urban hierarchy—which is the customary process in the more advanced countries—is faltering.[30]

Nevertheless, some penetration of the countryside is occurring, and our peasantry and small-town populations in the periphery are beginning to share some of the characteristics of the urban population. In this respect, our country is not greatly dissimilar to certain South American nations, such as Peru, where a similar process has been under way for some time. I refer to an article by Professor Julio Cotler of the Institute for Peruvian Studies in Lima who writes:[31]

> "The close links which [urban] immigrants maintain with their places of origin encourage the diffusion of new social, cultural, and political customs in the rural areas, as well as diverse social innovations, propagating new technologies and styles of consumption, diffusing new mass communication media, and recreating identifications with the rural society by means of a new cultural style . . .
>
> Further, a new and autonomous group is being formed which incorporates modern values, without destroying all of the traditional patterns, encouraging a climate of civic equality which is creative in its form, and not just a copy of that part of the culture from which it has been borrowed."

Professor Cotler refers to the importance of social contacts between urban immigrants and their home communities, as well as the formation of new modernizing elites, as facilitating variables in the process of innovation diffusion. He goes on to mention other change-transmitting instruments: Christian missionaries from abroad, the Peace Corps, State-fostered social movements such as *Cooperación Popular,* public works programs, and the growth of urban markets. These mechanisms are not unlike those we find at home.

The probable consequences of this gradually accelerating modernization of the periphery are perhaps not entirely expected. Contrary to what Your Excellency implies, the modernization of the periphery is unlikely to lead to a diminution in the flow of migrants to cities. Rather, migration to the national core will tend to increase. This is clear from indirect evidence obtained by studying the social composition of migrant populations, both in our own country and abroad. For example, migrants to the Indian city of Bombay were found to have much higher educational and skill levels than those who stayed behind in their areas of origin.[32] Again, with reference to India, Professor N. V. Sovani suggests that the poorest sectors of the population, the agricultural laborers, send out the lowest proportion of migrants, and that unemployment in four out of five Indian cities has its heaviest incidence among the resident population rather than among the newly arrived.[33] Quite similar results were obtained in a detailed survey of migration to Santiago de Chile.[34] There, migrants moving from the periphery to the national center came, for the most part, from already urbanized areas and were themselves favorably selected from among their home populations, at least with respect to educational attainments.

From these various studies one may conclude that the more a population comes to share in urban culture, the greater will be its propensity to resettle in large urban concentrations. In fact, the extension of new highways into peripheral areas has generally encouraged outmigration, as the large centers became more readily accessible and contacts increased.

The other major consequence of the diffusion process is less convincingly documented. I refer to the impact of modernization on attitudes and aspirations and its roundabout effect on political organization and action. Diffusion of urban innovations appears to awaken in a subservient, quiescent population the desire to share in the material and cultural progress of the nation, to cast off its bonds of dependency, and to press for basic socio-economic and political change.[35] Urbanization as a city-forming process as well as a process of diffusion introduces powerful discordant elements into a relatively placid situation, elements which, if they are seized and turned to advantage, may help in the construction of a truly national society. As Professor Cotler declares, speaking of recent urbanization experience in Peru:[36]

> "The new urban proletariat and its Cholo counterpart in Mancha India are—each in its own way as well as in a combined form—breaking in upon the structural and normative dependence of the Indian mass by mobilizing it socially and politically. Such mobilization entails a change from the local-family Indian pattern to another with class connotations. The confrontation with the [repressive] mestizo system takes place especially in the political sphere as such, since this type of resource is at the very root of the mestizo class status. Therefore, the 'access crisis' of the peasantry presupposes the questioning of the legitimacy of the caste system and thereby of the economic, social, political, and cultural cleavage which it entails."

So far, Your Excellency, I have ignored the spatial dimension of innovation diffusion, except by treating diffusion as a process which generally flows outward from core regions into the country's periphery. But now we have to look more closely at certain aspects of this critical dimension.

The diffusion of urban innovations does not occur at equal rates to all parts of the country; futher, there is a definite selectivity in the kinds of innovations that are adopted from one place to another. Recent studies by Professor Poul Ove Pedersen of Copenhagen suggest a rather orderly pattern that governs the diffusion process and relates it both to levels of economic development and to urban networks.[37] I should like to summarize some of his tentative findings:

- at higher levels of economic development, innovations diffuse down the urban hierarchy, from large to small central places; at lower levels, however, the friction of distance appears to have greater influence on diffusion rates: cities further away from innovation-generating centers tend to receive a smaller number of innovations during a given time period, regardless of their hierarchical position, than more accessible places;
- the time required for the complete diffusion of an innovation over a given spatial system decreases with a rise in the level of economic development of that system;

- innovations having a low threshold of population size (and income), diffuse more rapidly through a spatial system than high-threshold innovations;
- higher rates of participation in communication processes that are linked to innovations permit an increase in the rate of diffusion; but if the rate of participation in the communication process systematically decreases with town size, the diffusion process will be retarded;
- a low frequency of entrepreneurs capable of translating relevant information into actual innovations will tend to retard the diffusion process; since entrepreneurship is a scarce resource in all industrializing countries, and especially in their peripheries, the process of innovation diffusion will tend to occur more gradually there than in countries that are further along in their economic development;
- because the cities with the greatest overall accessibility tend to be among the first to adopt an innovation and will, therefore, also grow the most, hierarchical and geographical distances will approach each other as more and more innovations are diffused through the spatial system: thus, in time, innovation diffusion will tend to proceed downward through the urban hierarchy.

These findings are of extraordinary importance for urbanization policies. They support a policy of deliberate urbanization that will augment the number of centers for information diffusion as well as increase the size of selected communities located in the periphery to levels that permit the diffusion of high-threshold innovations. They support public subsidies to reduce threshold levels in selected centers and thereby render the location of certain innovations in these centers economically more attractive. They support policies for the extension of transport and communication services into areas which have already been touched by modernization. They suggest policies for raising the educational levels of the entire population and for increasing their exposure to information about urban innovations. They suggest policies for fostering conditions that are favorable to the formation of individual and social entrepreneurship. All of these measures would be consistent with a general policy for encouraging the growth of subsidiary core regions in the periphery. But the criterion for establishing new core regions is primarily an economic one: only those areas should be considered that offer exceptional opportunities for novel resource combinations promoting economic growth. At any given time, the Government will be capable of concentrating its efforts on only one or two new core areas that are especially endowed by nature or history. But the diffusion of innovations can be considerably accelerated by identifying within the remaining peripheries those areas whose populations seem to have an unusually high *internal capacity for social initiative,*—in other words, for organizing themselves to achieve rapid economic growth despite an apparently limited resource base. These could be called the "active" peripheries in distinction to purely "passive" areas that lack this internal capacity. In the language of public policy, we propose to designate active peripheries as social development poles.[36] A policy to activate social development poles would constitute a deliberate effort to draw populations away from the more passive to the active areas of the periphery by building up in the latter what Professor Janusz Ziokowski of Poland has called the psycho-social infrastructure of cities.[39] Among its components might be the following:[40]

(1) *Regional consciousness,* or the desire and will of local populations to promote the economic growth of their area; a group-focused image of economic growth.
(2) *Information transfer,* or the openness of leading sectors of the population to information external to the region and their capacity to reinterpret and use this information in ways that will make it relevant to local economic growth.
(3) *Autonomy,* or the ability of local populations to make effective decisions relating significantly to the economic growth of the area.
(4) *Innovative propensity of the social system,* or the affinity of local leadership for patterns of social organization conducive to sustained innovation.
(5) *Organizational capacity,* or the ability of local populations to organize themselves effectively for the solution of common tasks related to economic growth.

Not all of these variables are equally subject to influence by public policy. The critical policy variables appear to be local autonomy over development decisions, the promotion of information transfer, and the capacity to mobilize both internal and external resources for economic growth or social entrepreneurship.

There remain major problems in the identification of social development poles and in the conception and coordination of appropriate policies and programs to stimulate their growth. But these are practical problems that can, in principle, be solved. On the other hand, the even spread of urban innovations across all settlements, however small and isolated, is not a feasible solution during the present period of industrial transition.

4. The Territorial Organization of Power

The Head of State speaks:

"How should the Government organize its functions across the territory of the nation in order to carry out the development of the entire country, its regions and its cities, with maximum effectiveness? Our municipalities are powerless, lacking not only financial and technical resources but also leadership: their impotence has become a national joke. But the situation at the regional level is a little better. Our provincial assemblies, despite their constitutional authority, have nothing of any consequence to decide; unable to raise substantial taxes of their own, they are almost completely dependent on the Ministry of Finance whose current level of provincial transfers, however, is only five percent of the national budget. At the same time, political pressures are mounting not only for decentralizing Government activities but for going beyond this to a drastic devolution of effective authority to lower governmental units. The people in the Eastern provinces, for instance, are clamoring for greater autonomy in the management of their affairs: they claim, perhaps with justice, that too much attention has been given in the past to problems of the central region. The recent National Conference of Mayors presented the Government with a litany of urgent grievances: heading their list was a demand for more extensive municipal controls over housing programs, public works projects, and the establishment of industrial parks. How is the Government to answer them? Should we not seriously question whether the existing high degree of centralization in governmental organization is the best possible arrangement? I am fully aware that the number of trained technicians and administrators we have is very small in relation to what is being asked of them; that the influence of the opposition is strong in local areas and appears to be getting stronger; that the holders of traditional authority are still entrenched in many of our regions; that we cannot afford to waste resources on non-productive projects that fail to conform to national priorities. On the other hand, greater popular participation in decisions is a legitimate concern, and the coordination of local and regional programs can no doubt be better accomplished by a measure of decentralization. Perhaps, then, the present degree of centralization has gone far enough; useful, perhaps necessary, during the first period of independence, centralization appears now to have become a hindrance to national development. A major reorganization of governmental powers may be in order; new development regions may have to be formed. Is it not possible, therefore, to state some principles by which the existing system of governance might be assessed and from which the outlines of a major plan for governmental reorganization with a view to speeding the processes of national integration might be derived?"

And the Minister of National Planning replies:

Your Excellency is referring to the formal pattern of decision-making among the territorial units of a society. I should like to call it the territorial organization of power. This is indeed one of the critical issues of urbanization policy. For policy cannot be meaningfully separated from the institutions which generate it nor from the instruments for its implementation. It is also one of the least-studied aspects of urbanization. My staff has for some time now been engaged in a tentative elaboration of criteria that might be adopted as the basis for a large-scale empirical analysis of our present system of territorial power. The results of such an analysis would permit us to recommend a politically feasible system of reorganization that, at the same time, would contribute to the spatial integration of the nation.

We have identified three general criteria and nine specific measures for the evaluation of the territorial organization of power (TOP):

(1) Effectiveness, or the degree to which TOP is:
- autonomous
- responsive
- adaptive
- innovative

(2) Political adequacy, or the degree to which TOP is:
- consistent with the basic values and assumptions of the political system
- contributing to social mobilization

(3) Efficiency, or the degree to which TOP produces responses that are:
- timely
- accurate
- economical

If Your Excellency permits, I should like briefly to discuss each of these criteria and component measures.

The *effectiveness criterion* must be measured along four major dimensions: autonomy, responsiveness, adaptability, and innovative capacity.

Autonomy reflects broadly to the capacity of a political system to set its own objectives and to pursue them effectively. It therefore refers to (a) the locus of choice among the relevant values and (b) the ability to mobilize sufficient resources to implement the choice of values.

The first of these relates directly to the question of problem closure. If problems can be bounded territorially, so that solutions to them do not create major spillover effects, it is reasonable to assume that the authority to choose among the relevant values—to set objectives with respect to this problem—should be located centrally within the relevant problem boundaries. This measure may be refined by differentiating among the several facets of a problem. Some may be internalized quite locally, others may require a higher order of problem solution.

The question is complicated, however, by the second aspect of resource mobilization. Effective problem solution may require the mobilization of resources beyond the boundaries of a given problem. A claim is therefore placed upon resources that are more generally available, putting the initial problem interest in competition with those originating somewhere else. Initial autonomy may therefore have to be circumscribed, leading to a sharing of potential power among appropriate sub-units of a system. However, the distinction between goal-setting autonomy and autonomy over the allocation of resources permits the problem to be separated: a decision may be made with regard to an overall allocation of resources at one level, but the decision how to apply these resources may be made at another level.

Failure to respect this distinction and to allocate values at the higher level for problems effectively bounded at a lower level will lead to inconsistencies in the system. For the values chosen at the higher level may not be those that would have been selected at the level of problem closure.

It seems to me, nevertheless, that where outside resources are claimed, the powers at that level have an obligation to insist on a review of the proposed uses of the resources they control in order to ensure that problem solutions are technically adequate and will not lead to repeated claims for the same purpose or in larger (but unanticipated) sugsequent claims. This review function may be experienced as onerous, but it is absolutely essential if other and competing claims are to be given equal and fair treatment.

The second measure of *responsiveness* refers to the capacity of a system of territorial power sufficiently to take into account the large variety of particularized interests, needs, and values of the population affected by its decisions. It does not mean to respond in a positive sense to each and every need or interest that is filtered through the system. The ability "to take into account" depends partly on the perception of expressed interests by the relevant decision makers (choice of goals, use of resources) and their ability and willingness to give them serious consideration along with other needs. But for this to occur, there must exist appropriate channels of communication that will allow the expression and representation of these interests at the relevant decision levels. The issue raised concerns the involvement of different organized population groups in problem-solving, decision-making and implementing processes.

The performance of TOP with regard to the measure of responsiveness depends on the following:

- access of different population groups to the relevant levels of planning in the political system;
- levels of information aggregation in the planning system;
- information load at different levels of planning; and
- the propensity of authorities at any given level of planning to give serious consideration to the needs and interests of different population groups and/or work out acceptable compromises among competing groups.

Summarizing the evidence on the behavioral characteristics of territorial systems of power, we may conclude that, other things being equal, a system will be responsive when it is non-hierarchical, or hierarchical but decentralized; when the socio-psychological distance of its principal planning centers to the relevant interest groups is small; and when the value orientations of planners are substantially similar to those of the relevant interest groups.

Less clear is the question of the optimum number of levels in a planning system. Because information is lost as it passes through many levels of aggregation, the fewer levels a system has, the more responsive is it likely to be. On the other hand, the probability of access to the system increases with the number of levels in the system. Here, a trade-off appears to be possible between information and access.

All of this argues for a high degree of social participation in a new form of "transactive" planning so that the needs of different population groups get formulated, communicated, recognized, and included in the transactive planning process.

Since at any given level of transactive planning many different interests may be included, and since these interests may succeed in fighting each other to a deadlock, a responsive system of territorial power has need for special methods of conflict resolution (mediation) and general policy direction. Unless these elements are present, systems of territorial power which have a complex structure, may not be able to perform at all under conditions of high social participation.

The third measure of *adaptability* refers to the capacity of a system of territorial power to make those structural adjustments which will ensure a close fit with changing environmental conditions. The decisive variable here is the conditions of the transactive planning environment. Under conditions of rapid growth, "turbulent" change, and imperfect knowledge, highly adaptive organizational forms are needed. These would be characterized by:

- fluid, viscuous structure; a certain degree of "mild chaos";
- tight feedback loops, yielding accurate information about the environment and leading into a consciously elaborated evaluation system with sufficient salience to force a periodic re-examination of organizational arrangements in light of new information;
- a high degree of redundancy in the system: redundancy in this case means lack of clear definition of functions and roles, diffused responsibilities, multiple channels of access, and high capacity channels for information handling; and
- operating rules of sufficient generality to permit their periodic re-interpretation without challenging the normative bases of the political system.

The fourth and final measure to gauge the effectiveness of systems of territorial power, is their *innovative capacity.* This refers to the capacity of a system to respond creatively to newly perceived problem situations.

I would be less than candid to say that we are well informed about the variables contributing to innovativeness in the territorial organization of power. Nevertheless, a few general suggestions may be tentatively advanced:

- some mix of hierarchical and non-hierarchical structure would seem to be more promising for innovation than a system conforming entirely to either extreme. Systems must have the capacity to ocercome both internal and external resistance to proposed innovations. Only strong centralized control would seem capable of this feat;
- an information environment of high density would seem to be conducive to innovation: but overload problems must be avoided, since extreme conditions of overload may encourage a ritualistic enforcement of bureaucratic rules. On the other hand, unless a great deal of relevant information is available, significant innovations are unlikely to occur with great frequency;
- loose, inter-system linkages are probably helpful in permitting the more rapid diffusion of innovations from one system to another; this, too, is basically a problem of information. However, face-to-face contact is needed for innovation to become adopted by this route;
- a large organizational memory (together with a fast and competent retrieval system) will facilitate search processes that are part of the more comprehensive innovation process; the memory should answer such questions as: what has been tried already? What successes and failures were scored, and why? What answers were scored, and why? What answers were proposed but rejected, and why?
- a higher intelligence function is needed for both search and invention. This suggests linkages to external knowledge systems that are capable of using an even larger memory base than the organization and are possibly less restrained in the search for and filtering of innovative solutions; and
- a benevolent internal "climate" for innovation, including tolerance for ambiguity, contradiction, challenge, and non-conventional approaches; a willingness to assume organizational risks; and rewards for innovative behavior.

A second major criterion, Your Excellency will recall, is that of *political adequacy.* This may be measured along two major dimensions, answering to the following two questions: is the organization of territorial power consistent with the basic values and assumptions of the political system? And does it contribute to social mobilization?

With regard to the first question, a major variable appears to be the size of the system which is to be organized. We may assume that the values and assumptions which provide the yardstick for the consistency test originated in a system that was smaller (rarely larger) than the one currently being evaluated. Are the older values capable of being realized in the larger political setting? Have internal organizational adjustments occurred in the recentralization (decentralization, devolution) of powers? Is there an optimal-size system for the realization of political values, such as democratic citizenship and intelligent participation in political processes and transactive planning?

Turning now to the second question, social mobilization may be thought of as the sum of (1) a sense of social inclusion, and (2) effective involvement in transactive planning. Major societal goals can probably be achieved only under conditions of high social mobilization, i.e., in the positive absence of large-scale disaffection and alienation. Both inclusion and participation have been discussed already, but it is important to raise the issue of social mobilization independently of the effectiveness criterion, since social mobilization—an active society—may represent a value in its own right, regardless of its "effectiveness." Under conditions of high social mobilization, a society may show a high degree of tolerance for inefficiency, error, a low level of performance, and even for misguidance—and hence also be more stable—than under conditions of widespread "alienation." Whenever the nature of the system becomes valued for its own sake, sacrifices will be made to preserve it.

I come now to *efficiency* which we have identified as the third major criterion for evaluating the territorial organization of power. The response characteristics of the system are important here, including timeliness, accuracy, and economy.

With regard to *timeliness,* it is well known that responses may arrive either too late or too early to be either effective or acceptable to the people for whom they were intended. Responses that consistently "lag" behind problems, not only permit problems to continue, but to spread and multiply, until a crisis level is reached. On the other hand, responses that arrive too early are likely to produce unintended consequences, possible creating new problems on top of the existing ones. Where a number of coordinated responses are required, failure to be timely will not only lead to inadequacy of the total response, but may actually produce conflict among different responses, working at cross-purposes rather than together, and making competing claims upon the same resources.

In part, the question of timeliness concerns the time needed to process incoming and outgoing information. The more direct the route (the shorter the effective distance to be traveled) between a problem and the capacity for an adequate response, the greater will be the probability that the response forthcoming will also be "on time."

Another aspect of the same issue has to do with the perception of the problem to be solved, its overall dimensions and internal rates of change. Response times may be delayed because information on which adequate responses must be based is either not available or may become so only after a considerable delay. This argues for the existence of efficient feedback and intelligence mechanisms that are built into the system in such a way that the information they produce is seriously taken into account in the process of transactive planning. Problems that are caught early will tend to cause little damage. Sensitive feedback devices will increase the timeliness of response. This is especially true for problem environments that are subject to rapid change or are internally unstable.

Feedback devices, however, are not sufficient, especially under conditions of rapid change in the environment. To respond in a timely way, the system must also possess the capacity to make conjectures about the future, and to bring this information to bear at appropriate times on the process of transactive planning. The counterpart to "feedback" devices are "lookout" devices, and they must be built into the structure of territorial power to permit the system to react in ways that are both appropriate and timely.

The locus for "feedback" in the territorial organization of power is, however, not necessarily the same as that for "lookout" devices. Where feedback should be close to the point of impact of system response, and effective distance should be relatively short, lookout stations are better located at some distance from the action, so that a more comprehensive perspective and a clearer reading of the movement in the critical variables might be gained. Responses to the future are likely to be structural and will be less concerned with immediate operations.

The *accuracy* of the response is independent of its timing. A doctor may prescribe the correct medicine, but the patient may have gone to consult him when his illness was already too advanced for the medicine to do its work. Accuracy of response depends on the accurate diagnosis of the symptoms of a problem. The right questions must be asked about a problem and the information so obtained must not be distorted as it is filtered through several layers of information aggregation and transactive planning. The information must be at once complete and pertinent, and it must be interpreted correctly. The territorial organization of power must, therefore, be such that is consistently obtains correct information and makes correct assessments about the nature of the problems to be solved.

Is the probability of accuracy in information processing enhanced by certain types of organizational form? The question can be answered only in relation to particular problems. Problems that can be internalized (bounded) at the local level, for instance, would appear to require local response capabilities in order to arrive at solutions that are reasonably accurate. But the problem sets whose boundaries are drawn with a wider compass need to be resolved at that level in order to guarantee an acceptable degree of accuracy. Discrepancies between problem and decision (transactive planning) levels will tend to produce inaccurate responses and will therefore be relatively inefficient.

The cost of producing both timely and accurate responses is the third measure of efficiency in the territorial organization of power. It offers a framework for comparing the costs of alternative organizational arrangements. The relevant costs are for capital, labor, and transport and communication (information processing). Cost estimates may be made for systems which presumably will yield improved performance measured in terms of timeliness and accuracy. Or, alternate systems can be compared against each other on the basis of cost, holding performance constant. The evaluation of organizations, however, cannot be made entirely on the basis of efficiency, as some economists would have it. Measures of autonomy, responsiveness, adaptibility, innovative capacity, consistency with basic political values, and social mobilization must be introduced as additional criteria. Although trade-offs among these value measures are necessary the final cost of the optimal arrangement may well be higher than the minimum cost measured in terms of organizational efficiency alone.

Your Excellency, I have reviewed the criteria and measures for evaluating an existing territorial organization of power. By now it must surely be evident that the system of territorial power has a direct bearing on the spatial integration of the nation. It will determine the directionality of basic control processes between core and peripheral regions and, in this way, influence the flow and composition of investment, the diffusion of innovation, and the volume and direction of migration. For these reasons, it must be regarded as a major variable in the formulation of a national urbanization policy.

THE NEED FOR AN URBANIZATION POLICY

I have tried to show the formative character of urbanization for the development of national societies undergoing industrialization. In doing so, I have implicitly adopted a comprehensive paradigm of urbanization, incorporating innovation, innovation diffusion, control processes, investment flows, and migration. The paradigm suggests the divisive influence of urbanization on the spatial structure of society; it also shows how this stress-inducing character is capable of generating powerful counter-pressures that will work for closer integration.

A society that has set its course on industrialization cannot avoid accelerating the several related processes of urbanization. Any other hopes would be misplaced.

The outstanding example of a country that, throughout an entire generation, attempted to carry out a contrary policy of deliberate ruralization is Italy.[41] In 1931 and again in 1939, the Fascist government passed legislation which restored what Professor Robert Fried has called "a lop-sided version of serfdom in Italy."[42] Rural workers were forbidden to leave the land for cities or towns which were either provincial capitals or of "noteworthy industrial importance," unless they were able to prove steady employment. Those who left their villages without governmental permission were not allowed to register and secure residence permits. Nor were they able to work legally because they were not allowed, without official permission, to register at the unemployment office.

These laws, as it turned out, could not be seriously enforced. Despite great efforts to improve conditions of life in the villages and the lack of attention given to problems of Italian cities after the second world war, urban migration increased dramatically in the forties and fifties. During the latter decade alone, a million peasants left for cities, mainly in the North, and this, as Professor Fried pointedly observes, occurred *before* the repeal of the anti-urbanization laws.[43]

Deliberate ruralization in Italy not only failed to produce the desired results but exacted a high price in economic backwardness and social instability. Beginning in the 1950's, an enormous development burden had to be assumed for the southern regions of Mezzogiorno. Although in recent years the Government has adopted an active growth center policy, the tensions dividing core and periphery have been so great as to threaten national unity itself. The struggle for the recovery of the southern regions was a test of national survival.[44]

If national policy is unable to roll back urbanization during periods of active industrialization, the only alternative left is to direct this process into channels that will promote national development. Even this, however, poses extraordinary challenges. Most governments do little more than alleviate the most critical adverse effects of urbanization, such as maintaining minimum acceptable levels of public services in major cities. But the futility of this essentially negative approach is finally beginning to be known. Venezuela, Peru, Chile, and Brazil have started systematic efforts to overcome the growing imbalances between their national core areas and the periphery.[45] And a recent meeting on East Africa offers one of the most comprehensive analyses of national urbanization policies in any major region of the world.[46]

An excellent instance of how urbanization policies may be employed to promote the spatial integration of a society is afforded by Turkey.[47] The story begins with Kemal Ataturk who, in 1923, ratified his country's entry into the modern world by building a new administrative and cultural capital in the dusty plains of Anatolia. Thus displaced into the country's periphery, and a long way from the traditional seat of the Ottoman empire, the rulers of the young Republic extended a complex railway network across Asia Minor which had Ankara at its hub. The rail lines helped to link the central bureaucracy with major provincial centers. Young Turkey's modernizing elite needed these centers to stage their attack on what they saw as the economic and social backwardness of the nation.

By 1950, many of the early objectives towards which these policies were aimed had been realized. National unity had been attained to the extent that in the country's first free elections, the peasantry was able to vote the opposition party into power without engendering a major conflict. Westernization had perhaps exerted a more limited appeal, being principally confined to major cities. But industrial development had extended beyond the traditional centers, and the trans-Bosporus regions had been reclaimed for the nation.

Forced urbanization, however, had taken its toll among the peasantry which still comprised the vast majority of the population. Having voted the opposition party into office, the Turkish peasant now expected a policy of rural betterment to follow. The government not only complied; its forthright agricultural emphasis was accompanied by a monumental highway program that, although it served in part a military purpose, brought bus service within the reach of the remotest hamlet.

The results of these two policies were unexpected, however. Undoubtedly the policies had helped improve the conditions of life for the country's peasant population, but they had also triggered migration to the larger cities on a scale that had been hitherto unknown. Against its original intentions, therefore, the government was compelled to consider anew the problems of its rapidly expanding metropolitan centers. And by the end of the decade, the need for a more balanced approach to regional development was coming to be recognized. Appropriate institutions were created, and a policy of "concentrated decentralization" was initiated in economically promising areas such as Zonguldak and Cukorova.

Two observations based on this experience may be offered. First, it is interesting to note that during the first three decades of Turkey's modern period, a policy of deliberate urbanization was followed which emphasized the extension of central bureaucratic controls in a staged effort to promote the more rapid diffusion of urban (Western) innovations into the periphery. Turkey thus began its contemporary period with a concern for what I have called the spatial organization of modernization.[48]

This policy, pursued with single-minded concentration, eventually called forth a counter-movement to extend the benefits of economic development to the masses of the population who, in effect, had paid for deliberate urbanization during the earlier period. But the combined effects of modernization, agricultural development, and improved highway access created political pressures which led to a rethinking of spatial policies along the lines of a strategy for the encouragement of subsidiary growth centers. During this second period, then, Turkey's policy makers appeared to turn their attention to the spatial organization of the economy, endeavoring to bring about the integration of the entire society by the judicious application of resources in a few selected regions and their urban cores.

The second observation concerns the simple fact that, in order to be effective, urbanization policies must be pursued over very long periods of time. The first phase of Turkey's modernization lasted for a generation; the second period has been under way for twenty years. The effects of these policies on the spatial organization of the country have been cumulative. They have helped raise Turkey's per capita income to $300 and to achieve a substantial amount of internal cohesion and economic progress.

The commitment to urbanization policies carries with it a strong implication of accepting large-scale change in society not only as inevitable but also as desirable. Such change is unpredictable, however; in Professor Albert Hirschman's words, it is "a unique, non-repeatable, and *ex ante* highly improbable complex of events."[49] Blueprint planning is inappropriate for changes of such magnitude. What is needed instead is the creation of institutional capabilities adequate to cope with the enormous variety of urbanization issues from a national perspective. It would be illusory to believe that this can be accomplished smoothly. But by moving urbanization into the center of public discussion, the fundamental value of spatial integration can be explicitly raised as a question for public policy. It is because of this value that the dilemma for the industrializing countries is not a choice between rapid urbanization and national development but in how to combine these separate but complementary processes into the unifying framework of a dynamic order.

NOTES

1. For a particularly virulent attack on urbanization on the context of India's economic development, see Shanti Tangri, "Urbanization, Political Stability, and Economic Growth," in Roy Turner, ed., *India's Urban Future.* Berkeley, Calif.: University of California Press, 1962, pp. 192-212.

2. The idea of continuous innovation is basic to the concept of modernization which is currently in widespread use among social scientists. One of the earliest and best formulations of this concept is found in Manfred Halpern, "The Revolution of Modernization in National and International Society," in Carl J. Friedrich, ed., *Revolution.* Nomos, Vol. VIII. Yearbook of the American Society for Political and Legal Philosophy. New York: Atherton Press, 1966, pp. 178-216.

3. Glenn H. Beyer, ed., *The Urban Explosion in Latin America.* Ithaca, New York: Cornell University Press, 1967, p. vii.

4. See Chapter 3.

5. See Chapter 2.

6. See Chapter 5.

7. Joseph S. Szyliowca, *Political Change in Rural Turkey: Erdemli.* The Hague: Mouton and Co., 1966.

8. Edward W. Soja, *The Geography of Modernization in Kenya.* Syracuse: Syracuse University Press, 1968.

9. C.F.J. Whebbell, "Corridors: A Theory of Urban Systems," Annals of the Association of American Geographers, Vol. 59, No. 1 (March 1969), p. 4.

10. For a set of detailed propositions governing this conflict, see Chapter 3.

11. Edward W. Soja, op. cit., p. 112.

12. For a convenient summary of the current state of the art, see United Nations, *Urbanization: Development Policies and Planning.* International Social Development Review, No. 1. New York, 1968.

13. All of these questions are based on writing which purport to elaborate on the concept of "overurbanization." For a recent and critical review of this literature, see David R. Kamerschen, "Further Analysis of Overurbanization," Economic Development and

Cultural Change, Vol. 17, No. 2 (January 1969), pp. 235-253. The same questions with rather more attention to the policy implications of accelerated urbanization are discussed by Akin Mabongunje in two important articles: "The Economic Implications of the Pattern of Urbanization in Nigeria," Nigerian Journal of Economic and Social Studies, Vol. VII (March 1965), pp. 9-30 and "Urbanization in Nigeria: A Constraint on Economic Development," Economic Development and Cultural Change, Vol. 13, No. 4, Part 1 (July 1965), pp. 413-438.

14. John Friedmann, "An Information Model of Urbanization," Urban Affairs Quarterly Vol. IV, No. 2 (December 1968), pp. 235-244.

15. The relevant literature is discussed in an excellent recent paper by Joan Nelson, "The Urban Poor: Disruption or Political Integration in Third World Cities?" World Politics, Vol. 22, No. 3 (April 1970), pp. 393-414.

16. Clifford Geertz, *Peddlers and Princes: A Study of Two Indonesian Towns.* Chicago: University of Chicago Press, 1963. Geertz's concept of the bazaar economy has been generalized by W. R. Armstrong and T. G. McGee, "Revolutionary Change and the Third World City: A Theory of Urban Involution." Civilisations, Vol. 18, No. 3 (1968), pp. 353-378. Evidence that a similar phenomenon is occurring in Latin America is adduced by Lisa Peattie, *The View from the Barrio.* Ann Arbor: University of Michigan Press, 1968. And for Africa, see Peter C. W. Gutkind, "The African Urban Milieu: A Force in Rapid Change," Civilisations, Vol. 12, No. 2 (1962), pp. 167-191, and "African Urbanism, Mobility and the Social Network," in Gerald Breese, ed., op. cit., pp. 389-400.

17. A distinction is necessary here between individual and collective unemployment. In the new industrial city of Guayana, Venezuela, for instance, one-third of the migrants remain without a job one month after their arrival, and one tenth for over six months. After the first few months, however, recent migrants have about the same unemployment experience as the whole population, or about 12 percent. The evidence suggests further that migrants to Guayana are heavily recruited from among unemployed population groups. See the survey results reported by John Stuart MacDonald, "Migration and the Population of Ciudad Guayana," in Lloyd Rodwin et al., *Planning Urban Growth and Regional Development.* Cambridge, Mass.: The M.I.T. Press, 1969, chapter 5.

18. The reassessment of housing priorities among the poor is largely the work of John Turner. A number of publications have resulted from his researches which, starting in Peru, have spread out to a number of different countries throughout the world. One fairly recent summary is found in his contribution to the United Nations publication referred to in note 12, "Uncontrolled Urban Settlement: Problems and Policies," pp. 107-129.

19. Arnold S. Linsky, "Some Generalizations Concerning Primate Cities," in Gerald W. Breese, ed., op. cit., pp. 288-294.

20. Brian J. L. Berry, "City Size and Economic Development: Conceptual Synthesis and Policy Problems, with Special Reference to South and South-East Asia," in Leo Jakobson and Ved Prakash, eds., *Urbanization and National Development,* Vol. I, South and Southeast Asia Affairs Annuals. Beverly Hills: Sage Publications, 1971, pp. 111-155.

21. Some of the relevant literature includes: G. M. Neutze, *Economic Policy and the Size of Cities.* Canberra: The Australian National University, 1965; Alessandro Busca and Salvatore Cafiero, "Costo Social del Asentamiento," Cuadernos de la Sociedad Venezolana de Planificación, Vol. IV, No. 8 (August 1966), pp. 35-45; and Stanford Research Institute, *Costs of Urban Infrastructure for Industry as Related to City Size in Developing Countries:* India Case Study. Menlo Park, 1969; the chapter in the United Nations *International Social Development Review,* No. 1 (op. cit.), entitled: "Economic Aspects of Urbanization," pp.

71-78, where some of the relevant literature is reviewed; and most recently, William Alonso, "The Economics of Urban Size," Working Paper 138, Center for Planning and Development Research, University of California at Berkeley, November 1970. Alonso's is the first attempt to link city size to economic performance. In general, his argument sustains my conclusions.

22. William Alonso, "Industrial Location and Regional Policy in Economic Development," Working Paper N. 74, Center for Planning and Development Research, University of California at Berkeley, February 1968.

23. Ibid., pp. 39-45.

24. The French literature on *pôles de croissance* is critically assessed by Niles M. Hansen, *French Regional Planning.* Bloomington, Ind.: Indiana University Press, 1968; other general review articles, all of them quite critical of the concept, include: Heiko Körner, "Industrielle Entwicklungspole als Instrumente de Regionalpolitik in Entwicklungsländern," Kyklos, Vol. XX, No. 3 (1967), pp. 684-708, and David F. Darwent, "Growth Pole and Growth Center Concepts: A Review," Environment and Planning, Vol. I, No. 1 (1969), pp. 5-31. Important research contributions have been made by J. R. Lasuén, "On Growth Poles," Urban Studies (England), Vol. 6, No. 2 (June 1969), pp. 137-161 and "A Generalization of the Growth Pole Notion," Paper presented at the I.G.V. Colloquium on Regional Inequalities of Development (April 1971); Vida Nichols, "Growth Poles: An Investigation of their Potential as a Tool for Regional Development," Discussion Paper Series No. 30, Regional Science Research Institute (GPO Box 8776, Philadelphia, Pa., 19101), May 1969, and "Growth Poles: An Evaluation of their Propulsive Effect," Environment and Planning, Vol. I (1969), pp. 193-208; Leo H. Klaassen, "Growth Poles: An Economic View," Netherlands Economic Institute, March 1969. T. Hermansen, "Growth Poles and Growth Centres in National and Regional Development: A Synthetic Approach," Geneva: United Nations Research Institute for Social Development, 1969. This same Institute is about to publish a series of national studies of growth pole development. The series will be published by Mouton and Co.

25. William H. Nicholls, "The Transformation of Agriculture in a Semi-Industrialized Country: The Case of Brazil," unpublished paper, National Bureau of Economic Research, Inc., December 1967.

26. See, for instance, the chapters by Roberto Alamo Blanco and Alexander Ganz in Lloyd Rodwin and Associates, op. cit.

27. Richard L. Meier, *Developmental Planning.* New York: McGraw-Hill, 1965, p. 176.

28. Bert F. Hoselitz, *Sociological Aspects of Economic Growth.* New York: The Free Press, 1960, p. 154.

29. Jane Jacobs, *The Economy of Cities.* New York: Random House, 1969, chapter 2.

30. Poul Ove Pedersen, "Innovation Diffusion Within and Between National Urban Systems," Geographic Analysis Vol. II (1970), pp. 203-254.

31. Julio Cotler, "The Mechanics of Internal Domination and Social Change in Peru," in Irving Louis Horowitz, ed., *Masses in Latin America.* New York: Oxford University Press, 1970, p. 434.

32. K. C. Zachariah, "Bombay Migration Study: A Pilot Analysis of Migration to an Asian Metropolis," in Gerald Breese, ed., op. cit., pp. 360-375.

33. N. V. Sovani, *Urbanization and Urban India.* New York: Asia Publishing House, 1966, pp. 142-155.

34. Bruce M. Herrick, *Urban Migration and Economic Development in Chile.* Cambridge, Mass.: The M.I.T. Press, 1965.

35. This thesis is closely associated with the name of Karl W. Deutsch. See his "The

Growth of Nations: Some Recurrent Patterns of Political and Social Integration," World Politics, Vol. IV (1953), pp. 168-195, and "Social Mobilization and Political Development," American Political Science Review, Vol. LV (1961), pp. 493-514. A similar explanation may help explain recent violent outbursts among black inner city residents in the United States.

36. Julio Cotler, op. cit., p. 440.

37. Poul Ove Pedersen, op. cit. and Brian J. L. Berry, op. cit.

38. See Chapter 6.

39. Janusz Ziolkowski, "Sociological Problems of Regional Development," Proceedings of the First Scandinavian-Polish Regional Science Seminar. Warsaw: PWN–Polish Scientific Publisher, 1967, pp. 19-32.

40. See Chapter 6.

41. Robert C. Fried, "Urbanization and Italian Politics," The Journal of Politics, Vol. 29 (1967), pp. 507-534.

42. Ibid., p. 509.

43. Ibid., p. 519.

44. For accounts of Italy's regional policies since 1950, see U.S. Department of Commerce, Area Redevelopment Administration. *Area Redevelopment Policies in Britain and the Countries of the Common Market.* Washington: USGPO, January 1965; Joseph La Palombara, Italy: *The Politics of Planning.* Syracuse, New York: Syracuse University Press, 1966; and Robert C. Fried, "Administrative Pluralism and Italian Planning," Public Administration (Winter 1968), pp. 375-391.

45. Lowdon Wingo, Jr., "Latin American Urbanization: Plan or Process," in Bernard J. Frieden and William W. Nash, Jr., eds., *Shaping an Urban Future.* Cambridge, Mass.: The M.I.T. Press, 1969.

46. Michael Safier, ed., *The Role of Urban and Regional Planning in National Development of East Africa.* Kampala: Milton Obote Foundation Adult Education Centre, 1970.

47. The following account is taken from Malcolm D. Rivkin's detailed monograph, *Area Development for National Growth.* The Turkish Precedent. New York: Frederick A. Praeger, 1965.

48. See Chapter 4.

49. Albert O. Hirschman, "The Search for Paradigms as a Hindrance to Understanding," World Politics, Vol. 22, No. 3 (April 1970), p. 343.

Chapter 10

PROBLEMS OF SPATIAL DEVELOPMENT IN THE CAPITAL REGION OF CHILE*

I. SPATIAL ORGANIZATION IN NATIONAL DEVELOPMENT

Regions are formed by the joint action of nature and man. A river carves a valley through the mountains; man settles there, cultivates the soil, builds villages. He harnesses the river to his purposes for irrigating land and generating electricity. Pipelines carry the water of the river to distant cities. Highways connect the cities, and the flow of movement across them helps to define areas of attraction and relative subordination, creating a complex pattern of centers and regions. And so, nature and man work in harmony with each other. The works of man eventually predominate, transforming the original landscape beyond all recognition.

The processes by which man gives form to his environment may be subsumed under the general concept of urbanization. The direction and volume of urbanization processes influence the general trends of national development by establishing differential spatial patterns of economic growth and levels of living. They will affect the political "climate" of the country, define some of the crucial issues for public policy, and bear directly on the capacity for expansion of the national economy as a whole. In general, urbanization processes encourage the formation of socio-political, economic, and cultural relations among urban centers. Over long periods of time, however, they also dissolve and transform these relations. Formerly isolated places are joined together, old regions stagnate and decline, dynamic core regions are created, and the "friction

*I am indebted to Andrés Necochea for valuable assistance in preparing this chapter.

Map 1. THE PACIFIC-ATLANTIC AXIS
Future Access to International Markets

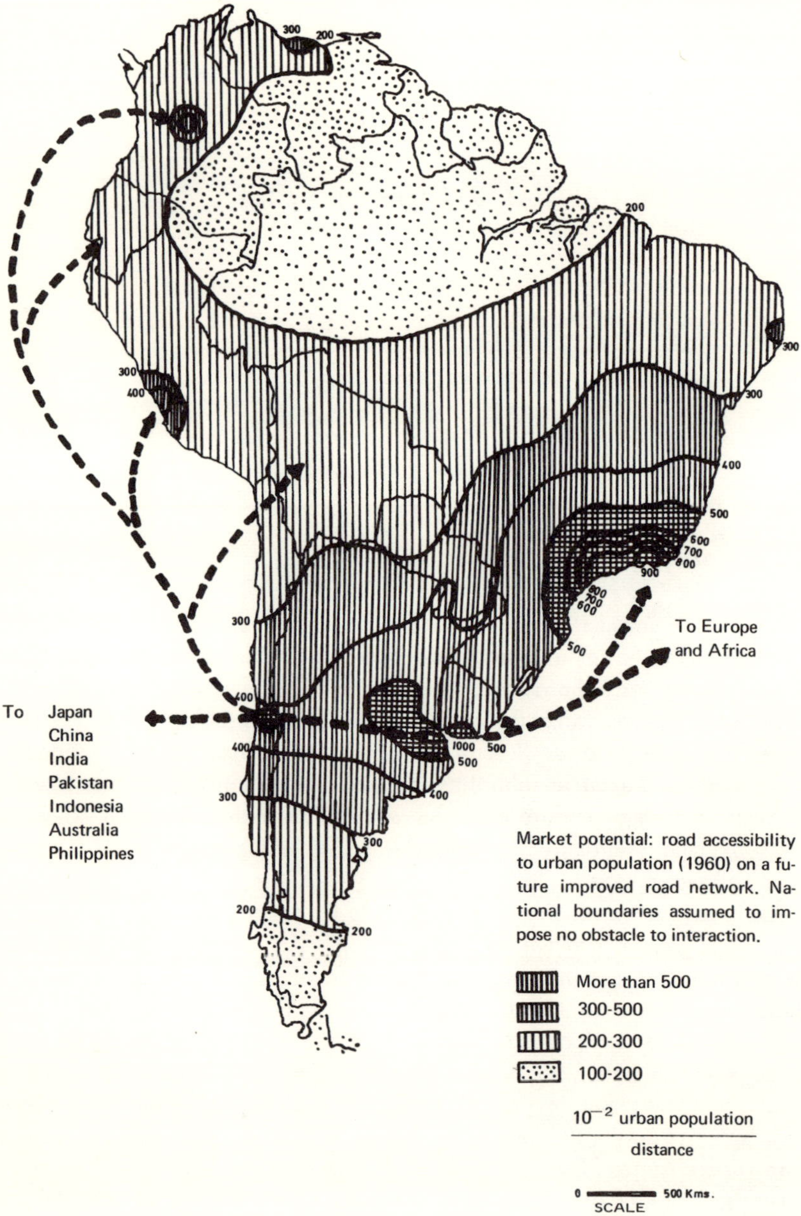

Source: Pedersen and Stöhr in American Behavioral Scientist, Vol. XII, No. 5 (May-June 1968) p. 7.

of space" is steadily diminished through improvements in transport and communications technology.

Most of these changes in spatial organization happen unintentionally; they are not planned. It has become increasingly evident, however, that patterns of urbanization are themselves subject to planned intervention. Instead of being treated as a mere byproduct of other developments, urbanization has in recent years become an object of public policy in its own right. By controlling the relevant processes, policy attempts to produce the spatial organization which best contributes to vital social objectives, such as employment, income, and general well being.

This chapter deals with the spatial organization of the Capital Region of Chile. Centered upon Santiago, the Region extends for about two hours' driving distance to the North, West, and South, its eastern boundary being defined by the cordillera of the Andes which rises abruptly behind Santiago, a spectacular backdrop to its busy life. Within this radius are found more than one-third of Chile's population, two-thirds of its urban population, and 70 percent of its industrial employment.

The spatial structure of this Region has evolved over centuries without conscious guidance or direction. It is our purpose to examine this structure, evaluate its impact on life in the Region and in the nation as a whole, and propose a set of coherent policies for its future development.

II. SOME ELEMENTS OF SPATIAL STRUCTURE

The Pattern of Settlement

The Capital Region of Chile occupies roughly the geographic center of the country. Astride one of the most important access routes connecting the Pacific with the Atlantic Ocean (Map 1), its broken topography has created a series of natural barriers which have shaped the basic character of its settlement pattern.

The eastern boundary of the Region is formed by the Andean cordillera. Practically an uninhabited area, its principal uses are for the extraction of copper (El Teniente Mine near Rancagua and La Andina Mine near Los Andes) and the pleasures of skiing at the fashionable winter resort of Portillo as well as at the more modest, but increasingly popular, metropolitan skiing areas of Farellones and Lagunillas. A highway and railroad connect the Capital Region with Argentina (Mendoza) across a high-altitude pass at Caracoles.

The Valley of Aconcagua defines the northern boundary of the Region. It extends in an east-westerly direction from the edge of the cordillera near Los Andes, to Llay-Llay on the Panamerican Highway and thence, via Calera-Quillota, to Concón near Valparaíso. Except for the immediate coastal area with its large concentration of metropolitan population, the economy of the Aconcagua

Map 2. CAPITAL REGION:
Settlement Patterns and Urban Subsystems

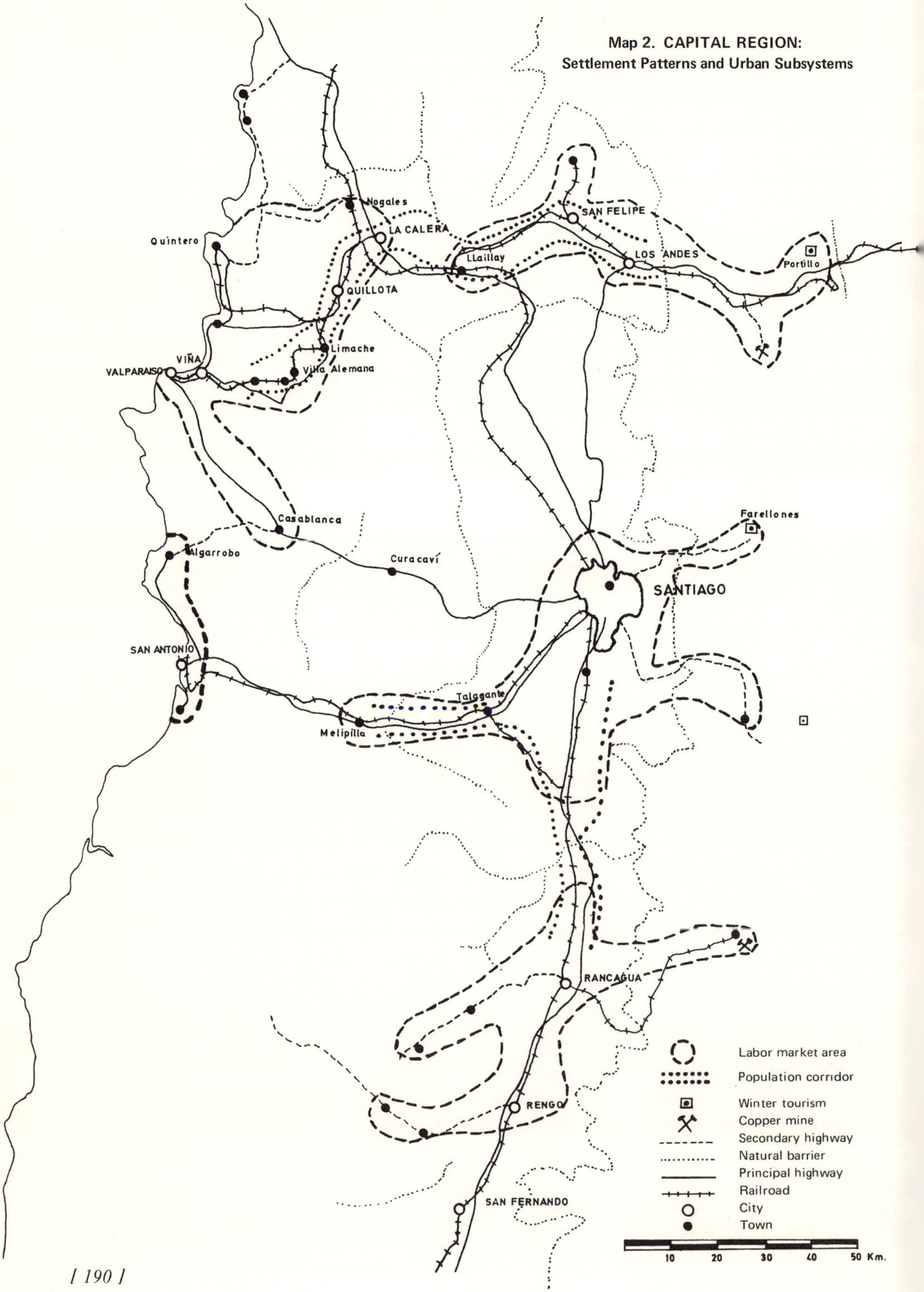

Valley is primarily given over to the intensive cultivation of semi-tropical fruits and vegetables (Map 7).

The Central Valley of Chile extends directly to the South of the San Felipe-Los Andes area and is focused on Santiago. Its fertile soils define a second agricultural zone immediately surrounding the national capital. Reaching southward to Rancagua, San Fernando, and beyond, it is punctuated at fairly regular intervals by a series of small agricultural service towns and provincial capitals (Map 7). At Santiago, the Valley widens into a basin which is closed by a narrows at La Angostura de Paine some 60 kilometers to the South. Immediately beyond the *angostura,* however the valley opens up again into the Basin of Rancagua.

Central Valley and coast are separated from each other by an extensive and largely unpopulated mountain area which is transversed by two major transport routes. The first connects Santiago with Valparaíso-Viña del Mar via Curacaví and Casablanca, while the other joins Santiago (and Rancagua) with the seaport of San Antonio.

The coastal area displays a variety of characteristics. At the extreme North of the Capital Region, Papudo and Zapallar are traditional summer resort areas. Further to the South, the metropolitan area of Valparaíso-Viña occupies a narrow strip of land along the coast, but has spilled over the adjoining hills, mesas, and valleys overlooking the Pacific Ocean. Still further South, and as yet unconnected with Valparaíso, commences a third area of urban settlements, favored by the population of Santiago for seaside recreation, especially during the summer months. Its northernmost point is given by Algarrobo, while its southern limit terminates at Rocas de Santo Domingo, near San Antonio.

A look at the settlement pattern (Map 2) clearly reveals how these topographical features have influenced structure of the Region. Most of the population lives either in the two metropolitan areas of the Capital Region (Santiago, Valparaíso-Viña) or in agro-industrial "corridors" which extend in two broad bands, the first from the coast along the Aconcagua Valley to Los Andes, the second southwards from Santiago, filling up the Central Valley portion of the Region to nearly its natural limits. Except for the small coastal strip between San Antonio and Algarrobo, the rest of the Region is only sparsely populated.

Accessibility Characteristics

Accessibility is an important characteristic for determining actual as well as potential areas of interaction. Map 3 was constructed on the basis of average driving speeds over the existing road network centered on Santiago, Valparaíso, and Quintero—a city which is likely to gain appreciably in importance during the next ten years. Accepting one hour as the maximum commuting distance, the map clearly reveals two major potential labor market areas which are physically separated today. Santiago extends halfway to the Aconcagua Valley in the

Map 3. CAPITAL REGION: Accessibility Characteristics

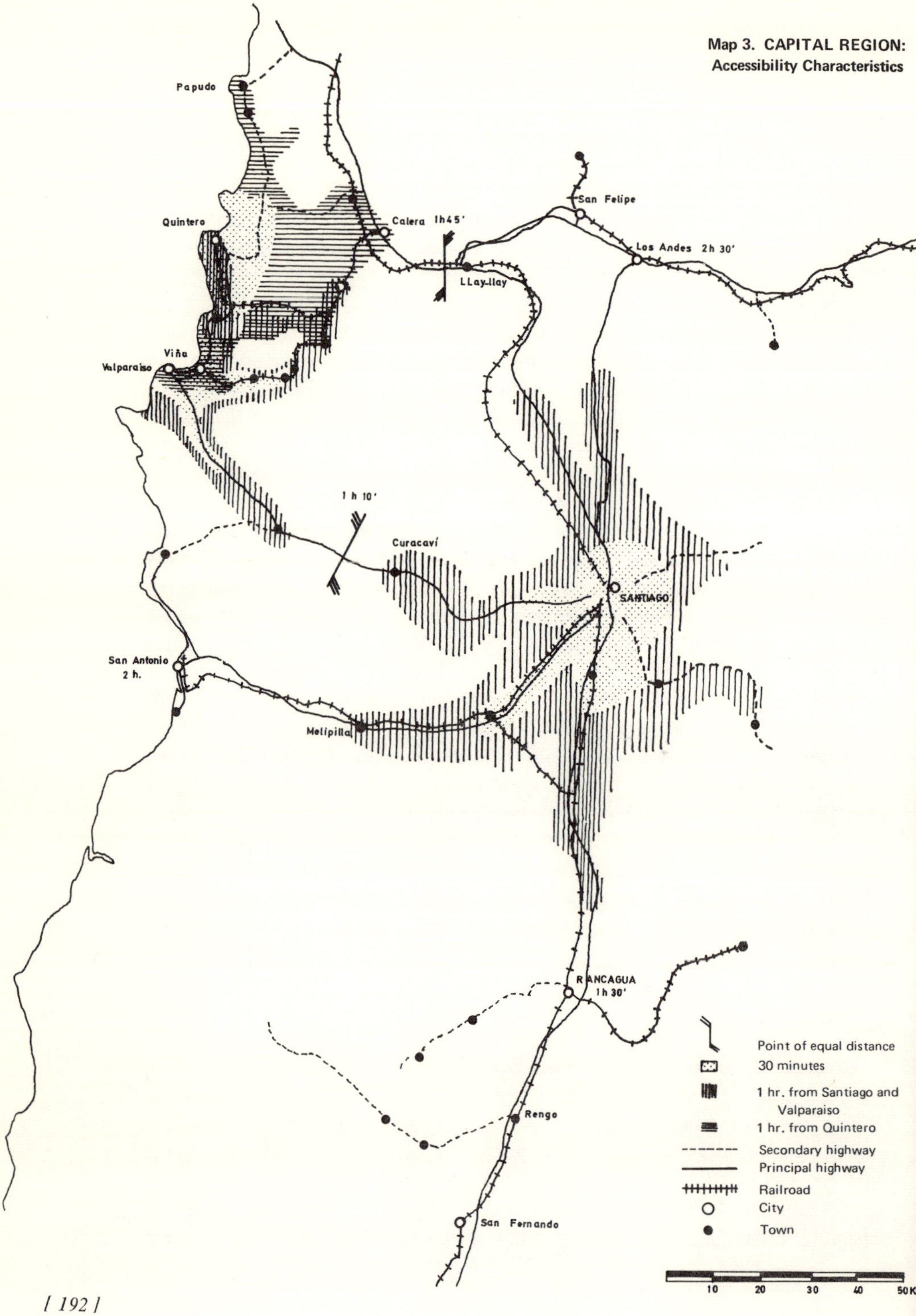

North, to Curacaví and Melipilla to the West, and about two-thirds of the way to Rancagua in the South. Valparaíso, on the other hand, ends abruptly to the South, but reaches eastward as far as Casablanca, northeastwards to Quillota, and terminates at Quintero in the North. By shifting the center of gravity to Quintero, however, Calera as well as Zapallar and Papudo are also included.

A more narrow definition of intense urban interaction is given by the thirty-minute isochrome.

Of some interest is the position of the Aconcagua Valley, especially its geographical center at San Felipe-Los Andes. This area is today equally accessible from Valparaíso and Santiago, though its natural configuration would suggest a closer relationship to the former. Lying just beyond the two-hour limit from either metropolitan center, its present economic orientation is ambiguous. Because of the greater attractiveness of Santiago as both a market and service center, its actual ties appear to be stronger with the national capital than with the port of Valparaíso. To the extent that its local economy turns increasingly to export markets, however, the relation of this region with ports along the coast will be accentuated.

The map also reveals the relative inaccessibility of the mountainous zone which separates the Central Valley from the coast. Most of its area is accessible only by dirt road and lies at distances of two to four hours from the major metropolitan centers.

Urban Subsystems

In Map 2, an attempt has been made to sketch the probable boundaries of future labor market areas which are centered on each of the five major urban agglomerations in the Region. This was done on the basis of known patterns of local interaction and commuting, ecological criteria, existing population distributions, and the accessibility characteristics discussed in the preceding paragraphs. These *urban subsystems* may be classified according to the degree of their dependency on external controls:

1. *Dominant:* Santiago
2. *Subdominant:* Valparaíso-Viña del Mar
3. *Dependent:* a. San Antonio
   b. Rancagua
   c. Aconcagua (San Felipe-Los Andes)

Each of these subsystems, except for San Antonio, is also the administrative center of a Province. They are arranged in the form of a wheel around the hub of Santiago to which they are connected by four spokes of transportation. The area of low population density which is excluded from this definition of urban subsystems, we shall call the *Intermetropolitan Periphery.*

A few economic indicators important for understanding subregional characteristics are summarized in Map 4. These indicators dramatically point up the

Map 4. CAPITAL REGION:
Economic and Social Indicators of Subregions

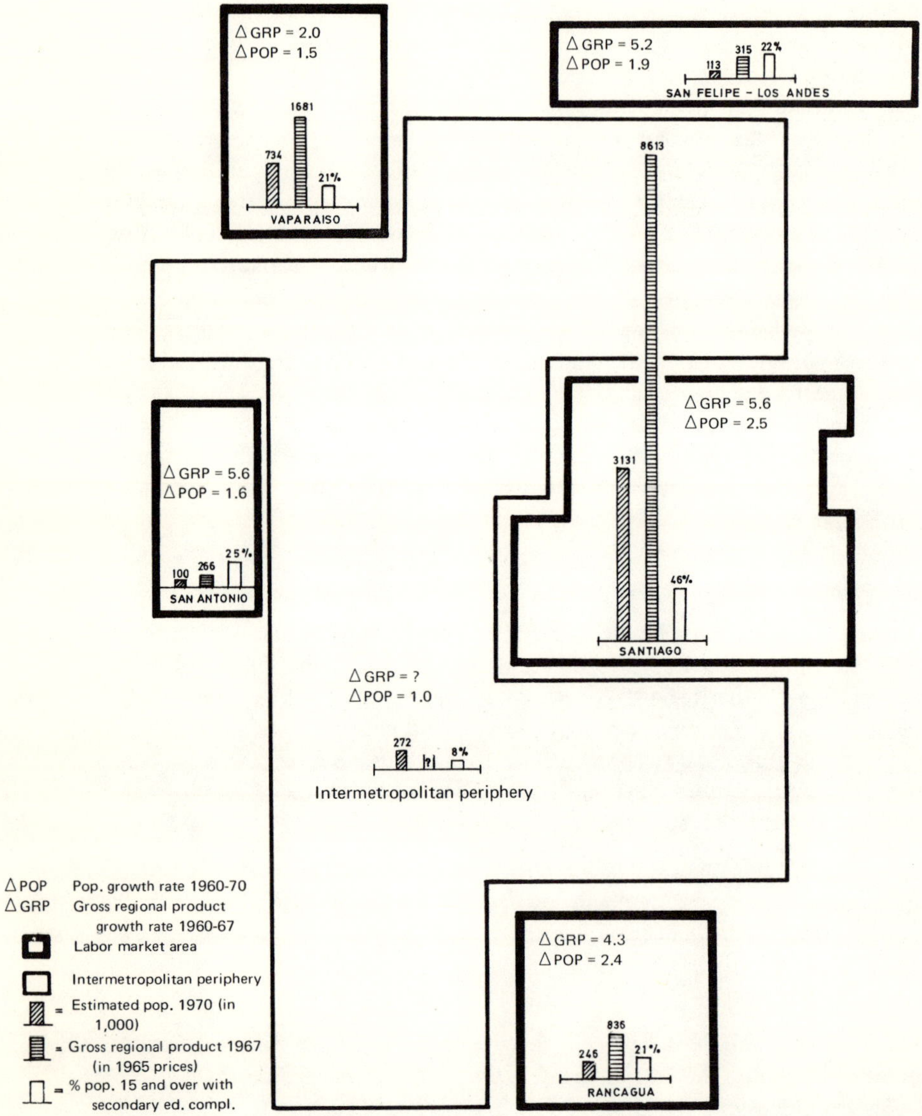

Note: Indicated subregions are not drawn proportional to size. However, they maintain their relative geographical position.

overwhelming importance of Santiago and the physically large but economically insignificant weight of the intermetropolitan periphery.

Movement Patterns

The urban subsystems of the region are, in turn, integrated amongst themselves by flows of communication, and the movements of passengers and freight. The relevant information is graphically presented in a series of diagrams on Map 5.

The patterns shown are consistent with our initial hypothesis regarding the definition of urban subsystems. The most complex pattern is for telephone calls. The linkage between the Quillota-Calera and the Valparaíso-Viña metropolis supports the inclusion of the former twin cities in the subsystem, but also reveals the existence of a weak bond between the Aconcagua-Valparaíso and Rancagua-San Antonio subsystems.

Passenger and freight movements are shown for only the principal routes. As in the case of telephone calls, the intensity of interaction is greatest between the Santiago-Valparaíso subsystems. Of special note is the substantial net-movement of freight *into* Santiago which amounts to two million tons out of a total movement of five million.[1]

As crude as these data are, they indicate very clearly that the Capital Region is articulated and dominated by Santiago. Intersystem movements which bypass Santiago are almost non-existent. This not only results in a separation of the northern and southern halves of the Region whose only link is with Santiago, but also suggests that subsystem complementarities are with Santiago rather than among the subsystems themselves. Aconcagua and Racagua are both agricultural and mining centers. Agricultural produce supplies principally the metropolitan market of Santiago, while copper revenues are the critical magnitude in the preparation of the national capital budget; the copper itself is exported via San Antonio and Valparaíso-Viña are both specialized ports (copper and grain; general merchandise) serving Santiago, as well as important centers of tourist Antonio and Valparaíso-Viña are both specialized ports (copper and grain: general merchandise), serving Santiago, as well as important centers of tourist attraction. For its own part, Santiago is the center of powerful institutions that govern not only the Capital Region but the entire country. It is the only subsystem in Chile which has successfully sustained a development process of its own. All other subsystems, with the partial exception of Valparaíso, are almost entirely dependent for their well-being on public and private bureaucracies in the nation's capital.

Map 5. CAPITAL REGION:
Major Interaction Patterns

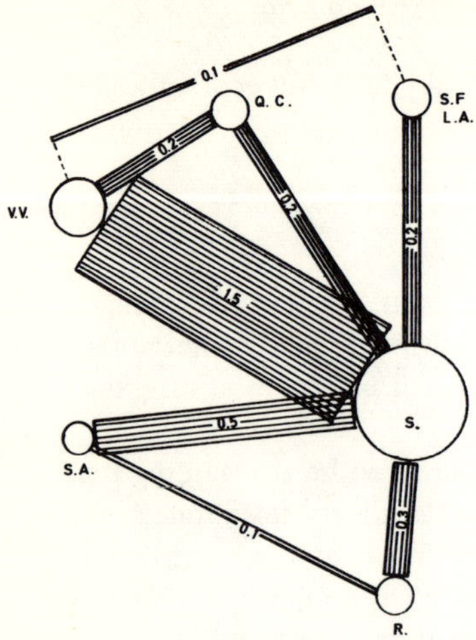

Telephone calls, 1967 (in 1,000's)
Source: ODEPLAN Kardex

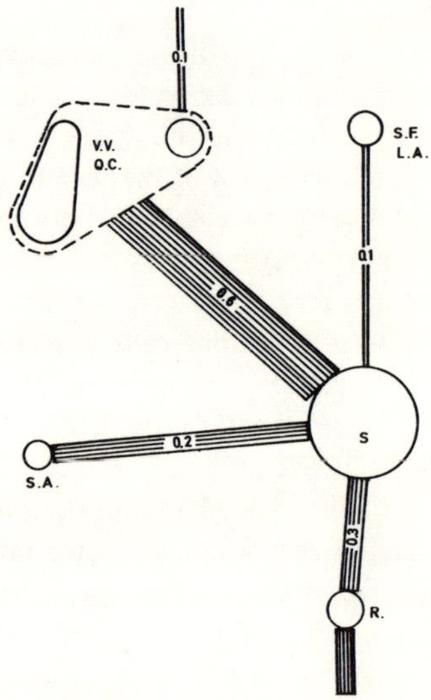

Passenger flows in public carriers, 1968 (millions)
Source: CIDU

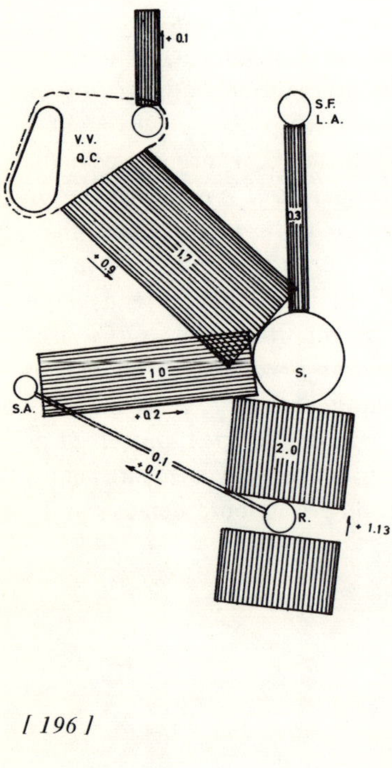

Freight movements, 1968 (million tons)
+0.2 positive balance in direction indicated
Source: CIDU

III. PRELIMINARIES TO A POLICY FOR SPATIAL DEVELOPMENT

The Region in an International Context

The Capital Region of Chile occupies a unique geo-political position on the continent of South America (Map 1).

(1) It is the most powerful center of economic activities on the West Coast of South America, South of Lima.
(2) It has extensive port facilities in San Antonio (copper and grains), Valparaíso (general merchandise), and more recently also in Quintero (petroleum, copper and copper products, general merchandise).
(3) Through this system of ports, the Region's industries are potentially able to produce for export to the Andean Market countries of Perú, Bolivia, Ecuador, and Colombia, as well as for the emerging markets of the Pacific Basin (Japan, China, India, Pakistan, Indonesia, Phillippines, Australia).
(4) It is the westernmost extension of the emerging South American system of cities, which extends from Rio de Janeiro in Brazil through São Paulo, Porto Alegre, Montevideo, Buenos Aires, and Mendoza to Santiago and Valparaíso. A direct linkage to this potential South American market exists today by way of the newly constructed International Highway from Valparaíso to Mendoza, as well as by rail. Although the present volume of trade across this route is still quite small, it may be expected to increase with time. By the end of this century, the Santiago-Buenos Aires axis may well have evolved into one of the important continental trade routes, linking the Atlantic coast of South America with the Pacific, and from there in either direction, to European-African markets, on the one hand, and Far Eastern markets, on the other. Neither Arica at the northernmost point of Chile, nor Lima, Perú, offer quite this array of locational advantages.

During the past thirty years, Chile's industrialization program has been following an explicit policy of import substitution. But the continuation of this policy is not likely to provide the country with the kind of economic expansion that is generally regarded as desirable (i.e., a secular rate of growth of between 6 and 7 percent). The long-term prospects for copper—today the country's leading export product—are only moderately good. World prices are unlikely to be maintained at present levels for another generation, and substitutes for copper are beginning to be introduced. At the same time, the rate of population growth in Chile has sharply declined to less than 2 percent a year; American aid prospects are uncertain; and the internal market for industrial products is expanding at only modest rates. The future of Chile's economy must, therefore, be seen as being tied increasingly to the country's ability to export industrial and agricultural products. This trend is already under way.

Although the country's current export markets are chiefly in the United States, Chile's trade with other nations will probably be intensified during the next decade. As an institutional arrangement, the Andean Market is already a reality. Active negotiations to establish a Latin American Common Market are continuing. And in 1970, the first conference which brought together academicians and public figures from over 15 Pacific Basin countries for a joint review of current economic and political issues gathered in Viña del Mar. These important pioneering efforts may well bear fruit a decade or two hence.

Given this outlook on Chile's future economic relations with the rest of the world, the geographical position of the Capital Region assumes special significance and assigns a distinctive role within the Region to the ports of Valparaíso and Quintero. Their volume of trade is likely to experience a very substantial increase, and many new industrial establishments whose major orientation will be to export markets and/or to raw material obtained from overseas may be attracted there. Other ports in Chile, such as Arica or Concepción, will share in this general expansion, but not nearly to the same extent, since they lack both a direct connection to the Atlantic "megalopolis" and an industrial establishment that is comparable in size and diversity to that of the Capital Region.

The Region in a National Context

To a major extent, the importance of the Capital Region in the Chilean economy can be ascribed to Santiago itself. All other urban subsystems of the Region are not only small by comparison but have rates of economic growth that are only one-third to two-thirds of those for Santiago (approximately 7 percent per year in regional product). Valparaíso, for example, is the principal port and seacoast center for the population of Santiago. Although its base in manufacturing is large compared to other Chilean cities, there has been little industrial expansion in recent years. The city of Valparaíso and, indeed, the entire labor market area which it integrates, is but a pale reflection of the booming economy of Santiago.

In general, Santiago has been absorbing a good part of the country's productive potential, with the result that its overall economic power has been steadily rising. This can be seen from the comparisons in Table 1. Between 1907 and 1970, Santiago increased its percentage of total population from 10 to 34 and, between 1960-1967, its share of gross national product from 40 to 45 percent.[2]

TABLE 1
PERCENTAGE SHARES OF POPULATION AND INDUSTRIAL PRODUCTION

	Population		*Regional Product*	
	1907	*1970*	*1960*	*1967*
Chile	100.0	100.0	100.0	100.0
Capital Region	27.0	46.8	56.8	59.3
Santiago	10.3	33.5	40.0	44.9

As sketchy as these trends are—and they could be more elaborately documented—they permit one to draw a number of conclusions concerning their effects on regional economic growth outside Santiago. The population of the nation's capital has to a considerable extent been supplied by the provinces of Chile.[3] This migration has been selective not only for age, since it is chiefly young men and women of working age who migrate, but also for education and technical abilities: those who remain in the provinces are generally less educated and less skilled than those who leave. As a result of continued migration, provincial economies have remained quite small relative to Santiago. In 1970, for instance, one-third of the provinces had less than 150,000 population, and no province, except for Valparaiso and Concepción, had more than a half-million inhabitants. On the other hand, only three provinces had population gains during the recent census period in excess of their rates of natural increase (Tarapacá, Atacama, and Aisén for a combined gain of 100,000), while 14 out of 25 provinces had gains of less than 1.5 percent a year and consequently experience out-migration on a very substantial scale. Most of the migrants, it may be inferred, eventually came to Santiago.

In short, the provinces have been exporting their most valuable asset to Santiago, educated young people. In addition, there has probably been a considerable net transfer of private capital funds from the periphery to Santiago. Viewed in a cold light, it might therefore be said that the principal function of the provinces in Chile is to supply the nation's capital with cheap labor, capital, food, and industrial raw materials, in return for which they receive just enough services (and an occasional manufacturing plan) to satisfy the minimum demands of their populations. Since provinces are organized as administrative and not political entities, they carry very little weight in the counsels of the central government and depend on the "power structure" in Santiago for nearly all decisions affecting the lives of their populations. Since they are dependent, they are unable to initiate and maintain a development process of their own.

It may be of some interest to speculate what the continuation of these trends would mean for Chile. The relevant data are summarized in Table 2.

By 1990, the Capital Region would generate two-thirds of Chile's national product and account for about one-half on the country's population. Most of the new growth would continue to be concentrated in Santiago.

TABLE 2
HYPOTHETICAL PROJECTION OF TRENDS, 1970-1990

	Population			*Regional Product*		
	Rate	*1970*	*1990*	*Rate*	*1967*	*1990*
Chile	2.0	100.0	100.0	5.5	100.0	100.0
Capital Region	2.7	48.0	51.0	6.1	59.3	67.6
Santiago	2.5	33.0	36.0	6.5	44.9	53.0

These trends do not represent an equilibrium solution for regional development. Unless drastic measures are adopted to contain, if not reverse, the tendency towards increasing centralization, Chile's economy will be virtually reduced, at least in geographic terms, to the economy of the Capital Region.

The Centralization of Power and its Consequences

It is currently popular in Chile to blame the increasing concentration of population and economic activities on the perfidities of the private sector. But a much simpler explanation would be to say that the attractiveness of Santiago as a location is due primarily to the centralization of the decision-making powers of government. A location in Santiago facilitates access to the power structure and this, under Chilean conditions, is probably the single most important condition for assuring the success of a new enterprise or the expansion of an existing one. There are subsidiary reasons, such as the size of the national market which lies within easy reach of Santiago, the relatively high level of specialized services in the city, and external economies—all of which contribute to make Santiago extremely attractive to business. But no enterprise can expect to prosper in Chile without the active support of government officials, and these can only be reached in Santiago.

In the statist economies of South America, this condition for the location of new enterprise is probably quite general. It also affects the ordinary person who, though he does not expect to transact any business of his own, nevertheless seeks out the national capital because the possibilities for obtaining a job, adequate housing, and schooling for his children, are vastly superior there than in the province or his origin. Indirectly, he is enjoying the fruits of centralized power.

The extreme form of centralization of power in Chile has had two important consequences. First, it has drained the will to organize for economic development in the Provinces. The bureaucratic obstacles to self-starting local efforts are so great that few people are willing even to try. This leaves the local economies almost entirely at the mercy of central decision makers. The second consequence is related to the first: regional development in Chile is not a continuous process of structural growth but a random and discontinuous process in which from time to time projects are imported from Santiago at the pleasure of central bureaucrats. Some of these, such as schools or public housing, correspond to an immediate local need. But industrial or agricultural projects are implanted according to project- rather than region-related locational criteria and constitute virtual enclaves of central power, without engendering significant multiplier effects in the local economy. After a short period of accelerated growth, the Province sinks back into its usual stupor.[4] Outside Santiago, there are only two provincial centers in Chile which are capable of generating even a small measure of development of their own: Valparaíso and Concepción. And both of these are urban concentrations of over half a million.[5]

None of this would matter a great deal if the bureaucratic center were, in fact, able to make better decisions than local people. Although it is widely assumed that this is so, there are strong reasons to think that the failure to decentralize often leads to wrong decisions, unjustifiable delays, and the neglect of potentially important development possibilities.

Central bureaucrats have an imperfect perception of local opportunities for development

Their knowledge of local areas tends to be rather abstract and superficial. Opportunities for investment are judged more in a national perspective (what does Chile need?) than as a response to local problems. The national perspective is certainly needed, but a local view of development is not necessarily contradictory to it. In fact, a country which is limping along on a rather slow rate of growth, such as Chile, might be well advised to promote *any* project that appears to be economically sound. The National Development Corporation (CORFO), for instance, tends to regard industrial development outside Santiago primarily in relation to natural resources, such as fisheries, horticulture, forestry, and sugar beets. It sees the country's periphery as a tremendous resource base for sustaining the market-oriented industrial plant of Santiago. The effects of its investments on local communities are considered only in passing. This approach may have considerable merit, but it leaves about two-thirds of the country's population without effective means to improve their well being except by moving to Santiago. In most of the industrially advanced countries, natural resources play only a minor role in economic growth; the chief contributing factors rather appear to be the technical skills and organizing abilities of their populations. This view of the development process is foreign to the engineers of CORFO.

Central bureaucrats are slow to respond to local problems

This is partly a result of the extremely cumbersome process of channeling information through the maze of central offices. The files accumulate, the necessary approvals multiply, documents get mislaid, study groups meet but fail to arrive at an agreement, and procedures must be obeyed according to the accepted norms of the bureaucratic behavior. Generally there is a long list of projects awaiting some kind of decision. Problems that might be resolved within minutes on the spot, because all of the relevant information is locally available, may require months of painstaking bureaucratic review at the center.

Central bureaucrats have a limited capacity for undertaking projects

In the central bureaucracy there are many capable people, but their abilities to act as national entrepreneurs are limited. The country's growth potential is therefore restricted to the working capacity of a few hundred individuals who occupy strategic positions in the Santiago hierarchy. For this reason, large projects, though time-consuming and risky, are preferred over smaller projects having primarily a local character. These are not only discouraged by institutional obstacles at the center; they are often ignored altogether.

The extreme centralization of power in Chile has therefore led to the internal colonization of the periphery by Santiago. Except for a very small number of activists, most of the country's population has been reduced to a status of utter dependency. They are prevented from exercising any initiative of their own and are expected patiently to wait for benefits that may ultimately flow from central wisdom. The idea that Chile's population might contribute anything at all to the development process other than their skills as docile "working ants" is not yet widely shared. Lip service is being paid to so-called popular participation, but this is conceived primarily as a way for peacefully petitioning the national authorities concerning local problems.[6] It is not understood as a way to mobilize the population for national development

The Image of the Region

The preceding discussion may help to throw some light on a vexing question. Should the Capital Region be considered as an emerging urban field in which the friction of space is no longer a decisive factor in the location of people and activities–all points of the "field" having relatively equal access to all other points–or rather as a *congeries of urban subsystems,* where each subsystem constitutes a separate unit for development.[7] The latter concept is usually applied to the study of national territories, while the former has been used to describe conditions of growing locational freedom in and around major metropolitan areas in advanced industrialized and post-industrial societies.

The implications for planning of using either one or the other concept differ markedly. If we think of the Capital Region as an "urban field," we would probably advocate centralized control and coordination of all the relevant development activities. The Capital Region would appear as an unbroken, continuous space, whose functionally specialized parts form a single whole. This implies extraordinary freedom in location, probably a policy of industrial and residential dispersal, and centrally managed networks of transport and public utilities. The "urban field" is the equivalent of a single large city and should be administered as such.

But if we think of the Capital Region as a "congeries of urban subsystems," where each subsystem constitutes a potential–if not an actual–labor market area, the primary focus of planning would be on the individual subsystems and only in the second instance on their systemwide integration. Instead of a single administrative authority for the entire urban field, selective decentralization might be encouraged as being more appropriate.

The "urban field" view basically argues that Chile's Capital Region is following the development pattern of large metropolitan systems in the United States and Western Europe. The Region which currently has about 4.7 million inhabitants and a per capita product of perhaps $700, can be expected to double both by the end of the century (assuming present trends continue). If only one out of every four families would own a car, the number of private automobiles would rise from a present figure of about 150,000 to nearly a half-million. Telephone communications would be vastly improved with the installation of automatic dialing. Time-distances would shrink to half or less than what they are at present or to about one hour. At the same time, the population would be much more highly educated than today, would have become more socially mobile, and consequently would begin to be geographically mobile as well, changing jobs according to minor differentials in pay and, in general, acting so as to maximize individual income.

These arguments, however, overlook a number of rigidities in the Chilean pattern of spatial development which, in a general way, support the conclusion that the more relevant concept for the Capital Region at the present time is that of a congeries of urban subsystems.

(1) Despite the rapid increase in automobility, at least three-quarters of the population will continue to be dependent on public means of transportation.

(2) Unless a very drastic redistribution of incomes should occur, most families will be able to spend less than one dollar for daily commuting for each working member, or about 10 percent of their combined incomes. This would leave very little for any other trips throughout the Capital Region. Geographic mobility has a high cost. To the extent that people are still living close to margins of minimum comfort, transportation is not likely to be ranked as high as other forms of personal consumption.

(3) Job mobility in Chile is traditionally low. A series of non-transferrable advantages, such as housing and social services, are tied to each job, and legislation protects the worker's right to these advantages. It is, of course, conceivable that this pattern will change over the next two or three decades, but this appears a rather unlikely prospect at present.

(4) Residential mobility is low in Chile. Once a family obtains a house of its own, it rarely changes to another one. The parents live there until they die, and their children will inherit it. Again, it is conceivable that this pattern will eventually change; it is not likely to do so, however, so long as Chile continues to be plagued by a serious shortage of housing.

(5) Even if the friction of distance could be reduced for everyone in the society, most people would still prefer to live in Santiago. Schools, social clubs, family entertainment, business lunches are all part of the requirements for social advancement and economic success. The potential freedom of the "urban field" is in large part an illusion.

In addition to these rigidities, a number of other reasons must be mentioned which cast doubt on the validity of the "urban field" as a planning concept for the Capital Region. The currently prevailing image of the region is one of relatively unrelated point locations. Urban labor market areas are not yet fully integrated. Distances are still long; a round-trip journey to any of the major urban subsystems from the center of Santiago, for instance, requires a whole day for completion. The regional image is somewhat clearer for the upper-middle class, but even professionals who presumably have a greater facility for conceptual thinking than other people, have difficulty in visualizing the Capital Region as their habitat. They are accustomed to use certain portions of it—especially for recreation—but, by and large, most of their lives are spent within the narrow radius defined by work place and their home.

The outlying subsystems of the Capital Region—Valparaíso, Aconcagua, Rancagua, and San Antonio—share with other parts of the Santiago periphery their almost total dependence on central power. As a result, their potential capabilities for development are not being used to their full capacity. Only a small part of the agricultural potential of the area has been brought into cultivation; touristic resources are waiting to be tapped; deep-sea fisheries need more scientific methods of exploitation; inter-regional complementarities have scarcely been identified. Each of the major subsystems presents its own configuration of planning problems awaiting solution. It is not at all certain that the kind of centralization of power which the "urban field" view implies would lead to a more efficient system of spatial organization conceived in terms of the interests of people where they now live. All that might result from this approach is the further concentration of population in Santiago to the detriment of its periphery.

For all these reasons, therefore, we conclude that the image of the region as an "urban field"—*una gran ciudad*—is not the appropriate one at present. But neither should it be entirely abandoned. What, in fact, is needed is a conception that will harmonize both the "congeries of urban subsystems" and "urban field" images of the Capital Region.

Development policy must always start with what is given in the present situation. In the case of the Capital Region, this suggests an initial focus on individual subsystems. At the same time, the Capital Region will increasingly come to behave as a single environmental system. And this suggests that public policy should strive to articulate the development of individual subsystems within a larger context that has as its long-term objective the creation of a spatially dispersed *gran ciudad.*

This compound and dynamic image of the Capital Region suggests the following six criteria for a policy of spatial development:

(1) Decentralization of power to selected subsystems of greater than ordinary prospects for development.

(2) Functional specialization of each subsystem within the context of the Capital Region.

(3) Acceleration of the rate of economic and demographic growth in the peripheral subsystems of the Capital Region.

(4) Integration of each subsystem as a labor market area, relying primarily on public means of transportation for commuting to work.

(5) Preservation of existing open spaces to define the boundaries of each subsystem.
(6) Functional integration among the five existing urban subsystems of the Region, especially of Valparaíso with the dependent subsystems of Aconcagua, San Antonio, and Rancagua.

IV. OBJECTIVES FOR SPATIAL DEVELOPMENT IN THE CAPITAL REGION

An Approach to Policies Planning

Traditionally, spatial planning has involved the preparation of detailed designs for urban areas in which each major activity was assigned a specific location. Once completed, this master plan for land and circulation was implemented primarily through a series of legal instruments, the most important of which are zoning ordinances. This approach, however, rarely yielded satisfactory results. On the one hand, the detail in which the plan was specified was too refined; on the other hand, the legal instruments available to planners were not sufficiently powerful to compel locational behavior consistent with the plan.

Rejecting the static master plan approach, we shall propose policies that are aimed at altering the rate and/or direction of six spatial effects of development (see Table 3). These effects are already operating in the Capital Region, as indeed, they are elsewhere, as the result of prior economic development. It is by acting upon the underlying processes that determine particular effects (the basic variables of Table 3) that are more adequate spatial organization may be given to the region.

The six effects identified are closely interrelated. Drainage effects, for instance, are the inverse result of polarization. Diffusion effects are heavily influenced by patterns of interaction. Dispersion effects are the almost inevitable by-product of rapid polarization. The reason for listing them separately—for they could easily be collapsed into a smaller number on purely logical grounds—is that that each effect points to a different area of potential intervention. The discriminating criterion, therefore, is to be found in the underlying variables (see column 3).

Spatial development policies will seek to influence the behavior of these effects, so that vital objectives of regional and/or national development may be advanced. It is extremely important to be aware, however, that spatial effects of development are produced even without explicit policy intervention. Policies may be devised that will accelerate, decelerate, contain, and redirect these effects. But because they are already in operation, and because policy can influence only the *rates* at which they occur at different locations, any attempt to bring about significant structural changes in the spatial organization of an area requires the steady application of a set of policies over extended periods of time. For this reason, the most practical way to achieve the purposes of spatial planning is through permanent changes in the institutional arrangements which

TABLE 3
SPATIAL EFFECTS OF DEVELOPMENT: DEFINITIONS, VARIABLES, POLICIES

Spatial Effects	*Definition*	*Basic Variable*	*Possible Policy Objectives*
1. Dominance	effective control by a core region of decisions affecting development in a dependent periphery	distribution of institutionalized decision-making authority and effective power	increase, decrease selectively
2. Polarization	concentration in space of population and economic activities	location decisions of individual family units and firms	accelerate, decelerate contain; establish new polarizing centers
3. Interaction	flows of information, trade, and people between two areas	functional specialization and channels of communication	intensify, diversify, restructure, stabilize
4. Innovation Diffusion	transmission of new ideas, customs, artifacts, and institutional forms across geographic space	communication processes	accelerate, channel, select
5. Dispersion	physical spread of the city into its immediate surroundings	occupance of land for urban uses	channel, order, contain
6. Drainage	net-withdrawal of production factors from peripheral regions	labor migration, capital movement, quantity and quality of natural resources	accelerate, decelerate, reverse, redirect

will ultimately be responsible for generating the effects desired. For instance, if access to power is the critical factor influencing the behavior of polarization, the territorial organization of power may need to be changed to bring about a different pattern in the spatial incidence of polarization. This basic approach to spatial development policy may need to be complemented by other measures, subordinate to the main purposes of institutional innovation.

Objectives of Spatial Development

The definition of objectives for the Capital Region is, strictly speaking, the responsibility of the political branches of government. It must reflect a continuing concern with the norms and values of spatial development. At the present moment, however, the very concept of the Capital Region is familiar to only a few dozen individuals in Chile and there is neither an administrative nor political framework of decision-making capable of formulating valid objectives for the Region and committing the necessary resources. Consequently, the possibility of clarifying already established objectives or of simply accepting them as the basis for a more detailed specification does not exist.

The following discussion of objectives has has its major purpose the clarification of the value premises underlying the proposals which we intend to make. It will also be useful in focusing attention on those aspects of development in the Capital Region which we believe to be of critical importance to the future of its population. Because of the Region's preponderant importance in the economy of Chile, it is evident that whatever objectives may be identified must be consistent with the general objectives for the development of the country as a whole. We shall not attempt to demonstrate this consistency, however. Our purpose is rather to suggest a way of thinking about regional objectives and to give enough detail to make the subsequent discussion of policy more intelligible.

We shall propose four major sets of policy objective which focus respectively on problems of Environment, Economic Development, Participation in the Management of Regional Development, and Levels of Living.

1. The Region as an Environment for Living. For an increasing number of people, the Capital Region is becoming a relevant living space or habitat. As population and incomes rise, and as geographic mobility increases, the question of physically ordering this space for greater amenity and the protection of its natural advantages will assume substantially greater importance than it has had up to now. Competition for land will become more intense; the destruction of natural assets will become an ever more real threat; and the urban structure of the Region may not adequately serve the needs of its population. Environmental objectives should respond to these concerns. They may therefore include the following:

(a) Preservation of resource amenities, with particular attention to the quality of the air, water, and the scenic and touristic resources of the Region, especially along the coast.

(b) Reduction of the "friction of space" among the urban centers of the Region, and gradual equalization of access to central services, tourist and recreational resources, and places of employment.

(c) Ordering the broad land use patterns of the Region to ensure their mutual compatibility and optimum use within the context of an increasingly interdependent and integrated regional system ("urban field").

(d) Densification of population and economic activities in the potential labor market areas (subsystems) of the Region, and preservation of existing agricultural and other open spaces as natural "green belts" to separate and define subsystems and to prevent their coalescence into a continuously built-up urbanized area.[8]

2. Economic Development. The Capital Region is the only area in Chile which is able to sustain an autonomous process of economic growth. Taking the region as a whole, its problems are primarily concerned with the spatial ordering—the *aménagement*—of its territorial economy, and not with accelerating its overall rate of growth. On the other hand, looking more closely at its present structure, one finds that the only subsystem primarily responsible for the propulsive quality of its economic growth is Santiago; the rest of the Region, as, indeed, the rest of the country, is peripheral to this core area and economically dependent upon it.

Two criteria appear to be relevant to an objective of economic development for the Capital Region. The first is *social efficiency.* This refers specifically to a requirement that future developments in the Region be carried out at the lowest possible social cost. This objective, though clear in principle, is difficult to apply in practice, especially where it refers to patterns of urbanization. If some method could be found that would fairly allocate the social costs of urbanization to the private users of the city, it is possible that further concentration in Santiago would be decelerated, with some of the smaller centers in the Region absorbing "spillover" growth. This is unlikely to happen, however, even if an "urban field" structure could be fully realized. The social efficiency criterion is, nevertheless, a useful one for evaluating alternative investment choices in the Region and, despite the obvious difficulties of measurement, should be retained.

The second criterion is the *optimum use of productive resources* in the Region. This criterion must be applied on a comparative basis and implies that in every case, the social productivity of investments in the Capital Region be judged in relation to investment opportunities in other parts of the country. There is reason to believe that subsystems peripheral to Santiago have not fully brought their existing resources into production. This is clearly the case with agriculture, where the present trend is toward an intensification of production by shifting cultivation to fruits and vegetables. It is also true with respect to the Region's unusual opportunities for the development of both national and international tourism. It appears to be true for fishery resources. And finally it is probably true for the locational resources of the Region, its strategic position with respect to the emerging South American "megalopolis" on the Atlantic coast, the Andean Market, and the vast Pacific markets which promise to become increasingly important in the future.

Any policy for the spatial development of the Capital Region should therefore be so conceived that it will encourage the more efficient use of these resources.

3. Participation in the Management of Regional Development. The idea that people should have direct participation in the processes of governance has become widely accepted in Chile during the last few years. Although the emphasis has been placed on the participation of the lower strata of society, the principle may be extended to other classes as well, since so much of the effective decision power rests in the hands of central government officials who are nearly immune to public pressures short of outright confrontation. Access to this power structure is easiest in Santiago and has been one of the chief contributing forces to the accelerated growth of the capital city. Even in Santiago, however, participation has in practice been restricted to a very small percentage of the population which, for reasons of birth, education, or economic power has been able to influence decisions in ways that are favorable to themselves. Populations outside Santiago have worked under an even greater handicap and for any substantial improvement in their condition have had to rely chiefly on the good will and benevolence of central bureaucrats.

Given the present structure of the Capital Region, with its population centered on five major labor market areas, increased and more effective participation in management decisions will require a substantial measure of administrative decentralization to those areas which demonstrate an unusual capacity for accelerated growth. The decentralization of power therefore appears not only as a means for widening the base of participation and increasing the responsiveness of government agencies to local needs, but also as a method for shifting part of the gravitational force responsible for accelerated growth in Santiago to other urban centers.

4. Levels of Living. The propensity of the Chilean economic system to concentrate the wealth of the country in Santiago has led to results which make the implementation of redistributive policies extraordinarily difficult. On the one hand, it has tended to increase differentials in living levels between Santiago and other parts of the country and has thus contributed to regional inequities. On the other hand, it has tended to increase differentials in living levels among the population of Santiago itself, and has thus contributed to social inequities.

Given the distribution of population in the Capital Region, with more than 70 percent included in the Santiago subsystem along, it is undoubtedly true that the majority of the disadvantaged sectors are also living there. This fact would suggest an incomes policy favoring disadvantaged persons, rather than areas; a policy, in other words, that would attack primarily social inequities. But the geographic mobility of the population is far from perfect in Chile, and in absolute numbers two or three million people may continue to live in the periphery. Moreover, the more that redistribution policies favor Santiago to the exclusion of other areas, migration will be accelerated with the result that most of the initial gains will be wiped out or that the government will be compelled to concentrate on redistributive policies at the expense of policies for growth and spatial development.

If this argument is correct, it would suggest that redistribution policies should address themselves simultaneously to problems of social and regional inequity. This would imply a general scheme for redistribution applied to persons regardless of where they live; improvements in social service levels within major subsystems of the Capital Region; and a policy for creation of new employment possibilities in subsystems peripheral to Santiago. The further rapid growth of the nation's capital provides no assurance that the economic situation of migrants to the center will substantially improve. On the contrary, the more rapid the growth of Santiago's economy, the larger will be the number of poor people moving to the center, and the more will peripheral economies be weakened.

TABLE 4
PROBLEMS AND POLICIES OF SPATIAL DEVELOPMENT IN THE CAPITAL REGION OF CHILE

1. *Dominance effect:*	effective control of Santiago over development decisions in Valparaíso
Major problems:	economic stagnation and reduced growth potential of Valparaíso; essentially passive local responses to development opportunities
Policy objectives:	decrease Santiago's dominance over Valparaíso
Policy direction:	alter the balance of institutional decision-making authority and effective power between the two cities by creating appropriate decision mechanism for Valparaíso that will increase the area's autonomy
2. *Polarization effect:*	continuing rapid concentration of population and economic activities in the Santiago subsystem of the Capital Region
Major problems:	high urban costs for maintenance and service; failure of populations living in peripheral subsystems to share fully in the benefits of regional development
Policy objectives:	decelerate the process of further concentration at Santiago and accelerate polarization at selected subcenters of the region
Policy direction:	change the pattern of location decisions by creating conditions favorable to rapid economic expansion in the Valparaíso and Rancagua subsystems and by direct public investments in these areas
3. *Interaction effects:*	lack of interaction among major spatial subsystems of the Capital Region
Major problems:	excessive dependence of subsystems on a single center; failure to develop spatially articulated complementarities in economic function
Policy objectives:	restructure interaction patterns by intensifying economic relations among spatial subsystems
Policy direction:	promote the functional specialization of spatial subsystems and improve communication facilities among them, increasing their accessibility with respect to each other
4. *Innovation Diffusion effects:*	slow diffusion of urban innovations to Aconcagua and Rancagua subsystems
Major problems:	economic and cultural backwardness of populations living in these areas
Policy objectives:	accelerate the process of innovation diffusion to strategic subsystems of the region
Policy direction:	improve communications between (1) Aconcagua and Valparaíso and (2) Santiago and Rancagua by promoting closer economic and social relations between these two sets of area
5. *Dispersion effect:*	rapid physical expansion of Santiago, at low densities, into agricultural areas surrounding the city
Major problems:	high cost of urbanization services; conversion of prime agricultural land to urban use; difficult access of urban population to major outdoor recreation areas; reduced efficiency
Policy objectives:	contain further physical expansion of Santiago
Policy direction:	gradually reduce the occupance of agricultural land for urban uses by increasing central city densities and by siphoning urban growth into satellite communities

TABLE 4 (Continued)

6. *Drainage effect:*	net-withdrawal of labor resources from Santiago's periphery to the center
Major problems:	excessive polarization at Santiago; reduced potential for economic growth at subsidiary regional centers
Policy objectives:	divert drainage effects from Santiago to subsidiary centers in the Capital Region
Policy direction:	redirect labor migration to the subdominant core region of Valparaíso and the industrial satellite city of Rancagua

Implications of Objectives for Spatial Development

The foregoing set of objectives must now be translated into policies for spatial development. For this stage of the analysis, the six major spatial effects of economic development discussed at the beginning of this section must again be considered. Table 4 lists each major effect, the principal social and economic problems which may be causally attributed to it, a concise statement of the broad policy objective to be achieved, and the principal policy directions for advancing toward the objective (without, however, at this stage, specifying the operational means for their implementation). This systematic accounting of major policies for spatial development is summarized in Map 6.

In the following section, the basic guidelines for an urbanization policy in the Capital Region will be elaborated in greater detail and supplemented by specific proposals intended to reinforce the basic policies framework described as well as satisfy objectives that would remain relatively unaffected by a purely spatial approach (e.g., the quality of the physical environment).

V. A POSSIBLE FUTURE: POLICY PROPOSALS FOR THE SPATIAL DEVELOPMENT OF THE CAPITAL REGION

Specific Policy Proposals

More detailed proposals for the spatial development of the Capital Region will be discussed under three general policies.

The first concerns the allocation of functional roles among each of the five urban subsystems of the Region, together with their Intermetropolitan Periphery. Through a policy of *territorial functional specialization,* we hope to achieve a pattern of complementarities among the principal labor market areas of the Region that will lead to an improved level of systemwide integration and the acceleration of polarization effects at Valparaíso and Rancagua.

The second policy deals with the role of *transportation* in spatial development. Our suggestions for specific projects in this sector are derived from the preceding policy for functional specialization and intended to support such a policy. At present, however, they should be regarded as simply a set of

Map 6. CAPITAL REGION:
Conceptual Framework for Spatial Development

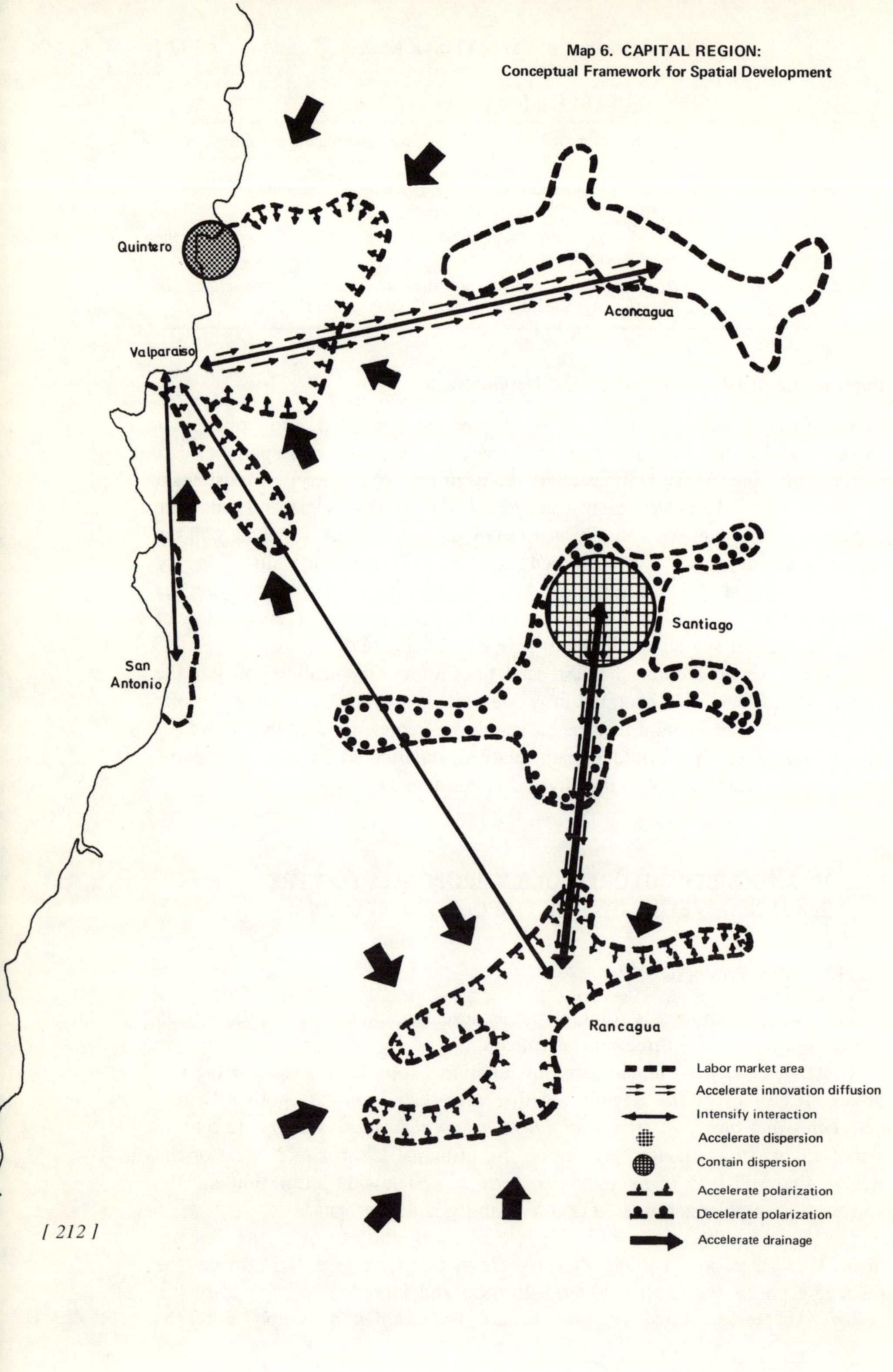

preliminary ideas. Each project will require careful and detailed study prior to its development.

The third policy focuses on *environmental protection.* This will be seen to have important implications for the spatial organization of the Capital Region. A policy addressed to this topic is necessary both to maintain the existing quality of environmental resources and to promote the Capital Region as an "environment for living" that will effectively serve the needs of a rapidly growing and urbanizing population.

Map 7 summarizes these three sets of policy proposals in graphic form.

1. Territorial Functional Specialization. Each urban subsystem offers specialized advantages which must form the basis for any program of spatial development in the Capital Region. They include:

a. For Valparaíso
- Export-oriented manufacturing industry, to be located primarily in the Quintero-Las Ventanas area, with a view to an eventual integration of the Region with Andean, Pacific, and Atlantic markets.
- Urban tourism, expecially at Viña del Mar.
- Regional administration, centered on Valparaíso.
- Higher education, centered on Valparaíso and Viña del Mar.
- Intensive agriculture (fruits and vegetables primarily) in the Quillota-Calera area.
- Modern fishery industry, centered on Quintero.
- National Defense (Naval Base at Valparaíso, with an extension at Quintero.

The development of these functions in the Valparaíso sybsystem will require, among other things:

(1) Creation of a Metropolitan Development Corporation for the Valparaíso-Aconcagua subsystems.

(2) Concentration of central government offices related to the promotion of export/import trade at Valparaíso.

(3) Expansion of port facilities at Quintero.

(4) Construction of an international airport at Quintero, alternative to the present installations at Pudahuel near Santiago.

(5) Development of export-oriented manufacturing industry at Quintero.

(6) Development of a modern fisheries industry, together with specialized installations, including research, at Quintero.

(7) Intensification of urban development at Quintero.

(8) Installation of a hydrofoil transport service between Quintero and Valparaíso-Viña del Mar.

(9) Development of the Calera-Quintero-Valparaíso railroad for rapid, inter-urban commuting.

(10) Expansion of university facilities at Valparaíso and Viña del Mar, as part of a national plan for higher education.

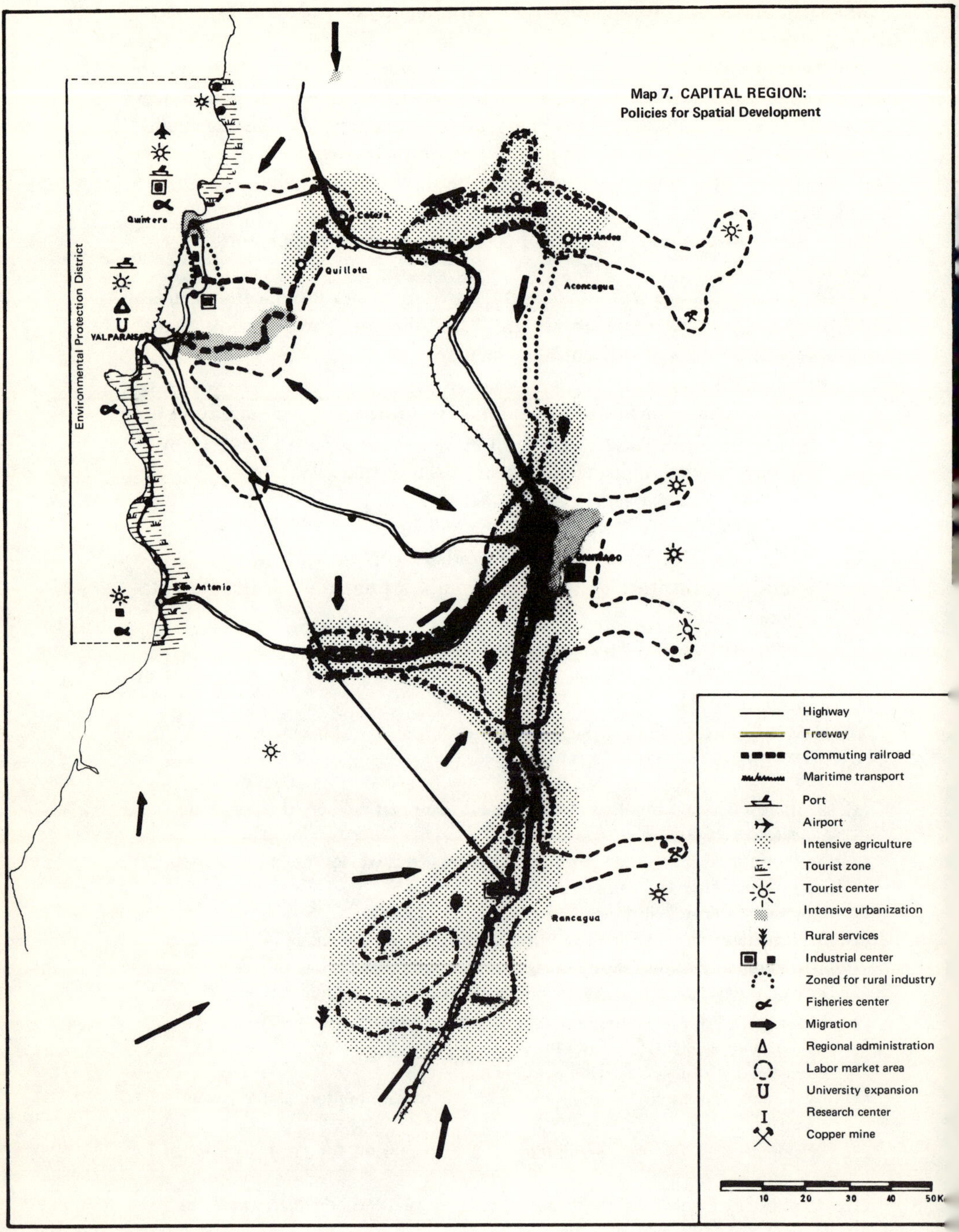
Map 7. CAPITAL REGION:
Policies for Spatial Development
Environmental Protection District
Quintero
Quillota
Aconcagua
VALPARAISO
San Antonio
SANTIAGO
Rancagua
Highway
Freeway
Commuting railroad
Maritime transport
Port
Airport
Intensive agriculture
Tourist zone
Tourist center
Intensive urbanization
Rural services
Industrial center
Zoned for rural industry
Fisheries center
Migration
Regional administration
Labor market area
University expansion
Research center
Copper mine
10 20 30 40 50

b. For Aconcagua

– Intensive agriculture (fruits and vegetables, with special emphasis on production for export).
– Agriculture-related processing industry.
– Copper mining (La Andina Mine).
– National and international tourism at the ski resort of Portillo and at thermal baths near San Felipe-Los Andes.
– Medium-sized, non-toxic industry located in the San Felipe-Los Andes area. Exclusion of all other manufacturing industry from the Aconcagua subsystem.

All of these developments are currently under way and will not require special implementing measures, except for the legal designation of those areas in the Capital Region where industrial locations would be restricted.

c. For Santiago

– Industrial expansion in areas of present concentration (San Bernardo, Puente Alto, and along the Panamerican Highway to the North).
– Further development of all economic, administrative, and cultural functions, except as these may be modified by developments in peripheral subsystems.

The functional specialization of Santiago will continue largely as in the past. In order to regulate its internal spatial structure, however, and further integrate its labor market area, we propose to:

(1) Intensify commercial developments in the center of the city at the major points of intersection of the regional transport system, possibly near the Central Railway Station.
(2) Integrate outlying urban areas, such as Melipilla and the city of Rancagua, through an improved system of rapid railroad commuting.
(3) Complete, as soon as possible, the general development plan for Santiago which is currently under study, and create an institutional mechanism capable of implementing its proposals and of providing continuing guidance for urban development.

d. For Rancagua

– Major center of industrial expansion for Santiago.
– Copper mining (El Teniente Mine).
– Research center, especially related to agriculture, mining, and other natural resources.
– Exclusion of all but agricultural processing industries from the "corridor" connecting Santiago with the Rancagua subsystem.

Rancagua is the leading candidate for becoming Santiago's first major industrial satellite city. Four measures are proposed to implement this policy:

(1) The official designation of Rancagua as an industrial satellite for Santiago.
(2) Strengthening the existing Development Corporation for the Province of O'Higgins (CODO) by decentralizing CODO's Technical Secretariat to Rancagua.

(3) Strengthening existing offices of central government ministries in the city.
(4) Establishment of one or more major research centers related to the development of regional resources.

e. For San Antonio
– A specialized port for copper exports and grain imports.
– Tourism and popular recreation along the entire coastal area.
– Port-related manufacturing industry at San Antonio.
– Development of modern fisheries and related industries at San Antonio.
– Exclusion of manufacturing industry from all other areas.

San Antonio's principal role will be as center for tourism along the entire coast, from Rocas de Santo Domingo to Algarrobo. This will require careful planning for the preservation of its natural attractions (beaches, pure water, open space) as well as the provision of specialized services (hotels, summer camps, tourist information, etc.). In line with our proposals, the present coastal road will eventually connect with Valparaíso to the North, permitting the integrated development of this exceptional natural resource for the use of Capital Region populations. Present reforestation programs in the area should be continued on an accelerated basis.

f. For the Intermetropolitan Periphery

The economy of this extensive area is still insufficiently understood. In 1970, it had slightly more than 5 percent of the total population of the Capital Region (or approximately 270,000 inhabitants) whose principal activity was in subsistence agriculture. Its population is extremely dispersed, relatively inaccessible to major urban subsystems, and very poor. The intermetropolitan periphery tends to have a large volume of migration to the principal centers of employment, specifically to Santiago.

The future of this area requires more study than we have been able to give to it. In general, however, we are inclined to believe that its potentialities for development are quite restricted, and that public policy should continue to encourage the emigration of its population. This might be directed by concentrating major urban services in the more viable centers of the periphery and by a careful planning of its road and bus network which would increase the accessibility of these centers to their hinterland and to the major labor market areas of the Capital Region. At the same time, studies should be undertaken to explore the possibilities of converting the present land to more extensive use, such as forestry and grazing.

2. Transportation. The transport sector tends to be developed independently of regional considerations along purely functional lines. We propose, therefore, that plans for the expansion and improvements of the transport system in the Capital Region be subordinated to a general policy for spatial development. In line with this thinking, we are able to identify three types of programs which should be given priority consideration.

a. Ports and airports
   - Expansion of port facilities for general merchandise, petroleum, and fisheries at Quintero.
   - Construction of an alternate international airport to Pudahuel at Quintero, using, if possible, the existing facilities of the naval airbase located in the area.

Both measures would provide the basis for the proposed urban-industrial expansion at the site, would lead to the more efficient use of existing port installations at Valparaíso, and provide a direct air connection for the Valparaíso subsystem to the rest of the country, to Argentina, and to the countries of the Pacific Basin.

b. Highways and roads
   - Improved road connection between Quintero and Nogales on the Panamerican Highway to facilitate the spatial integration of the Aconcagua subsystem with the new port city at Quintero.
   - Construction of a bypass road leading from Rancagua to Melipilla and Casablanca, providing a direct connection between the southern and northern halves of the Capital Region.
   - Construction of a tourist route from Valparaíso to Algarrobo in the South, opening a new tourist zone along the coast which connects the Valparaíso subsystem with that of San Antonio.
   - Completion of the limited access highway which is currently under construction between Santiago and San Antonio to facilitate metropolitan access to the tourist areas along the coast.
   - Conversion of the Panamerican Highway between Santiago and Nogales to four lanes, whenever projected traffic volume suggests the desirability of this investment.
   - Improvement of the limited access, four-lane highway between Santiago and Rancagua to further reduce travel time between the metropolis and its satellite city and to accommodate the increasing traffic volume over this route.
   - Conversion of the existing highway between Santiago and Valparaíso, via Casablanca, to four lanes and limited access whenever projected traffic volume suggests the desirability of this investment.

c. Railroads
   - Conversion of existing railroad lines between Calera and Quintero/Valparaíso; Melipilla and Santiago; and Rancagua and Santiago to rapid inter-urban commuting. This would help to integrate each of the affected labor market areas as well as provide a better connection to Santiago's industrial satellite city. It would also tend to concentrate economic activities and population at either terminal of the commuting line, as well as at intermediate stations (see Appendix C at back of book for supporting arguments).
   - Improvement of the line between Santiago and Valparaíso, via Calera, as

an alternative to a new high-speed system being proposed to provide a direct connection between the two metropolitan centers.

3. Environmental Protection. Important coastal sectors are currently being treated by major sources of air and water pollution. A partial listing of these sources include:

(a) Sewage disposal from Limache-Quilpué-Villa Alemana into the bay at Viña del Mar endangering the use of the urban beaches in the area.
(b) Heavy air pollution from existing copper refining plants (ENAMI at Las Ventanas, near Quintero). Although the area is sparsely populated today, future growth in industries and population may seriously impair its livability. The present extent of water pollution caused by these plants is not known.
(c) Danger of water and beach pollution from vastly increased volume in crude oil transport to storage facilities at Quintero.
(d) Serious air and water pollution from the oil refinery at Concón, currently destroying one of the most important tourist and recreational assets in the area.
(e) Air pollution dangerous to health at La Calera where the existing cement plant reputedly deposits one ton of cement dust per day upon the surrounding city.
(f) Sewage disposal from Valparaíso and Viña del Mar which is deposited in the bay without substantial treatment, with unknown effects on marine ecology.

With growing industrialization, this already serious situation is likely to get worse. A legal and institutional mechanism is therefore proposed that would bring the entire coastal area, from San Antonio to Papudo, under the control of an Environmental Protection District. A similar district might be established for Santiago, although we do not advocate this at the present time. Some details relating to this proposal will be given in Section VI.

Probable Consequences of Proposed Policies

The joint effects of these proposals are expected to bring about those changes in the spatial organization of the Capital Region which we identified as desirable in Sections III and IV. They would:

– decelerate growth at Santiago by directing economic activities (and population) towards the Valparaíso and Rancagua labor market areas;
– integrate each of these areas through improved systems of transportation;
– encourage the emergence of an "urban field" concept of the region through a system of territorial functional specialization and a transport policy subordinated to the demands for spatial development;
– encourage the bilateral integration of the Aconcagua-Valparaíso, Valparaíso-San Antonio, Santiago-Rancagua, and Valparaíso-Rancagua subsystems;
– provide for adequate environmental protection along the critical coastal area;
– provide for a concentrated pattern of industrial location, with consequent-

ly greater possibilities for internalizing multiplier effects in the respective labor market areas; and

– delimit each urban subsystem by large areas devoted to extensive use for agriculture and recreation.

Some of the implications of the policy proposals for spatial development are quantified in Table 5, using a series of alternative assumptions as to possible rates of growth in regional product and population. The most "optimistic" set of assumptions for future growth is shown in Map 8.

For the next twenty years, projections show Santiago's population rising from 3.1 to 5.1 million and regional product increasing by a factor of four. By

TABLE 5
ALTERNATIVE PROJECTIONS OF REGIONAL PRODUCT AND POPULATION, 1970-1990

A. REGIONAL PRODUCT

	Hypothetical Growth Rates (percent per annum)			*Regional Product (million escudos in 1965 prices)*			
				1990			*1967*
Area / Growth Rate	–	–	–	4.5	5.5	6.5	–
Chile	4.5	5.5	6.5	54,393	67,715	84,140	19,765
Chile less C.R.	1.8	4.7	4.5	12,137	21,990	21,164	8,054
Capital Region	5.8	6.1	7.4	12,839	45,726	60,487	11,711
Aconcagua	2.0	3.0	4.0	497	622	776	315
Valparaiso	4.5	5.5	6.5	4,628	5,792	7,156	1,681
Santiago	6.0	6.5	7.5	32,899	36,659	44,572	8,613
San Antonio	3.5	4.0	4.5	567	656	732	266
O'Higgins	5.0	6.0	7.0	2,061	3,139	4,273	836

B. POPULATION

	Hypothetical Growth Rates (percent per annum)			*Population (in thousands)*			
				1990			*1967*
Area / Growth Rate	–	–	–	1.0	1.5	2.0	–
Chile	1.0	1.5	2.0	11,672	13,675	14,214	9,566
Chile less C.R.	0.0	0.7	0.8	4,971	5,543	6,300	4,971
Capital Region	1.8	2.1	2.7	4,763	6,957	7,914	4,595
Aconcagua	0.6	1.0	1.5	126	139	152	113
Valparaiso	2.2	2.8	3.5	1,134	1,274	1,461	734
Santiago	1.6	2.0	2.5	4,302	4,652	5,130	3,131
San Antonio	1.4	2.6	3.0	106	162	179	99
O'Higgins	3.5	3.8	4.0	489	518	539	246

SOURCE: 1967 data from Office of National Planning.

Map 8. CAPITAL REGION:
Projections of Population and Regional Product, 1970-90

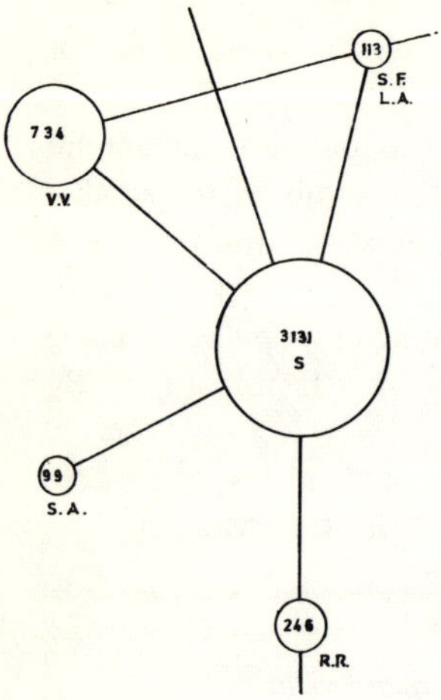

1970 Population (thousands)

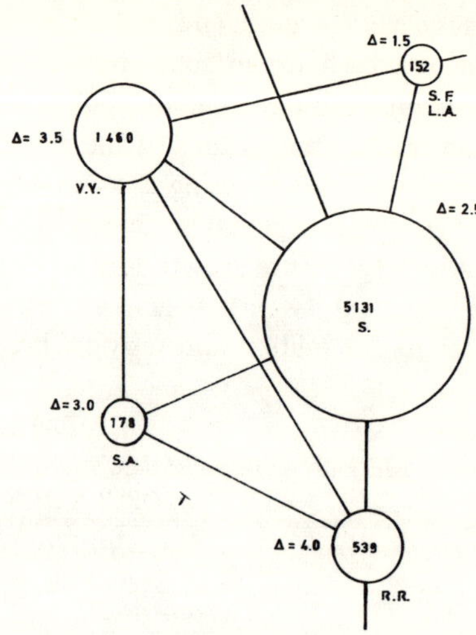

1990 Population (thousands)

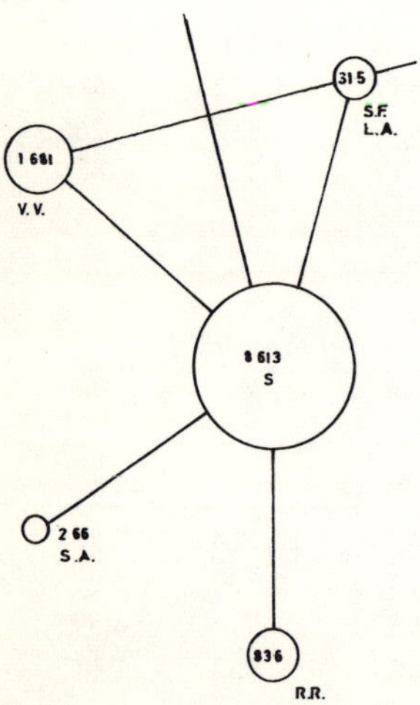

1967 Gross Regional Product
(million escudos)

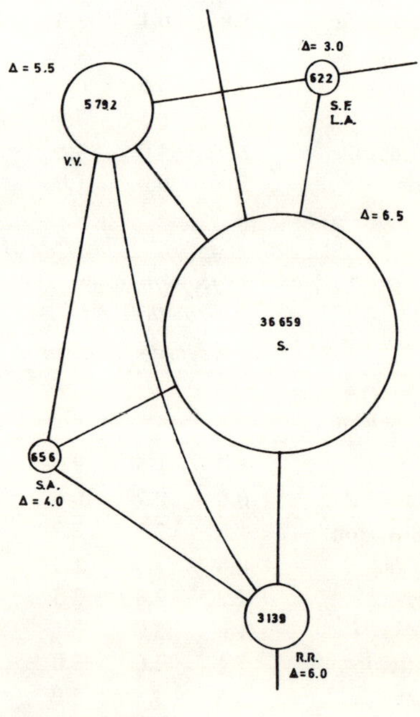

1990 Gross Regional Product
(million escudos)

1990, Valparaíso's population would have reached a threshold size of 1.5 million, while its regional product would be over three times as large as at present. Although this would raise the per capita product in both areas, it would substantially increase the existing differences in the levels of per capita product between the two major metropolitan subsystems of the Region. The projection, therefore, does not appear to be a stable one. Population growth in Valparaíso will either have to occur at less than the projected rate, or the increase in regional product would have to be greater.

There is no basis for making "firm" projections of the future. But if polarization effects should be accelerated at Valparaíso, as we have argued, its per capita regional product would have to grow more rapidly than that of Santiago.

The relationships seem to be more in balance for Rancagua, which shows slightly more than a doubling of population and a product increase similar to that for Santiago. Given Rancagua's "satellite" position vis-à-vis Santiago, this projection appears to be consistent with our policies. The two remaining subsystems would continue to be small in size, with less than 200,000 population each.

Our policy proposals are merely illustrative for the moment and will require continuing study to ascertain their economic feasibility and timing. But if the image of a possible future is to be brought nearer to the present, institutional innovations will have to be made that will provide a framework for more detailed study and the ultimate implementation of policies for spatial development. Our positive recommendations are directed primarily to this point. We are not prepared to defend particular transportation proposals, for instance, beyond the point that we have carried the discussion. But we firmly advocate the following institutional arrangements as necessary to a spatial development policy which seeks to follow the objectives outlined in Section IV.

VI. INSTITUTIONAL INNOVATIONS FOR SPATIAL DEVELOPMENT

Designation of a National Capital District

Of all the regions in Chile, the Capital Region is the most difficult to administer. It appears as both, a "congeries of urban subregions" and an emerging "urban field." As the former, it requires a localized approach to development; as the latter, it calls for regionwide planning and coordination.

The administrative structure of the Region consists of four Provinces (Valparaíso, Aconcagua, Santiago, O'Higgins-Colchagua); and a Metropolitan Zone (Santiago). Our analysis of the spatial organization of its economy has further identified five potential labor market areas which are separated from each other by an Intermetropolitan Periphery.

No present institutions correspond to any of these divisions. Development programs are the responsibility of different national Ministries and semi-autonomous Corporations, the most important of which are the Ministries of Public Works, Housing and Urban Affairs, Education and Health; the Agrarian Reform Corporation; and the National Development Corporation. Each of these agencies of Government proceeds in large measure independently of the others. Their programs are conceived separately and do not correspond to any general strategy for the spatial development of the Capital Region. Indeed, it is doubtful whether the image of a Capital Region has penetrated very far into the thinking of the responsible officials.

The conditions of development on the Capital Region are of supreme national concern. The Region is the seat of National Government. It is the country's most densely populated area. It has a majority of the industries, the largest port installations, and the country's major international airport. It is the principal node of communication with the outside world and the cultural capital of Chile. It is also the only center of banking and finance.

For all these reasons, the Capital Region calls for special treatment in the management of its spatial development processes. We therefore propose the creation of a National Capital District which would include but not replace, the four Provinces of the Region.

The National Capital District is conceived as an administrative territorial division for achieving a better coordination of development programs. An Inter-Ministerial District Council would be directly responsible for approving all policies, plans, programs, and projects affecting its development. The Council would be chaired by the Head of the National Planning Office whose staff would also provide the services of a Technical Secretariat to the Council.

Creation of a Metropolitan Development Corporation for Valparaíso-Aconcagua

The National District Council corresponds to the "urban field" concept of the Region. But our basic policies framework also proposes a special concentration of efforts in the Valparaíso subsystem. The complexity of this task, as well as the need to obtain the continuous and direct participation of local populations in the management of regional development (see statement of objectives in Section IV), suggests the desirability of creating an autonomous corporate body that would guide and actively promote developments in the area.

We therefore propose a Metropolitan Development Corporation that would be located in the city of Valparaíso but whose authority would extend to both the Provinces of Valparaíso and Aconcagua. A Corporation of this type could be organized in several ways, one of which is illustrated in some detail in Appendix D at back of book.

Basically, the Metropolitan Development Corporation would be empowered to draw up plans and programs for regional development, undertake certain projects of its own, and make development loans consistent with its basic objectives to business firms, municipalities, *juntas de vecinos,* and others. Its annual program budget would require review and approval by both the National Planning Office and the National District Council, and would subsequently be incorporated in the national budget as a separate item for submission to the Congress. In this way, coordination would be automatically achieved, and the consistency of the Corporation's programs with national and overall regional objectives would be assured.

At the same time, in its decision processes, the Corporation would formally assign a major role to neighborhood councils, municipalities, the universities, and the principal productive sectors of the regional economy. The Corporation would thus be made more responsive to the needs of the population (for further details, see Appendix D at back of book).

The transfer of effective power to make and carry out development decisions to two of the major urban subsystems of the Capital Region will tend not only to strengthen their attractiveness as possible locations for industry and other economic activity, but promote the closer integration of the Aconcagua subsystem with that of Valparaíso.

Decentralization efforts of this kind seem to have had favorable results in Chile wherever they have been seriously tried, especially at Arica. There, in collaboration with a technical staff furnished by the National Planning Office, a *junta de adelanto* has been highly successful in promoting the economic growth of the area as a center of international tourism and industrial production, with a long range perspective of becoming one of the principal exporting centers to the Andean market countries, especially Bolivia and Perú. In Valparaíso itself, the recent experiment in administrative decentralization, with the establishment of a Zonal Delegation of the Housing and Urban Affairs Ministry, has apparently met with widespread approval among the local populations who feel that their needs are now being served more effectively than before.

Establishment of a Free Trade Zone at Quintero

In order to promote the location of export-oriented industry at Quintero, the establishment of a *Free Trade Zone* is proposed that would permit the importation of duty-free raw and semi-processed materials and their manufacture into finished products exclusively for export to other countries.

A Special Planning and Development Authority for Quintero

The projected rapid growth of Quintero will pose some of the most intricate problems of town planning and development in Chile. A number of critical functions will have to be accommodated to each other in an environment of extraordinary natural endowment. These functions include:

- expansion of general port facilities
- construction of a new international airport

– expansion of petroleum storage facilities to supply the refinery at Concón, with adjacent port facilities to permit the entry of giant tankers
– expansion of heavy metallurgical industry
– establishment of a Free Trade Zone
– establishment of a major fisheries port and related processing facilities
– development of facilities for recreation and tourism along the coast
– construction of port facilities for rapid hydrofoil service to Valparaíso and Viña del Mar
– increase of population to between 100,000 and 200,000 within the next twenty years

The possibility exists for constructing Quintero as a symbol of Chile's new determination to become one of Latin America's major industrial powers. The combination of locational advantage and scenic resources would permit the design of a city of striking beauty. However, the uncoordinated, sectoral development of its many functions could only result in their mutual interference and very likely, also in the destruction of its environmental advantages.

To assure the harmonious development of the New City of Quintero, we propose the establishment of a special Planning and Development Authority which would be directly responsible to the Minister of Housing and Urbanism. This Authority would have powers sufficient to coordinate all urban and port-related developments of the new city.

The existing Intercommunal Planning Office of the Ministry of Housing in Valparaíso is unable to carry out this task. Moreover, it is essential that the Planning and Development Authority be physically located at Quintero and have sufficient resources of its own to undertake essential investments in the basic facilities and infrastructure of the city.

An Environmental Protection District

The exceptional resource advantages of the coastal area of the Capital Region are in danger of partial destruction from environmental pollution on a large scale. Some instances of present pollution of the air, water, and beaches in this area were cited earlier in Section V. There is every reason to expect that pollution will increase very substantially with the expansion of industries, port facilities, tourism, and urban population in the area.

To prevent this from happening and to reduce existing levels of pollution, the establishment of an Environmental Protection District is proposed that would extend its authority from Papudo in the North to San Antonio in the South and specifically include the entire urban subsystem of Valparaíso. This District would be established by a law of the National Congress and would be administered by a District Commission, responsible to the Ministry of Health, with headquarters in Valparaíso.

The District Commission would establish maximum permissible levels of pollution as well as other appropriate standards to regulate the discharge of waste products into the physical environment. For this purpose, it would have at its disposal a Research Unit to study the effects of pollution on the ecology of the area, to establish scientific standards of pollution control, and to monitor current pollution levels. The Commission would be empowered by law to require compliance with its rulings and to levy fines where compliance is not obtained.

Designation of Rancagua as an Industrial Satellite City of Santiago

In order to demonstrate its determination to actively seek and promote the development of Rancagua, the Government should *officially designate it as an Industrial Satellite City of Santiago.*

In addition, the Development Corporation for O'Higgins (CODO) should be strengthened by establishing a qualified Technical Secretariat to CODO in Rancagua, transferring this function from its present location in the National Development Corporation offices (CORFO) in Santiago. This Secretariat should be housed in the same building together with an expanded staff of the Ministry of Housing's Communal Programming Office in Rancagua. A general development plan for the area, together with medium-term and annual investment programs should be prepared jointly by CODO and the Communal Programming Office as the basis for a concerted development effort.

Designation of Agro-Industrial Corridors and Tourist Zones for the Regulation of Industrial Location

One of the important policies for the spatial organization of the Capital Region is the use of areas devoted primarily to agriculture and recreation, limiting the physical spread of its major urban subsystems. Coincident with this traditional urbanistic concept of "green belts" is a theory, more fully explicated in Appendix C (at back of book), that a location pattern of industrial concentration is preferable to one of dispersal. We have therefore argued in favor of excluding industry—except for agricultural processing—from designated agro-industrial corridors and tourist zones. This policy would have the additional benefit of protecting large portions of the natural environment from air and water pollution.

The implementation of this proposal would require the legal designation of those areas in which, with specified exceptions, no manufacturing plants may be located. A legal instrument of this kind would represent a start towards a broad, Region-wide Zoning Plan.

NOTES

1. The two million tons of freight movement shown between Rancagua and Santiago includes all shipments to and from the South of Chile and could not be separately calculated for these two subsystems.

2. For further detail on the increasing concentration of economic power in Santiago, see Sergio Boisier, *Algunas Hipótesis sobre un Modelo de Desarrollo de la Zona Metropolitana.* Santiago: CIDU, 1970.

3. Surveys made in 1962 and 1963 show that approximately 35 percent of Santiago's population was born elsewhere. (CELADE, "Encuesta sobre inmigración en el Gran Santiago." Santiago, 1965).

4. For this reason, the National Planning Office has stopped making regional product projections that extend for more than three or four years. Recent trends were found to be entirely misleading as indicators of the future. On the question of the initiative for local development, see Chapter 6.

5. Size of city may have something to do with the degree of external dependence of its economy. This hypothesis is briefly discussed in Appendix B at back of book.

6. Germán Castro, *Manual de Participación de la Comunidad en la Programación DEL DESARROLLO LOCAL.* Santiago: Ministerio de Vivienda y Urbanismo, 1970.

7. This question has been incisively posed by G. Geisse and J. L. Coraggio in an unpublished paper, "Areas Metropolitanas y Desarrollo Nacional," Revista Latinoamericana de Estudios Urbano Regionales, Vol. I, No. 1 (October 1970), pp. 51-62. The authors refer to urban field as "una gran ciudad" and to the concept of a congeries of urban subsystems as "un pequeño gran espacio." Their definitions, however, are consistent with those we have used here.

8. Yet another reason for "green belts" is the effect they would have on the concentration of manufacturing industry in centers especially suitable for this purpose. Industrial concentration will tend to accelerate local multiplier effects and bring other benefits which are explained in greater detail in Appendix C.

Part Three

PRACTICE

Chapter 11

FROM THE WORKBENCH:

ASPECTS OF PLANNING PRACTICE

This and the following chapters of Part III concern the formulation and carrying out of urbanization policies. Logically, these chapters follow from the preceding discussion of theory and policy. Indeed, the progression from one part to the other would seem to be a gradual one and, if feedback loops are added, the sequence appears to be a tightly knit and integrated process.

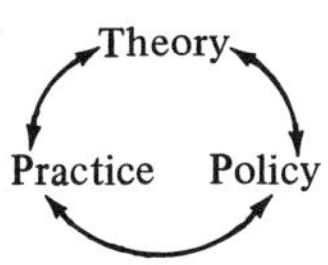

But the facts of the matter suggest another conclusion. Taking only the two extremes, a yawning gap of incomprehension separates the theory and practice of urbanization policy. The reasons for this are inherent in the conditions of theorizing and practice. Because this fact goes generally unnoticed, and because I believe that theory ought to be more closely joined to practice, I should like to devote the following pages to examine the principal sources of this separation.

1. Theorists and practitioners address themselves to different questions, with the result that what looms important to one is of small consequence to the other. Migration, for instance, is a favorite subject of theory in the social sciences, but planning practitioners, though interested in obtaining estimates of expected migration to specific areas, are much more attentive to the problems arising from migration such as housing urban migrants at their destinations. The practical question of how to provide adequate shelter for those who need it has not attracted more than a handful of theoretical minds.[1]

Location theory might be cited as yet another example. One of the topics traditionally of interest to economists and geographers, location theory has produced a literature of impressive proportions. But spatial planners make only minimal uses of this literature (see Chapter 13); instead, they frequently devote themselves to the related but quite different issue of how to reduce unemployment in specific localities. Questions of optimal economic location are of secondary interest to them and rarely arise in the context of urban-regional development. On the other hand, unemployment results in part from insufficient economic opportunities. Although different explanations have been tendered to account for local imbalances between workers and jobs, no theory has yet appeared capable of enlightening planners about practical measures for reducing such unemployment, especially in labor surplus economies, where Keynesian theories do not appear to apply.

2. *Normative questions are inseparable from planning practice; they are either eliminated or treated by way of formal assumptions in theoretical work.* The examples just given also help to illustrate this point. In migration theory, considerations of social value are simply irrelevant. Propositions about migration behavior are true or false regardless of the personal preferences of the investigator concerning geographic mobility or its direction. Questions of housing policy, on the other hand, cannot be discussed at all without considering questions of equity, sources and forms of financing, and standards of adequacy as central aspects of the argument. This conclusion may be put even more emphatically. For any given policy, the question of whether it is true or false does simply not apply. The critical test of any policy instead is a pragmatic one; the policy will either accomplish what it sets out to do or not. Most of the controversy surrounding policy proposals turns on central questions of value.

Economic location theory—which for the most part is deductive—does manage to incorporate a value judgment by positing the criterion of efficiency (and, in certain cases, profit). The theory is therefore able to produce "solutions" that are optimal with respect to this specific value. But once the value has been laid down, everything else is derived in logical sequence from it. Discussions of local employment policy, on the other hand, are permeated with normative questions.

3. *The constraints of theory are artificially established, but planning practice is hemmed in by a large number of conditions that appear to occur at random and are not capable of being dealt with by theory.* Both types of constraint are so well known that they require little further explication. Theory, typically, assumes a frictionless environment (perfect knowledge, an undifferentiated plain), assumptions which may be gradually relaxed but which can never be

abandoned without forfeiting the possibility of theoretical explanation. The constraints on practice, however, form integrally a part of the problem to be solved. The realities of the political and administrative situation suggest what might be feasible or not. The time available for analysis and the costs of gathering and processing information limit the extent to which policy can be grounded in theoretical analysis. Existing institutional arrangements will further influence the type of policy that is evolved, while the resources available together with competing claims for them set limitations to what can be accomplished, regardless of theoretical optima. For these reasons, conclusions that are theoretically valid may not be capable of realization, whereas the constraints on practice are irrelevant for theory.

4. *Practitioners typically consider a much larger number of variables than theorists. For this reason, their requirements for data processing are also greater.* The purpose of theory is to explain phenomena; the more simple the theory (i.e., the fewer the number of variables needed in explanation), the more "powerful" it is said to be. A powerful theory peels back all extraneous variables in order to concentrate on the phenomena to be explained. Although its demands for data will certainly be large, they are confined to what is absolutely essential to the argument. Quite the reverse is true of planning, where the demand for information is seemingly inexhaustible, though the constraints of time and resources may impose severe limits on how much information can be brought together to arrive at a decision. A criterion of relevance applies to both theory and practice, except that what is relevant for practice is much more difficult to determine. Information about the conditions of implementation or the acceptance of a particular course of action by all groups likely to be affected may be of crucial importance to planners. The success or failure of particular policy proposals may hinge on having an accurate sense of timing, knowing the right channels for introducing new ideas, and a detailed understanding of the current political situation. Theorists need to know none of these things; the contents of their theory is independent of political moods.

5. *The language used in conveying theoretical ideas is different from the language of everyday practice.* The former is abstract, conceptual, mathematical; the latter is concrete, discursive, and primarily verbal. Theorists generally employ the jargon of their discipline when addressing themselves to other specialists. But practitioners may variously make use of political rhetoric, diplomatic equivocations, administrative jargon, or the language of the public information media. Most practitioners will not be schooled in the languages of science nor, for that matter, in the related processes of reasoning. This communications gap is the ultimate reason why theorists may be said to inhabit quite different worlds, and why practitioners avail themselves so little of what is known in theoretical science. The principal media of communication in the sciences, such as preprints or journals, are simply not read by those who might

extract from them some practical meaning. On the other hand, the failure of practitioners to generate a demand for the products of theoretical learning contributes to making theory increasingly esoteric.

These problems have no simple solution. A style of planning may have to be evolved which draws experts and clients more closely together in a process of mutual learning.[2] Professional education might itself have to be recast to enhance the scientific understanding of practitioners (Chapter 16). Closer collaboration between practitioners and scientists on problems of practical impact might have a salutory effect on both efforts (Chapter 15). Theory may be more directly inspired by problems of public policy (Chapter 1). But it is evident that the structural differences between the worlds of theoretical research and practice are not likely to disappear.[3]

The diversity and richness of the practitioner's environment makes it rather difficult to write coherently about practice. Most accounts tend towards the anecdotal and episodic. For this reason, the chapters which follow may seem less of one piece than those which make up the earlier parts of this book.

Chapter 12 focuses on the transition between policy and practice. It is a review of the principal measures for implementing regional development policy. The framework for the discussion is given by a set of principles of policy implementation taken directly from practice.

Chapter 13, written jointly with Walter Stöhr, presents a practitioner's view of theory and is based on experience in Chile where the authors assisted in the establishment of an institutional system for regional planning.

Chapter 14 deals with the implementation of a major regional development program in the Guayana region of Venezuela. After pointing out how approaches to regional development may differ according to the character of the region and the objectives for its development, the chapter proceeds to analyze a series of policy choices that confronted the planners of Guayana. The current popularity of core-region strategies lends special interest to this chapter where the pros and cons of such a strategy are discussed on the basis of a historical instance of policy application.

The remaining two chapters both concern education for urban-regional development planning. Chapter 15 points to some of the inadequacies of American planning education for work in the developing countries. It explores the question of whether U.S.-trained planners working abroad may be considered agents of social change to a greater degree than they are at home, and concludes with a set of specific proposals for closer collaboration between American planning schools and overseas centers for urban-regional development.

The final chapter (16) presents the elements of a graduate course of studies for urban-regional planners who will work primarily in developing countries.

NOTES

1. The principal housing theorists since World War II include Catherine Bauer Wurster, Charles Abrams, John Turner, Leland Burns and Charles Frankenhoff. Of these, only the last three are still living. A much larger number of experts have contributed to the analysis of housing policy-technical libraries around the world are choking on low-income housing studies—but a theory of housing has not emerged from any of these efforts nor is it ever likely to do so. The housing question is too intimately bound up with the realities of particular policy situations to lend itself to more than very low-level generalizations.

2. This planning style is discussed in great detail in John Friedmann, *Retracking America: A Theory of Transactive Planning.* New York: Doubleday, 1973.

3. The tension between theory and practice is apparently less severe in communist societies, but this advantage is gained at the price of abandoning theoretical social science, with its ahistorical bias, as it is currently understood in the West.

Chapter 12

THE IMPLEMENTATION OF REGIONAL DEVELOPMENT POLICIES: LESSONS OF EXPERIENCE*

To plan for development is to practice the art of the improbable. Although we have come a long way in our ability to prescribe policies for regional development, we now need to consider the specific methods of implementation. I shall comment on the relevant experience under four major principles.

1. Planning must be joined to effective power for implementation.
2. Political commitment to a policy must be sustained.
3. The use of instruments for regional development must be concerted.
4. A national balance in interregional development must be maintained.

Failure to observe these principles in practice increases the likelihood that regional planning will remain ineffective.

I. JOINING PLANNING TO EFFECTIVE POWER

In most countries, planning for specific sectors of the public economy, such as transport, housing, education, electric power, or agriculture is considerably more successful than either coordinated intersectoral or regional planning at the

*I should like to thank Arie Shachar and Barbara Stuckey for many helpful comments in the preparation of this paper.

national level. The reason is that sectoral planning is done by the very agencies that are in charge of making the programmed investments. For instance, a Department of Transportation plans for the use of its own resources, including its regulatory powers, to achieve intended, if usually quite limited, effects. A Rural Housing Agency does likewise; the allocator of its investments is not the "unseen hand" of a competitive market but an explicit decision process in which technical factors are weighed alongside other considerations. Planning in these cases is used to guide the daily work of the agency. It provides the necessary information and analysis, produces future projections, furnishes evaluations of costs and benefits, and in a multiplicity of other ways is tied into the formulation of policies, program development, and project design.

Planning can be called effective in this sense because it is intimately joined to the exercise of power, that is, to control over the legal and monetary resources to carry out intended actions. Where planning is so integrated with organized actions, it must operate under a set of realistic constraints that make its contributions immediately relevant to the managers of power. Plans are usually formulated in terms of the available instruments for implementing programs; questions posed by the potential use of these instruments are the problems to which planning will generally respond.[1] Sectoral planning has, therefore, a close fit with the available instruments of power.

At the national level, both intersectoral and regional (or spatial) planning are further removed from the management of effective power than is sectoral planning. In most countries, the role of national planners is to advise on policy. Yet, being divorced from the instruments of action, national planners have no specific means for getting their advice accepted. Indeed, their advice is often judged irrelevant to the immediate needs of resource managers. This point is made, time and again, in the new literature on regional planning.[2] Planners' advice, separated from the uses of effective power, accomplishes little. Concluding his lengthy study of regional policy in Britain, Gavin McCrone writes: "Regional planning cannot be said to have had much impact on the regional problem so far and in many respects it has still to find its role. Much depends on whether the regional bodies are to remain advisory or are to be given some executive power."[3]

Regional development authorities, having a corporate structure, are the leading exception to this condition of relative impotence. Established as semi-autonomous agencies of the national government, they operate within the confines of a specific geographic area and generally control resources sufficient to carry out a variety of closely coordinated investment programs. The prototype of all these agencies is the Tennessee Valley Authority, established in 1933. The Guayana Corporation of Venezuela, the Cauca Valley Corporation in Colombia, and the Superintendency for the Development of the Northeast (SUDENE–Brazil) are more recent and equally well-known examples. At the urban scale, the Foundation for the new capital of Brazil, the New York Port Authority, and a proposed Metropolitan Development Corporation for Valparaíso (Chile) embody the same principle of fusing planning with effective power.[4] In all these cases, planning focuses attention on the short-run future and is likely to be more concerned with strategies, programs, projects, and even operational details than with long-range plans that endeavor to portray an image of the finished work.

The corporate approach to regional development has yielded impressive results whenever the powers of the development agency were commensurate with its mission. But these results have had to be bought at a cost. The authority of a regional agency extends over a defined geographic area that is only a small part of the national territory. Since few industrializing countries have adequate financial and technical resources to undertake more than one or two concerted efforts of this kind at the same time, use of the corporate approach has meant, in practice, that certain regions were favored at the expense of other areas.[5] The justification for such a policy of *imbalanced* development has nearly always been based on arguments that stressed the overriding national importance of the regional scheme. Even so, the demonstration effect of a successful regional enterprise is often difficult to control. Regions that consider themselves neglected frequently bring pressure on the government to mount similar efforts in their own areas. Although strenuously resisted by national planners who fear a geographic dispersion of resources and, more generally, a gradual attrition of their own powers to guide the overall course of national development, these pressures may lead to a multiplication of regional agencies and, consequently, to a dilution of resources for promoting development in any one area. Returning to the example of South America, both Venezuela and Colombia have six regional development agencies in operation, each with some resources of its own; Perú has blanketed the entire country with a series of Departmental Development Corporations; and Brazil has established, in addition to the powerful SUDENE, three regional superintendencies, the Foundation for the Development of Brasilia, and several other intensive programs of regional development.

The apprehensions of national planners appear to be well founded. Development Authorities are potent institutions. Being regionally based, they often succeed in capturing the support of local populations and so become increasingly immune to central control and direction.[7] The technical staffs attracted to these agencies—not least because of the higher salaries and the elan of their mission—often provide it with superior technical talents than those available to central authorities. In a sense, one might say that where planning is closely linked to effective power, the true danger lies in its becoming excessively effective. Regional authorities, for example, are extraordinarily skilled in mobilizing external resources. They become the pampered clients of international lending agencies and star exhibits of national progress. They may even start to act as quasi-governments for their own regions, replacing central authorities in important respects. This behavior may raise not only questions of political representation and accountability but, particularly in multi-national states, the grave issue of regional separatism. National governments are jealous of their prerogatives. The drive of regional authorities for territorial supremacy is, therefore, likely to be checked before it can pose a threat to national unity. On the other hand, once they are institutionally entrenched, regional development agencies are difficult to eliminate and become more or less permanent features of planning administration.

II. SUSTAINING POLITICAL COMMITMENT

Rudimentary patterns of urbanization and regional social structure, established quite early in a country's history, tend to maintain themselves over long periods of time. Subsequent flows of controlling decisions, innovation diffusion, migration, and economic location will tend to reinforce this pattern. The most probable future pattern will therefore resemble that of the past. Planned changes may require decades and even generations of counter-intuitive effort to alter this structural pattern in a significant way.

By way of illustration, the following data may be cited (Table 1). Despite a century of technological change and population growth, the relative proportions of major regions in the United States have changed by only a few percentage points, and these changes reflect primarily the opening of the new areas West of the Mississippi to settlement.

TABLE 1
GEOGRAPHIC DISTRIBUTION OF POPULATION IN THE UNITED STATES, 1850-1960 (in percentages)

Year	*North*	*South*	*West*	*U.S.*
1850	57.0	34.5	8.5	100.0
1900	62.0	32.0	6.0	100.0
1920	60.0	31.0	9.0	100.0
1940	57.5	31.5	11.0	100.0
1960	54.0	30.5	15.5	100.0

SOURCE: Jerome P. Pickard, "Trends and Projections of Future Population Growth in the United States, With Special Data on Large Urban Regions and Major Metropolitan Areas, for the Period 1970-2000." U.S. Department of Housing and Urban Development, Technical Paper No. 4, July 22, 1969.

Numerous other examples might be adduced, such as the relative stability in the rank-size order of cities down to approximately the tenth rank.[8] In any event, this built-in tendency of any spatial pattern to maintain itself can be counteracted only by controlling the basic rates of flow through the system and by changing their direction in ways that are highly improbable. These improbable rates of change must themselves be maintained over long periods of time. They require sustained public intervention.

Let me give a simple illustration. Suppose that the capital city of a country has established primacy over all other cities by a factor of 5 : 1, so that the largest city has five times the population of the next three largest cities combined. This is the case of Lima, Perú. In the foreseeable future, the most probably pattern of population migration will be one that will tend to maintain this ratio in the distribution of urban population.[9] If, for some reason, the national government should decide to reduce the primacy of the capital to a ratio of, say, 3 : 1, it would have to adopt policies for accelerating migration towards provincial cities while decreasing the flow of population to the national capital. This change in the direction of migration would represent a significant departure from the normal, system-maintaining pattern. In order to accomplish its objective, the government would have to sustain its commitment to this policy over a period of several decades. Furthermore, the policy change would have to be sufficiently bold in conception to ensure that non-planned, random events, such as changes in industrial technology, would not cancel any possible gains achieved through application of the policy.[10]

Spain, for instance, has adopted a kind of infant industry policy for encouraging the industrial development of four growth pole centers. The location subsidy is paid to private firms for a maximum period of seven years, after which it is to be withdrawn. Commenting on this feature of the policy, H. W. Richardson asks some pertinent questions:[11]

> "What will be the long-run consequences of a short benefit period strategy? What happens when the designation period for the four new poles expires? Will a new set of poles be created for another five or six years, and then another set, and so on? In this case, the long-run effect would be that a large number of centres would have experienced modest expansions, and that Spain would have a hierarchical distribution similar (relatively if not absolutely) to that which exists at present."

In other words, Richardson is implicitly arguing for a sustained policy of giving favored treatment to selected areas having a high potential for industrial expansion. Frequent shifts in the objects of this policy may not lead to any system-wide change whatever.

The practical difficulties of sustaining major policy commitments should not be underestimated. If the period of desirable commitment should last for several decades, it will have to cut across several periods of governmental change, becoming a supra-political policy capable of receiving endorsement from more than one political regime. Few policies are ever likely to enjoy such support. Above all, they must serve important long-term national requirements of development. Although this functional relationship is not always easy to prove, it must at least be widely believed to exist. In addition, experience suggests that such policies must conform to a number of pragmatic criteria.

The first criterion is that *the policy must be simple in concept and dramatic in its implications.* So conceived, a policy will set the popular imagination on fire and will tend to be elevated to a national commitment. Examples come readily to mind:

- the conquest of the desert (Israel)
- the harnessing of major river basin resources (Ghana, Egypt)
- the building of new towns (England)
- the building of a new national capital (Turkey, Brazil, Pakistan)
- the reclamation of new land from the sea (Holland)
- the construction of a transcontinental transport system (USA)
- the creation of a major industrial growth center (Venezuela)

These historical examples have at least one characteristic in common: they refer to projects whose primary impact is physical. Their appeal derives from man's emotional attachment to the land and from his immense pride in major works of physical construction. If the projects are sufficiently large to become national in scope and project an image of man pitting himself against the forces of nature (deserts, rivers, the sea) or, alternately, raising the towers of a city to heaven, they stand a good change of mobilizing national support.

Irrational motives are powerful incentives to action. Support for building the new capital of Brazil was mobilized, in part, because of the "almost mystical feeling that the city-bred Brazilian has for the interior; a feeling that the great wealth of the country is locked up there, far away from the coastal region where ninety percent of the people live."[12]

A "mystical" belief of a different order was the inspiration of Islamabad, the new capital of Pakistan. "The capital of a country is the focus and centre of the people's ambitions and desires," wrote General Ayub Khan, former President of Pakistan. "It is wrong to put them in an existing city. The city must have a colour of its own and character of its own. And that character is the sum total of the aspirations, the life and the ambitions of the people of the whole of Pakistan. With the two provinces of Pakistan, separated as they are from each other, you want to bring the people on a common platform. The thing to do was to take them to a new place altogether."[13]

Yet another epic theme—that man should dominate nature—lies at the root of much regional planning in the Soviet Union where it has been used for more than a generation as a basis for sustaining policy commitments. In a textbook for school children published in 1929, the authors summon up truly poetic feelings that reach far beyond a purely rational examination of costs and benefits. They write:[14]

> "We must discover and conquer the country in which we live. It is a tremendous country, but not yet entirely ours. Our steppe will truly become ours only when we come with columns of tractors and plows to break the thousand-year-old virgin soil. On a far-flung front we must wage war. We must burrow into the earth, break rocks, dig mines, construct houses. We must take from the earth. . . ."

This appeal to protean feelings, however, is not without its hidden costs. For the ultimate objective is not a physical artifact (a flowering desert, a new capital city, a transcontinental railroad, the conquest of the steppe), but some specific improvement in the national "welfare," as expressed in relevant social and economic terms, such as an increase in production, improved social conditions, or national integration. The physical project, although intended as a means towards these ends, often replaces the ultimate objective and assumes the character of an end in its own right. The desert *must* be irrigated, the dam *must* be built, the land *must* be reclaimed, the highway *must* be built, the steppe *must* be made fertile—regardless of the costs imposed or even the ultimate (and sometimes doubtful) benefits. The project—like the moonshot—becomes a national obsession. A man *must* be placed on the moon; he *will* be placed. Alternatives to the man on the moon are *not* explored; alternative uses for the resources are *not* studied or seriously contemplated. The irrationality of the original appeal (which, I would claim, ensures a sustained national commitment to the undertaking) may also lead to a perversion of the ultimate objectives of the project expressed in welfare terms, as the unforeseen consequences of the project spillover to affect other areas of national life. Projects of this kind can be a country's ruin.

The very fact of national commitment suggests that the project will not be abandoned even if negative consequences should materialize. The project is undertaken, as it were, on trust in the "principle of the hiding hand" that any problems which might arise in the course of its implementation will somehow produce an adequate, creative response to deal with them.[15] Impending ruin, joined to the promise of great benefits (however vague), may call forth extraordinary efforts to avoid it. Only a fraction of the power available from the United Arab Republic's Asswan Dam is currently being used. The fact that the facilities are now available but wasted is leading to a redoubled effort to increase the country's industrial capacity, particularly of heavy power users. The destruction of fisheries in the Nile delta, due to changes wrought in the physical ecology of the delta by the construction of the dam, is adding impetus to the development of new fisheries in Lake Nasser. The changeover from periodic flooding to artificial fertilization in Upper Egypt is seriously disturbing an ecological balance that has been maintained successfully over thousands of years. It is also inducing a major effort to bring about structural changes in the organization of agriculture that promise to bring the population of the region into the twentieth century within the span of a few years.

The second criterion is that a policy, in order to enjoy long-term political commitment, *must be embodied in new institutions capable of outlasting periodic changes in the government.* Three forms of institutionalization may be tried. The first is *legal* and may be illustrated by the provision in the Brazilian Constitution of a fixed percentage of national revenues that must be applied each year to the development of particular regions.[16] This provision is mandatory on all governments (so long as it remains a part of the Constitution); it has the disadvantage of being excessively rigid and unrelated to the actual cost of realistic economic opportunities for investment. The total amount thus set aside, may be either too small or too large and can easily lead to a serious misallocation of national resources.

The second form of institutionalization is *administrative.* This may be illustrated by the proposed regionalization of public administration and development efforts in France.[17] To the extent that regionalization is successful, shifting the gravity of decision-making power away from the Paris region, this reorganization of the French administrative apparatus will be difficult to reverse. The obstacles to regionalization have proved to be formidable, however. Institutionalization by administrative decree may be too weak an instrument to bring about substantial changes in the probable distribution of development benefits among the different parts of a country.

The third form of institutionalization is *corporate.* An example would be the creation of the Guayana Corporation of Venezuela.[18] This corporate authority has ample financial resources; it has gradually built a political basis of support in the region of its primary impact; and it has managed to attract superior technical skills and leadership abilities. Its principal drawback has been the acquisition of too much power, a problem which already has been discussed. There is no question, however, about the effectiveness of this type of institutional arrangement. By now, the Guayana Corporation has become the largest single enterprise in Venezuela. In the future, its powers may be restricted, but its existence is reasonably ensured for a long time to come.

Finally, long-term policy commitments require that *the policy in question produce visible results quickly.* A policy is most vulnerable during the first few years of its application. If results are not timely, those opposing it may be more successful in their efforts to persuade the government to abandon it in favor of some other alternative. Yet frequent changes in policy for spatial development are counter-productive. They usually involve changes in objectives as well. Where this happens, the results of one policy will be counteracted by the results of alternate policies and so will tend to cancel each other.

The requirement of quick and visible results argues, once again, for works of physical construction. A dam is visible; the building of a new capital city is visible; a new subway is visible. The social and economic effects of works of physical construction, such as increased employment, higher incomes, more efficient administration, more efficient mass transit, are not visible in the same way. They are captured only in statistics and, in many cases, are difficult to attribute to a given policy. Indeed, quite aside from the possibility of their measurement, there is a deeply seated folk prejudice (not entirely unwarranted) against relying exclusively on statistical measurements for justifying the continuation of public policies. As nearly everyone knows, statistical results can be deceptive. Moreover, the effects of a given policy may take years to produce results that are statistically significant. The instruments of measurement are frequently insensitive to minor but significant variations in the phenomena observed. When a program of physical construction is a feasible alternative to a people-oriented program that will have intangible results, the physical project is likely to be chosen.

A land redistribution scheme, for instance, which leads to the quick reallocation of property to a landless peasantry is likely to be more popular with politicians than a policy seeking to encourage cooperatives, expand production credit, and create new marketing arrangements as elements of a general land reform program. The latter policy would almost certainly be slower in its impact, and the total amount of land passing to the peasants would be smaller. Politicians, no less than the public, are primarily interested in the symbolic gesture. Although willing to allocate millions of dollars for a quick scheme of land redistribution, they may be reluctant to support a more elaborate and complex program of reform whose results would be more gradual and, in any event, less visible and dramatic.

The sunken costs in the original investment may make the policy, once initiated, irreversible. President Kubitschek insisted on accelerating the building of Brasilia so that succeeding governments might find it impossible to back off from the original decision. A tremendous effort was therefore launched to complete the physical infrastructure of the city under his administration. The potential monstrosity of having a brand-new ghost capital sit unoccupied would force subsequent governments to make the city work. Adversity would thus call forth its own solutions.

III. CONCERTING THE USE OF INSTRUMENTS FOR REGIONAL DEVELOPMENT

The measures that have been applied to carry out specific regional policies are extraordinarily varied. Some order may be imposed on this jumbled profusion by classifying the principal policy measures in the following categories: physical land use controls, directed migration, public investments, and financial location incentives. Following a brief discussion of these measures, I shall argue that none of them can be relied upon to produce significant results unless a set of complementary measures is brought to bear in a coordinated and focused way on a limited geographic area.

Descriptive surveys, such as the study recently published by the Organization for Economic Cooperation and Development, are of little help to policy makers who are interested in knowing the relative effectiveness of particular measures in a variety of contexts.[19] Professor Rodwin's excellent review of national policies for urban and regional development does attempt to arrive at a balanced judgment concerning their success, but his evidence is, for the most part, impressionistic and only incidentally related to the instruments of implementation currently available.[20] What we need, and do not have, is a theoretical scheme for evaluating different methods for implementing regional policies in a variety of socio-economic settings. I do not propose to undertake this ambitious task in the present chapter. My comments will rather be directed to the question of inherent limits in a number of widely used practices. The larger issue of their relative effectiveness requires a more systematic study.

a. Physical land use controls

Land-use zoning applied to regions on a national scale has been tried in a number of European countries but not, to my knowledge, in any of the newly industrializing nations. A typical example is the Danish Zoning Plan which uses a four-part division of the country: urban and industrial development areas; areas potentially attractive for urban and industrial development; landscapes of recreational and cultural values; and agricultural areas.[21] A more elaborate classification emerges from the further subdivision of these categories and from stipulation of patterns of spatial juxtaposition.

Conceptually, Denmark's zoning plan appears to be a useful device for focusing public attention on critical problem areas. Its principal use is as a technical document for guiding public decisions. But, unless positive incentives exist for ensuring a desired pattern of location, the essentially negative controls of a zoning map are unlikely to be very effective. Urban experience in the United States suggests that land zoning is, at best, a procedure for slowing down the processes of changing land occupance.[22] Essentially a negative instrument intended to prevent certain undesirable actions from occurring, it generates no positive action of its own.

Despite these obvious shortcomings, a simple zoning plan that is easy to administer (through a system of licensing, for example) may be a useful complement to other, more positive measures, expecially in areas of relatively dense settlement. For the capital region of Chile, I have proposed the designation of agro-industrial corridors and tourist zones for the purpose of excluding manufacturing activities from these areas. The creation of an Environmental Protection District to control urban pollution was also suggested.[23] These measures, however, have not been enacted. Given the agenda of priorities in the newly industrializing countries, the political appeal of regional zoning legislation is likely to remain low.

b. Directed migration

Direct controls over migration have been attempted in several socialist countries.[24] Before a person can acquire legal residence in certain cities, he must be able to produce an official work permit or, at least, show evidence of having found a job and housing. This allows the planning authority to restrict migration to large cities and to facilitate a policy of industrial decentralization by assuring an adequate supply of labor in the smaller centers.

As with other controlling devices, this system of direct migration controls is basically a preventive measure. Even so, it is difficult to stop the flow of urban migrants altogether, and to hold the population of major cities to the limits ideally desired by the planners.

In a number of Western European countries, directed migration has taken the form of extending financial incentives to workers, without restricting labor mobility as such. Sweden has carried this concept of an active migration policy farthest. Allowances are given to workers who leave depressed areas as well as to those who move to growth centers within them. Since 1963, around 10,000 workers from the North have obtained special transfer grants each year to take up jobs in the central and southern areas. At the same time, over 6,000 workers per year have obtained such grants for moving from the southern and central regions to jobs in growth centers within the northern areas.[25]

Incentive programs such as Sweden's would, in principle, appear to be more suitable for mature industrial than for newly industrializing countries where migration to cities from rural areas occurs at a pace in excess of the capacity of the urban economy to absorb new workers. For these countries, the Socialist model of direct migration control would seem to be more appropriate. The enormous task of managing such controls must be recognized, however. The policy would also raise the inevitable question of whether individual mobility ought, in principle, to be restricted at all. The whole issue deserves more careful study than it has received.

Finally, the provision of housing and social amenities in selected urban areas may be mentioned as a possible instrument for attracting population to them. This policy has been applied in Great Britain and Israel as one of the principal inducements for population to move into new towns where job opportunities have tended to follow rather than precede an expansion of the labor force.[26] Occasionally, it has also been argued that the improvement of living conditions in rural areas might contribute to stemming the massive transfer of population to cities. This belief, however, has not been borne out in practice. Generally, it is the most "urbanized" part of the rural labor force that shows the highest propensity to migrate to cities. The closer the bus services, and the extension of mass media into rural areas, among other means—the more people can be expected to leave their villages to seek their fortune in cities.[27]

c. Direct public investment

This is the most widely used method for implementing regional development policies. The underlying notion is that the provision of critical facilities in designated districts, such as generating plants for electric power, transport facilities, housing and other social services, or large manufacturing plants in basic industries, will help establish conditions favorable to private investors and so set in motion a self-sustaining process of regional growth.[28] Such investments are best carried out in coordinated fashion according to a carefully worked out program of development. Isolated efforts, such as construction of an electric generating plant, are not likely to produce significant changes in the attractiveness of a region for private investors without supporting investments in other facilities. This consideration argues strongly for the corporate approach to regional development planning. Even a tightly coordinated approach, however, offers no guarantee that a cumulative growth cycle will be initiated. For instance, the potential growth center of Concepción in Chile has proved incapable of attracting private investments on a scale sufficient to allow this important provincial city to compete with Santiago as the country's principal center of manufacturing, and this despite the general excellence of Concepción's facilities, including a leading private university. The city's remoteness from national markets, its relatively high cost of living, the absence of a diversified structure of private services to industry, and the extreme centralization of decision-making power in the nation's capital combine to militate against the self-supporting economic growth of the region. Further overhead investments or even the enlargement of the government's steel plant in the area would do little to encourage private investors unless additional conditions preventing spontaneous growth were to change. The situation may be radically transformed under the present Popular Unity government of Chile as a result of both the increased public ownership of industry and a general restructuring of the final output of manufacturing towards a greater emphasis on capital goods and working class consumer products.

Examples of the failure of public investment programs to induce economic growth are probably more numerous, on balance, than the successful cases. Though public infrastructure may be necessary for regional development, it is by no means a sufficient condition. Additional incentives, especially of a financial nature, may have to be offered to bring the private calculus of costs and benefits into harmony with policy designs. The social response of local communities to government incentives is also relevant. Similar economic programs may not elicit similar responses on part of the community. In some cases, a program may lead to its mobilization around the objective of economic growth while in others it will meet with apathy or even passive resistance. This ability of the local community to organize itself for economic growth and development must be considered an important variable in the implementation of development policies.

Almost no empirical studies have been carried out to substantiate this point, and the evidence that does exist is chiefly based on personal impressions.[29] Nevertheless, it would seem that the social structures of communities do differ with respect to their ability to generate and absorb changes conducive to economic growth. Some towns are essentially "open" and dynamic systems, welcoming change and capable of rapidly adapting to new conditions of life; other towns are highly controlled by traditional elites and are much less receptive. The question deserves a great deal more study than it has received.[30]

d. Financial location incentives

These are among the most popular devices for promoting the development of designated areas in private or mixed enterprise economies. Principally, they include *tax incentives, import tariff reductions, location subsidies,* and *investment credits.* These measures are applied to individual firms in the hope of inducing them to invest their capital in the development areas. Experience has provided contradictory evidence concerning their effectiveness. Four major problems may be identified:

(1) It is politically difficult to restrict the geographic incidence of financial inducements to only a few areas. Yet, the multiplication of areas eligible for favored treatment will encourage an excessive spread of industrial locations, dissipating the potential effectiveness of the incentives. An interesting proposal for a regionally differentiated payroll tax has recently been made in England by Professor Colin Clark. This would involve varying rates which, at one extreme, would penalize most heavily locations in London and, at the other, give numerous benefits to firms operating in the Hebrides.[31]

(2) Unless financial subsidies make up a substantial proportion of total variable costs in production dependent on location, they may not lead the industry to locate in areas that, in other respects, are lacking in decisive economic advantage. The subsidy must be at least sufficient to compensate the producer for higher costs of transport of raw materials and/or finished products as well as for other costs he may incur by locating outside existing core regions. Yet the provision of an adequate level of subsidy may place an excessive burden on the public treasury and decelerate the overall process of national economic growth.

(3) Financial incentives may encourage industry in locations that are inefficient from the standpoint of private cost accounting and thus reduce the overall competitive position of national industry in foreign markets. Additional export subsidies may therefore be required. This, however, would further raise the costs of regional development to the nation.

(4) Withdrawal of financial incentives after the initial starting-up period may lead to a shift of industries out of the region and back to the core areas of the country. Should this happen, and the mere threat to do so would be sufficient, a policy of permanent subsidization might be required, resulting in a structural distortion of the resource allocation pattern that would be difficult to reverse.

The most celebrated instance of the use of financial incentives is the development program for the Brazilian Northeast.[32] Some 600 industrial projects resulted from this program between 1960 and 1968.[33] Even so, the creation of new jobs has apparently been unable to keep pace with population growth in the region, and unemployment is reported to be increasing. Moreover, the various financial incentives, such as the 50 percent relief in corporate income tax for an equivalent amount of investment in the region, have had, on balance, a regressive effect on the distribution of income.

The Northeast development effort is notable in that it combines a number of the "success" elements which have been discussed. It is focused on a single region; it has acquired an institutional form in the Superintendency for the Development of the Northeast (SUDENE); it has survived several changes of government by acquiring a strong political base in the region; and it has achieved a remarkable degree of coordination in the use of development instruments.

An insightful critique of this program has been made by Professor Walter Stöhr in his comprehensive survey of regional development policies in Latin America. His conclusions are worth citing at some length, because they demonstrate that even the very best programs may only partially succeed in what they set out to accomplish.[34]

> "The conviction that a direct strengthening of the economy and the improvement of social welfare levels was essential for obtaining a lasting effect led to the creation of SUDENE. Some 600 industrial projects were the result between 1960 and 1968. A road and communications system was established, a modern educational system was introduced, versatile professional opportunities were newly opened in the region, and a change in the style of regional politics and decision-making was initiated along with the opportunity for the states to participate in program formulation for the allocation of federal funds in the region. Of the original strategy components of SUDENE–intensification of industrial development, reorganization of agriculture, and relocation of population surpluses–only the first one was achieved to a major extent. Since the main incentives were for capital inputs, most of the new industries turned out to be highly capital-intensive, using modern technology; their contribution to the creation of regional income and the expansion of regional markets however, was relatively small. In effect, these new industries remained modern enclaves within a large and backward area. They received their capital and technology from the outside, mainly from the South-East, and shipped an increasing amount of their products to South-Eastern markets. The hoped for impact on the rural areas by absorbing manpower and creating additional income for rural population was not realized. Due to the lack of regional market expansion, these new industries created yet another problem for the nation: subsidized competition from the North-East emerged for some of the industries in the South-East, and in some sectors idle capacity arose due to the limited absorption possibilities of the national market.
>
> "A great part of the income created by this new industrialization flows back to the South-East. In addition, an instability has recently been created by pressure on the government to liberate the sale of newly established enterprises by their original tax-favored owners. This would indicate an inclination of investors in the South-East to withdraw their capital from the North-East as soon as they have secured their initial tax benefit.
>
> "The rural sector has been relatively little changed, first because practically all of the private capital under the tax savings legislation went into industrial rather than agricultural investment, and second, because the relocation of surplus agricultural labor as well as agricultural reform proceeded very slowly. Both have recently been de-emphasized even more because of fear that they would aggravate the unemployment problem still further.
>
> "In summary, three key problems have arisen in connection with the program for the North-East. The first lies in a failure to create a self-sustained process of regional growth with decreasing dependency on extra-regional inputs. To a large extent, this may be ascribed to the lack of a sufficient regional multiplier effect under which growth would increasingly be based on regional savings, and regional markets would be enlarged. The second problem results from a failure of the development process to diffuse from the few industrial enclaves to the rest of the region. The third problem is the insufficient compatibility between the development of the North-East and national development criteria."

IV. MAINTAINING A NATIONAL BALANCE

Discussion of the preceding three principles has, in various ways, stressed the importance of an approach focused on specific development requirements of a given region. It is now necessary to back off from this emphasis to attain a wider national perspective. The territory of a nation can be divided into major development areas each of which will present a different problem constellation to the planner. Regional development strategies must reflect these differences, but they must also strive to achieve a balance of all programs across the nation. I shall call this the balanced approach. A balanced approach is required, first, because the development issues of different spatial subsystems are interrelated among themselves and second, because the highest level of policy integration must be the nation. A regional development program conceived and carried out without regard for its impact on the rest of the country is likely to be harmful from a national viewpoint. The successful implementation of a regional policy, therefore, depends on the ability of planners to relate it in functional ways to the development of the nation as a whole.

Whereas the focused approach clearly favors some areas over others, the balanced approach requires a mechanism not only for adequately differentiating among strategies for regional development but also for assuring the consistency of these strategies (and the programs through which they are implemented) with the overall development policy for the nation. In other words, all areas of the nation will receive at least some separate attention; a few areas will receive concentrated attention; and the policies and programs for all areas will be coordinated at the national level through sectoral and intersectoral policies, programs, and budgets. Because this arrangement is fairly complicated, a diagram may help to clarify it further (Figure 1).

In the diagram, three levels are distinguished: Level I is essentially political and brings together the Council of Ministers as well as other political decision makers of national stature. Level II includes primarily the technical skills of the central guidance system (national planning, budget office, treasury, central bank, and technical officials from the ministries and other specialized agencies). Level III joins political and technical functions in the several regions and cities of the country.

Effective power in this system refers to the management of budgetary resources. Policy-making, planning, and investment programming are closely linked in the diagram with effective power. For ease of communication, it is advisable that planning and budgetary functions be closely related or even joined within a single agency or Planning Ministry. In addition to the financial budget, a separate accounting for manpower and resources is suggested. Few countries have so far extended their management systems to include the budgeting of critical skills and scarce physical resources, but it is desirable that they do so.

Spatial planning-budgeting is shown to relate closely to both the sectoral and

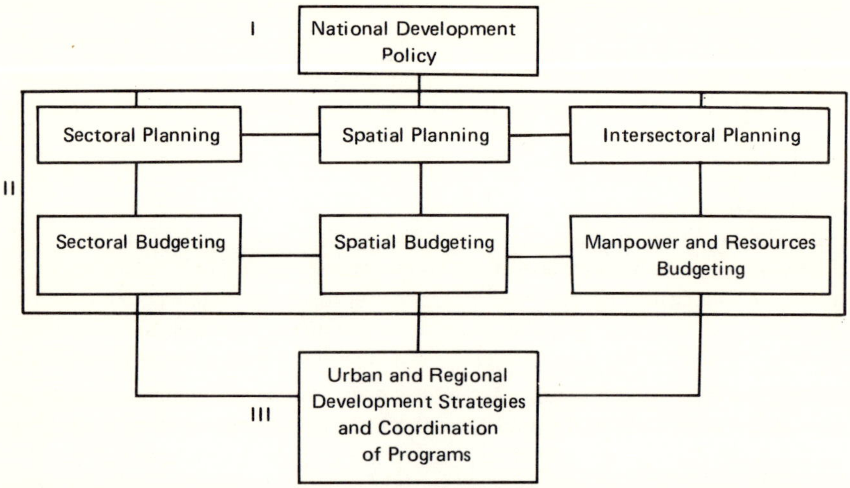

Figure 1: A Comprehensive System for National Planning, Programming, and Budgeting

intersectoral planning-budgeting process. This suggests that spatial planning represents a new level of synthesis which attempts to coordinate both sectoral and intersectoral programs according to the requirements of regional policy while, at the same time, assuring the consistency of these policies with the overall development policy for the nation.

To make this possible, Level III becomes a necessary component of the planning-budgeting system. Specialized planning offices must be established in each of the country's development areas (including the major cities), replicating organizations at the national level. In those regions which are undergoing intensive development, the Regional (or Metropolitan) Development Authority replaces the planning office as the focus for planning and coordination. In contrast to the ordinary planning office which commands no investment budget of its own, the Regional Authority has a corporate form and its own source of funds for carrying out its work.

An intricate web of communication brings these several elements into a related system. The administrative arrangements can get rather complicated and in every case must be adapted to local circumstances. A number of variants of this basic scheme are possible, depending on whether the country concerned is a federal or unitary state and the extent to which participation in policy planning is practiced by different groups and organized interests.

This model for a comprehensive system of planning for regional development is based on a prototype evolved in Chile between 1964 and 1969.[35] Similar arrangements have been put into practice in France[36] and Turkey.[37] Although no country has so far succeeded in perfecting a system of regional program budgeting, considerable progress has been made in Chile, where an inter-regional econometric model is used for achieving consistency between regional and national growth targets.[38] Each year, the National Planning Office provides regional offices with a current statement of policy, program guidelines, and budgetary ceilings. These are used to detail the components of the annual investment program in close collaboration with the representatives of the functional agencies in the area. A draft of this program is sent back to the national office for revision and for coordination with other regional programs, with overall national economic policies, with the Budget Directorate of the Ministry of Finance, and with the Council of Ministers. The original draft submitted by the regional office is thus subjected to several revisions with the active collaboration of the regional office until a satisfactory allocation of investment funds has been made. The regional (or spatial) budget for the country as a whole is then submitted, through the Minister of Finance, to the President for his approval. It is not, however, submitted to the Congress for legislative action, but remains an administrative document. Since representation in the Congress is on a territorial basis, explicit discussion of regional allocations would lead to needless controversy and would undoubtedly produce so many changes in the budget, as to render the entire budgetary process meaningless.

Even as an internal, administrative document, however, the regionalized national budget has many uses. It results in a good deal of intersectoral consultation; it helps to focus attention on the regional consistency of sectoral programs; it serves to highlight the requirements of certain projects that are of national importance; it heightens awareness of the systemic character of inter-regional relations; and it allows for tests of inter-regional balance in the programming of national investments. The regionalized national budget certainly does not lower the efficiency of resource use; quite possibly, it succeeds in raising it. All in all, it is a hopeful beginning.

CONCLUDING COMMENTS

Regional development planning, as a component of national planning, is a very recent innovation. Its extraordinary appeal is reflected in the frequent international conferences on the subject, the growing number of scholarly publications, and, most importantly, in the acceptances of this new technique by an increasing number of national governments. No one would deny that many problems of a conceptual and practical nature remain to be solved. But the base already exists for any country, wishing to achieve the maximum beneficial effects from its development programs, to undertake a regionalization of the development effort. An adequate body of theory exists; major policy issues for regional development have been identified; good analytical techniques can be applied to furnish a factual basis for spatial planning; there is a wide array of implementing tools; sufficient experience has been gathered to distill from it a number of principles for the implementation of regional policies; and the necessary technical expertise—thanks to the efforts of the United Nations and a number of development institutes in different parts of the world—is becoming available. Although the gap between intent and realization remains a large one, progress is visible on all fronts.

NOTES

1. This is clearly demonstrated in the case of city planning which, traditionally, has had to rely on zoning as its principal instrument of effectuation. The master plan for a city is chiefly a two-dimensional map of desirable land uses that can be easily translated into a zoning map for purposes of control. Lacking direct influence over municipal investments, city planners have done little so far to advance the art of capital budgeting for urban development. Their planning expresses primarily the instruments of control available to them.

2. Gavin McCrone, *Regional Policy in Britain.* London: George Allen & Unwin, 1969; Niles M. Hansen, *French Regional Planning.* Bloomington: Indiana University Press, 1968; Niles M. Hansen, *Rural Poverty and the Urban Crisis.* A Strategy for Regional Development. Bloomington: Indiana University Press, 1970; Lloyd Rodwin et al., *Planning Urban Growth and Regional Development.* The Experience of the Guayana Program of Venezuela. Cambridge: The M.I.T. Press, 1969; Lloyd Rodwin, *Nations and Cities.* A Comparison of Strategies of Urban Growth. Boston: Houghton Mifflin Co., 1970; Malcolm D. Rivkin, *Area Development for National Growth.* The Turkish Precedent. New York: Frederick A. Praeger, 1965; John Friedmann, *Urban and Regional Development in Chile.* A Case Study of Innovative Planning. Santiago: the Ford Foundation, 1969 (offset); Walter Stöhr, *Materials on Regional Development in Latin America: Experience and Prospects.* McMaster's University (Hamilton, Ontario), Department of Geography, 1970; U.S. Department of Commerce, *Area Redevelopment Policies in Britain and the Countries of the Common Market.* Washington, D.C.: U.S. Government Printing Office, 1965; and H. W. Richardson, "Regional Development Policy in Spain," Urban Studies, Vol. 8, No. 1 (February 1971), pp. 39-54. Gordon C. Cameron, *Regional Economic Development.* The Federal Role. Washington, D.C.: Resources for the Future, Inc. 1970; T.N. Brewis, *Regional Economic Policies in Canada.* Toronto: The Macmillan Co. of Canada, 1968.

3. Gavin McCrone, op. cit. p. 276.

4. See Appendix D at back of book.

5. The poverty of administrative skills is a more serious impediment to successful development planning than is normally realized. Jean Claude Garcia-Zamor has recently assessed the Brazilian situation in a short but insightful article, "An Ecological Approach to Administrative Reform: the Brazilian Case," Institute of Latin American Studies, The University of Texas at Austin, Offprint Series No. 98 (1960). An extended, more systematic treatment of the topic is found in Fred W. Riggs, *Administration in Developing Countries.* The Theory of Prismatic Society. Boston: Houghton Mifflin Co., 1964.

6. See Walter Stöhr, *Materials. . . .*, op. cit., Table 1.

7. Philip Selznick, *TVA and the Grass Roots,* Berkeley: University of California Press, 1949.

8. On the stability of the rank-size order of cities, see John Friedmann, *Regional Development Policy,* op. cit., p. 147; J. R. Lasuén, F. Wasservogel, and A. Montserrat,

"Some Traits of the Process of Growth of the System of Nations: Stability, Polarization, Diffusion." MS; and J. R. Lasuén, "Urbanization Hypotheses and Spain's Cities System in Evolution." The Hague, October 1967. Manuscript.

9. Professor Kingsley Davis has constructed a 4-city index of primacy covering the period 1950-1970 for a total of 38 countries. During these 20 years, only Mexico showed a marked decline in the index, from 2.44 to 1.22. All other countries changed their index in either direction by less than a single order of magnitude, and most countries displayed remarkable stability in the overall pattern of population distribution. Stability was greatest for countries which had a low rate or rural to urban migration. See Kingsley Davis, *World Urbanization 1950-1970.* Volume I: Basic Data for Cities, Countries, and Regions. Berkeley: University of California, Population Monograph Series, No. 4, 1969, Table G.

10. The probability of success of such a policy would also depend on the overall growth of population, the extent of prior urbanization, and the rate of rural-to-urban migration. See D.H.K. Amiran and A. Shachar, *Development Towns in Israel.* Jerusalem: The Hebrew University, 1969.

11. H. W. Richardson, "Regional Development Policy in Spain," Urban Studies, Vol. 8, No. 1 (February 1971), p. 50.

12. Glenn V. Stephenson, "Two Newly-Created Capitals: Islamabad and Brasilia," Town Planning Review, Vol. 41, No. 4 (October 1970), p. 321.

13. Quoted in Stephenson, op. cit., p. 323.

14. M. Ilin, *Velikii Plan.* Moscow: Government Publication, 1929. Cited in Albert E. Burke, "Influence of Man Upon Nature–the Russian View: A Case Study," in William L. Thomas, Jr., ed., *Man's Role in changing the Face of the Earth.* Chicago: University of Chicago Press, 1956, p. 1048.

15. The "Principle of the Hiding Hand" is discussed in Albert O. Hirschman, *Development Projects Observed.* Washington, D.C.: The Brookings Institution, 1967, chapter 1.

16. Antonio Cerqueira Antunes, "La Política de Industrialización del Nordeste Brasileño," Latin American Institute of Economic and Social Planning, November 1966 (mimeo).

17. Lloyd Rodwin, *Nations and Cities,* op. cit.; also Niles M. Hansen, *French Regional Planning,* op. cit.

18. Lloyd Rodwin and Associates, op. cit.

19. Organization for Economic Cooperation and Development, *The Regional Factor in Economic Development.* Policies in Fifteen Industrialized OECD Countries. Paris, 1970. See also Europaeische Wirtschaftsgemeinschaft, *Massnahmen der Regionalpolitik in den Mitgliedstaaten der EWG.* Brussels, March 24, 1964.

20. Lloyd Rodwin, *Nations and Cities,* op. cit.

21. *Zoneplan 1962 for Denmark.* Copenhagen: Landsplanudvalget, August 1962. In the United States, Hawaii, Oregon, and Maine have adopted statewide zoning controls, and Colorado is currently studying such legislation.

22. R. F. Babcock, *The Zoning Game.* Municipal Practices and Policies. Madison: University of Wisconson Press, 1966.

23. See Chapter 10.

24. See, for example, the account for Bulgaria by Philippe Bernard, "Growth Poles and Growth Centres as Instruments of Regional Development and Modernization, With Special Reference to Bulgaria and France." Vol. VIII of *Growth Poles and Growth Centres in Regional Development.* Geneva: United Nations Research Institute for Social Development,

1970 (UNRISD/70/C.25); and Kosta Mihailović, *Regional Development in Eastern Europe: Experiences and Prospects.* Geneva: United Nations Research Institute for Social Development, 1970 (UNRISD/70/c.49), pp. 93-101.

25. Organization for Economic Cooperation and Development, op. cit., p. 79.

26. "Urbanization: Development Policies and Planning," International Social Development Review, No. 1 (New York, 1968), pp. 89-95.

27. The evidence is widely scattered, but nearly all studies that have looked into this issue agree on this basic point. See, for instance, Joan Nelson, "The Urban Poor: Disruption or Political Integration in Third World Cities? " World Politics, Vol. 23, No. 3 (April 1970), pp. 393-414.

28. For a discussion of the role of overhead investments in the urbanization process, see Lowdon Wingo, Jr., "Latin American Urbanization: Plan or Process? " in Bernard J. Frieden and William W. Nash, Jr., *Shaping an Urban Future.* Essays in Memory of Catherine Bauer Wurster. Cambridge: The M.I.T. Press, 1969, pp. 62-64. A more exhaustive treatment is A. J. Youngson's *Overhead Capital.* A Study in Development Economics. Edinburgh: University Press, 1967.

29. The social response of communities to the challenge of economic development is extensively discussed in Chapter 6.

30. Mr. Chung-Tong Wu is currently preparing a doctoral dissertation on "the social conditions of economic progress in Hong Kong" at UCLA's Urban Planning Program. This study is expected to shed some light on the critical social variables in regional economic development.

31. Gavin McCrone, op. cit., pp. 187-188. Professor McCrone argues strongly in favor of the regionalization of fiscal policies. Unfortunately, we do not as yet have any experience on which to base a reasonable assessment of such policies.

32. Antonio Cerqueira Antunes, op cit.

33. Walter Stöhr, *Materials. . .*, op. cit., p. 62.

34. Ibid., pp. 62-62.

35. See Chapter 13. A more detailed evaluation has recently been made by Mariano Valle, "Planning Regional Development in Chile. Achievements and Perspectives." Cambridge, Mass.: Special Program on Urban and Regional Development (SPURS), 1969. Typescript.

36. Niles M. Hansen, *French Regional Planning,* op. cit.

37. Lloyd Rodwin, *Nations and Cities,* op. cit. See also Turan Ersoy, "Financing of Housing and Urban Development in Turkey," a Report. May 1970 (mimeo).

38. "A Model of Interregional Programming and Compatibility," National Planning Office (ODEPLAN), Santiago, 1968. This, in turn, is based on concepts developed in L.B.M. Mennes, Jan Tinbergen, and J. George Waardenburg, *The Element of Space in Development Planning.* Amsterdam: The North Holland Publishing Co., 1969. Dr. Mennes served for a time as consultant to the National Planning Office of Chile.

Chapter 13

THE USES OF REGIONAL SCIENCE: POLICY PLANNING IN CHILE*

THEORY

Every scientific endeavor aims in some measure to be useful, either by deepening man's knowledge of the world or by extending his control over the objects of his knowledge. Regional science is no exception to this rule. Its fundamental concern is the changing pattern over time of relations between terrestrial space and society.

We all know, of course, that the hoary distinction between the "basic" and "applied" sciences is badly misleading, especially when it suggests a hierarchical relation, with the basics peering down their noses upon an inferior species called planners or, with equal lack of logic, planners nodding condescendingly to "useless" basics. It is misleading because action itself generates knowledge and knowledge prompts action. Thus, an intimate relationship evolves between the two uses of science which, when it fails to be nourished, quickly leads to the debilitation of both. Action points to what is relevant; it insists on leading the scientists to the concrete problem itself. Science helps to define problems with greater precision and to choose the proper instruments for intervention.

We propose to speak here as practitioners and planners, and, as such, we are perhaps more conscious of the great difficulties posed in passing from mathematical models that describe partial or general equilibrium conditions

*This chapter was written jointly with Walter Stöhr. It reflects the situation in Chile up to 1966.

under highly artificial conditions to situations that are scientifically "impure," imperfectly known, subject to change, and often uncomfortably real. And because we are engaged in trying to solve problems that matter to people, we always work under the pressure of great urgency and within a highly charged field of social values in tension. We do not dispose over unlimited time and our answers are never morally neutral.

Nevertheless, the passage must be attempted. It would be fatal for regional science if its students resigned themselves to investigating the properties of abstract models—and damn the public and its problems! The theoretical and the practical must be joined in action. Should this require a change in the orientation of scientific behavior, let it be so. It is the way of all creative thought.

The aim of regional scientists is to study social behavior with respect to and contingent upon relationships in space. Webber has discussed the "elastic mile," and we know that measurement of space is one of our most difficult problems.[1] Space has many dimensions, whether we look upon it as objectively measurable or subjectively. But regional science would quickly dissipate the wide interest it has generated were its spaces capable of being measured unambiguously. Once we abandon the certainties of Euclidean space, there is no limit to the number of geometries that may be relevant to a given problem; it all depends on the problem.

The regional scientist is presumably interested in discovering general explanations (laws) for known spatial distributions of human activities, their historical evolution, and the transmission through space of this process. We do not propose this as a formal definition for regional science. But we do claim that most of the work in the field may be thus described. How will a gain or loss in production in region A affect production, income, employment, population, and settlement in regions B, C, . . . , N? This question is typical. We may also wish to know how and why this particular pattern has arisen, and whether it conforms to some known distribution capable of being explained by a model that relates spatial distributions generally to other leading variables.

But what about the other interest of our science, our deep involvement in questions such as the economic development of regions, urban development policy, and industrial location, whose far-reaching effects on social welfare are undisputed?[2] Can we profit from our codified knowledge—the theory—in answering urgent questions of program and policy, and can we intensively use techniques that have become the stock-in-trade of serious scientific work among students of space relationships? These are the crucial and perhaps embarrassing questions we must answer.

Before we come to them, let us examine the method and contents of regional science more closely. The early doubts and hesitations have been left behind; we now display far greater confidence than we did only a few years ago.[3] Still, we may ask: Is there really nothing more substantial holding us together than the ambiguous, magic words "space" and "region"; or do we have more solid common ground?

First, there is the bundle of analytic methods; the most important–by no means the only ones–have been competently examined and expounded by Walter Isard.[4] There is no need to list them; they are generally known. What is characteristic, however, of the regional science approach to the study of relationships in space is that these relationships are seen as a system endowed with mathematical properties. This is, indeed, one of the master ideas of regional science, not original in itself, but new in relation to the field of study to which it is being applied. Behavior is related over distance; cities and regions are not isolates, but together form stochastic energy systems that are subject, by extension, to the laws of entropy and information.[5]

But regional science is also propositional, and a body of hypotheses is emerging capable of being tested in the real world. There are, for instance, propositions about the spatial structure of economic activities, about the statistical distribution of city sizes, and about the pervasive effects of distance in the ordering of regions.[6] There are also propositions about the processes of spatial transformation which refer to the role of exports in regional economic growth, to regional multipliers and linkage effects, to core-periphery relations, to the relationship of changes in economic structure and location, and to the pattern of migration flows.[7]

The regional planner, then, has a wealth of method and a more or less well-ordered body of substantive knowledge. He must face up to the difficult problem of adapting this knowledge to historically real situations under circumstances that are often trying. In doing so, he must consider four main questions.

First, how generally applicable is the theory? Is it applicable to his particular case? Do we have a universal theory, valid through all space and time, or rather a particularized theory, a generalized explanation of a unique event?

Let us take the case of the famous Christaller hexagons.[8] One of two relevant propositions concerns the hierarchical ordering of settlements according to social and economic (mainly service) functions, the second relates to the efficiency of an hexagonal arrangement for the distribution of settlements under stated conditions (a perfect plain, even distribution of resources, etc). Evidence has been assembled in favor of these two propositions in a number of studies, but no critical test has been made to reject them. Perhaps none can ever be made, for hierarchy, like region, is a concept rather than a fact and, as for hexagons, we have no statistical test of fitness in situations where, as almost everywhere in the world about us, plains are not perfect nor evenly endowed with nature's bounties, history has left its mark, and the distribution of human capital is strongly skewed. But it seems increasingly difficult to hold to notions of hierarchy and nucleated settlement in highly mobile communications-oriented, post-industrial societies. In the United States the basic form of settlement is no longer the little *Kreisstadt* of southern Bavaria but the "urban field," transcending relations of dominance and dependency in a "pattern of point locations and connecting flows of people, information, money and commodities" that extends over an area whose space dimensions were undreamed-of in Christaller's scheme and its subsequent modifications.[9] Hierarchy may well be a nineteenth-century concept; at least in organization theory, it is being replaced now by looser, more flexible arrangements.[10]

Theories conform to cultural style. We found hierarchy perhaps because we looked for it; and we looked for it because hierarchical order conformed with our beliefs in a rational order. But we may have deluded ourselves. Our beliefs were perhaps no more than a residue from times when hierarchy was stylish because it was a useful and practical notion that seemed to be confirmed by social experience.[11] If this is so, then regional science theory is in large measure (perhaps totally) relativistic with respect to time and place. And this will have clear implications for the practitioner who is inevitably tied to a historical situation.

The second set of questions deals with the relevance of regional theory to policy. Does existing theory have any answers, in principle, to the concrete problems put by planners? Clearly, there is a dilemma here: the more specific the theory, the greater will be its probable relevance but the weaker also will be its powers of broader explanation. And more: specificity means not only orientation to the unique case (location behavior of American suburbanites, for instance) but often also a marked relaxation of constraints. In the attempt to be relevant, the ability to say something that might be useful *anywhere* is lost. Simulation exercises are typical examples of this dilemma.

There are two other aspects of theoretical relevance. The first has to do with the choice of theory which may or may not correspond to the range of phenomena with which planners have to deal. What theory, for instance, do we have that would help us to estimate the future potential for economic growth of any coordinate position within a given national space? One could suggest some tools, such as regional input-output studies, but the predictive powers of this analytical model have yet to be proved; moreover, its accuracy diminishes rapidly as the size of region is reduced. The planner who wishes to identify potential core regions for development, is, therefore, left to rely mainly on intuition and on what may be called broad professional judgment.

The second aspect concerns mainly the form of the theoretical model, particularly whether it is intended to explain change or equilibrium conditions. Planners who are pressed for solutions to immediate problems have little time to think about the possibilities of long-run equilibrium, static "best positions" or optima. The long run is, at best, five or six years away, but the real issue is what to do tomorrow that will not be regretted five years hence.

The third set of questions to consider is the relevance of analytical techniques. Even if the planner's knowledge is not very helpful—how helpful we shall see when we turn to Chilean experience in regional policy planning—perhaps he can be saved by sophisticated manipulation of data, developing lower-level theories along the way, thus arriving at powerfully based conclusions.

The main difficulties with using any given technique, particularly in the less developed countries, are these:

(1) scarcity of the necessary data;
(2) excessive time required to carry out analysis;
(3) high cost of investigation;
(4) the high degree of sophistication required relative to the available professional skills.

Continuously greater refinement and subtlety of technique may thus be self-defeating.

We now come to the fourth and last and undoubtedly the most difficult set of questions. In a practical situation, what is the nature of the relationship between the technical expert (who presumably will wish to draw on scientific knowledge) and the wielders of political influence? Altschuler has suggested that expert judgment is generally better received by others if it is clothed in the language, difficult for the layman, of hard science and mathematics. Engineering judgments are much less subject to question (e.g., road performance standards) than urbanistic judgments (e.g., zoning standards).[12] But the issue is more complicated. The nature of the relationship has not, to our knowledge, been deeply probed.[13]

The planner must accept the practical constraints of the decision process. If he wishes to prevail—and who would not?—he will have to play two distinct roles for two different audiences. In his first role, played out before his professional peers and professors in his parent discipline, he must hide behind a mask of scientific perfection. In his second role, played each day for the edification of his employers whom, for brevity's sake, we shall call the politicians, he must appear in shirt sleeves and join the rough talk of the builder, doer, and operator. He must be willing to cater, cajole, and simplify: above all simplify, so that all complexity may be reduced to maxims, may be learned and memorized, easily dramatized, and readily applied. And he must push for results, propose the second best or agree to it if the best cannot be had, and even condenscend to live with third, fourth, or fifth best choices if no others are available. He must be sensitive to changes in political currents, to the predilection of this or the other powerful ministry or institution and to the limits of political tolerance.

But if all this is true, what will be left of his professionalism or of his access to the body of theory and technique that supposedly distinguishes him from the run-of-the-mill *politico*? Or have we exaggerated, purposely biasing the presentation? We shall have to examine the evidence.

CHILE'S REGIONAL PLANNING TO THE PRESENT

The Chilean experience is pertinent evidence. Since 1964, the Government has supported a regional planning program that is having wide influence throughout the country and is designed to establish a link between national economic policies and programs of local development. The program envisions the possibility of establishing regional priorities for investment as a means of accelerating the process of national economic growth and furthering the integration of the national territory; it seeks effectively to relate all regions to the development process in Chile through direct participation in decisions; and it promotes the incorporation of additional resources into the system of national production by giving close and continuing attention to development possibilities at the local level that might otherwise be neglected.

These are the broad purposes. But the larger meaning of regional development planning in the sense of articulating local needs with national objectives is more ambitious. It is nothing less than an effort to achieve a closer integration of the economy in its territorial dimensions and to guide the resulting settlement pattern so as to create a national sense of identity cutting across all sectors, regions, and classes.

Antecedents of Regional Planning

Regional approaches to Chile's development had been discussed for several decades. This was a natural result of the country's geographic shape, extending over 2400 miles down the western coast of Latin America, from the hot desert in the North into the polar regions of Antarctica. Servicing the whole national territory from Santiago was difficult until the recent and marked improvement in communication. Regionalism thus grew up, rooted in landscape, economy, and social structure. Oversimplifying, mining characterized the North, commercial agriculture the Center, and sheep ranching and oil drilling the South. Of these three zones, the central region was the most populous, growing more rapidly than other parts of the country. In professional circles, especially among architects, the extraordinary growth of Santiago was usually lamented, and the decentralization of industrial development was advanced to become one of the leading ideas during the forties and fifties.[14] However, the instruments for accomplishing such an objective did not exist as yet, and the scientific basis for regional planning was practically unknown.

The Constitution of 1925 provided for the creation of political assemblies in each province, but no implementing action has ever been taken, and Chile's national administration continues to be one of the most centralized in Latin America. Branch offices of several Santiago ministries were established in the provinces, but were limited to minor administrative routines. The most effective of these provincial offices were those of CORFO, the National Development Corporation, created in 1939 to promote industrialization. CORFO's provincial offices were entrusted some responsibility for local development, such as the administration of local credits. In CORFO's central office, a small department was in charge of supervising the provincial offices and contracting for some regional economic research with the universities, in the hope that eventually the prospects for evolving a systematic development policy in the provinces might become more favorable. The Ten-Year Plan prepared by CORFO in 1961 was devoid of any explicit recognition of the regional element in national development.

The Phase of Initial Persuasion

Regional planning began seriously in June of 1964 when a team of Ford Foundation advisers arrived in Chile. The time was ripe for a change. A new Government was to be elected in September, and the economic advisers of the two principal contending parties—the Christian Democrats and the left-wing FRAP forces—were basically sympathetic to the regionalization of the national plan. This sympathy, however, had to be translated into specific proposals regarding institutions and procedures.

Initial contacts were made with the head of CORFO's Regional Planning Department. It was decided to propose to the economic policy planners of the Christian Democratic Party the creation of a project location group within the Bureau of the Budget (Ministry of Finance). The first reason was that Budget had already a small project evaluation office, and the consideration of location variables would appear as a logical extension of this activity. Second, and more importantly, it was thought that one of the main instruments of regional development policy at the national level was the locaction of infrastructure and directly productive investments. Since central Government decisions were decisive in this case, and since all projects would eventually have to obtain approval from Budget, the selection of that office seemed theoretically sound.

With the approval of the Budget Director, a small professional staff was assembled and began the task of gathering existing regional economic data from many dispersed sources and of compiling a complete documentation of regional research done in Chile up to that time. When the new Government took office early in 1966, the project location staff was incorporated into the National Planning Office (ODEPLAN) as a Division of Regional Planning.

Regional science clearly had little to contribute at this stage of Chile's planning system. It did, however, serve to provide a conceptual base for negotiations with the national economic planners. Location economics constituted a strong conceptual link between sectoral national planning and local development. It was suggested that a national policy for regional development could be made effective by influencing location choices for new project investments.

The Phase of Gaining Political Acceptance

To this point, only bureaucratic victories had been won: the recognition of a function, the establishment of a new office, and the recruitment of a core professional staff. To be effective, however, regional planning had also to prove its usefulness in contributing to the solution of national problems. These might be the comparative evaluation of alternate locations for a sugar refinery, proposing solutions for a persisting local unemployment situation, the elaboration of a strategy and a program for the development of a key region in the South of the country, or the regional allocation of sectoral investment targets in the national plan. To put it bluntly, the planners had to prove that they could improve the quality of decisions even if their data were imperfect. This meant foregoing long-range and more fundamental research in favor of more immediately useful studies. The problem was to find a balance between pressures of the moment and long-range studies that would yield an optimum in terms of actual decision results. Finding this optimal point was not always easy.

Five principal tasks came to preoccupy Chile's regional planners during the initial years: delimitation of planning regions, creation of an institutional framework for regional planning, formulation of regional investment priorities, construction of regional development indicators, and regional project identification. Each of these will be briefly discussed.

Although it is true that regional planners need regions, the question of how they should be drawn for practical purposes is a far less serious matter than many regional scientists might be willing to grant. In the barely inhabited peripheral zones of Chile's economy, where resources are very incompletely known, the tracing of exact boundaries is an essentially meaningless exercise. In the more densely settled areas, on the other hand, the shape of regions would vary with the purposes for which they were intended. In neither case could regions be delimited primarily on the basis of the retrospective and essentially static criteria which regional science offers. Thus, in the peripheral areas, regions had to be devised that would constitute an adequate framework for resource development and colonization. In the central zone, they had to be instrumental for introducing changes in the existing semi-feudal agrarian structure and, in addition, had to be oriented to the predominant urban economies in the area. In the end, a rather flexible overlapping system was devised consisting of core regions or *polos de crecimiento* (based on earlier studies available), provinces (as the fundamental administrative units of the national Government), groups of provinces (for more rational development programming), and special problem regions. In brief, every attempt was made to avoid regional boundaries that would, by their very rigidity, constrain social and economic change. From time to time, the existing pattern would be reassessed so that it might be adapted more properly to the currently prevailing conditions.[15] The techniques used for identifying regions were rudimentary, consisting mainly of analysis of data relating to regional characteristics (to calculate indices of similarity) and other data bearing on city-hinterland relationships for the more important centers. Practical feasibility for the administration of development was an additional consideration.

A second major task was to establish for each group of provinces, composing Chile's basic development regions, a special programming staff or ORPLAN (Oficina Regional de Planificacion), that would act as a local version of the national planning office. Many problems were encountered in carrying out this decision, but the creation of these offices was primarily an administrative measure decided on grounds quite independent of regional science.[16]

One of the main expectations from regional planning at the national level is the setting of investment priorities by geographic area.[17] Unfortunately, the prospects for formulating a satisfactory set of universal criteria for investment location are not very promising.[18] In Chile, a method has been evolved that avoids the logical problems inherent in the choice of a rationally consistent set of criteria. As a first step, this involved the selection of regions where development efforts should be relatively concentrated; the rest of the country would be treated as a residual category. Two types of areas were considered: (1) core regions with good prospects for rapid expansion in the near future, and (2) special problem areas. Again, regional science offered no practical guidelines for defining points of maximum development potential across the space of the national economy. .The concept of market potential seemed too narrow an approach, especially where much of the development within regions would initially be resource-oriented.

The second step involved the analysis in depth of the socio-economic situation in a few regions. In this analysis, no new ground was broken; reliance was placed instead on conventional methods. More sophisticated approaches are now being considered, however.[19]

The final step has to do with influencing actual investment choices. Here, two procedural instruments were involved. The first refers to the creation by the President of the Republic of an interministerial investment coordinating committee, with the Regional Planning Division of ODEPLAN serving as the secretariat. Initially, the purpose of this committee was to make recommendations, on the basis of regional information and other considerations, concerning relative investment priorities when budget cuts become inevitable. Of greater potential significance was a decision to present the national capital budget not only in administrative terms but also by major development sectors, regardless of the ministry to which the funds are finally allotted. This form of budget presentation greatly facilitated a regional breakdown of capital expenditures and eventually led to an integration of regional with sectoral investment programming.

One of the major problems encountered in setting regional priorities was the absence of a satisfactory regional accounts framework. Regional product estimates for Chile exist for 1960 and 1964; however, the probable margin of error for individual totals lies between 20 and 25 percent. With data so inaccurate, even relative levels of regional development are hard to assess, not to mention the meaning of annual fluctuations. Although work on accounts is continuing, explorations are currently under way to devise partial sets of development indicators for regional economies. Some of these indicators should be useful in predicting short-term changes in regional product, others would be more useful in anticipating secular trends.

The last major planning effort during the first few years was concerned with project identification.[20] The relative shortage in Chile is not capital but competently drawn-up investment projects. This shortage is felt especially at the level of medium-sized and small projects in the productive sectors. At the same time, it was indispensable that regional planning should move to practical results quickly by identifying projects to be undertaken in specified regions. For this, it was necessary to create a project-generating mechanism at the regional level with direct links to the executive agencies primarily responsible for formulating projects and for implementing or promoting them. This critically important function was built into the structure of the new regional planning offices from the start, and an ODEPLAN section at the national level was designated to work with these offices in the identification of investment opportunities, to bring them to the point where a ministry would begin to become interested in the proposal.

Summarizing, it can be seen that the major activities of regional planning in Chile during its starting-up period could make only limited use of the rigorous quantitative methods which are the hallmark of regional science. Absence of an intellectual tradition in this new field, at least within Chile, was one reason for this failure to use advanced methods. Pressure of time and inadequate data were others. But the most persuasive reason of all was the fact that the kind of questions investigated by regional planners were either of little interest to regional scientists (e.g., an institutional framework for regional planning) or had so far not yielded to their efforts (e.g., identification of points of maximum development potential).

The Phase of Gaining Professional Acceptance

Political acceptance of regional planning is not enough, however. An internal battle of survival must also be fought within the central planning agency where econometricians are arrayed against an interdisciplinary group of regional planners who work largely without formal models and who are as much oriented to implementation as they are to elaborating plans. The econometricians tend to regard the regionalists as amateurs. They themselves take an excessive pride in mathematical calculations (which, however, do not explicitly take into account the often substantial error in their basic data).[21] In point of fact, neither group can survive for long without some compromise between mathematical rigor and pragmatic approaches.

For the regionalists in Chile, a period is approaching in which professionalism must be proved in tough competition. Painstaking quantitative analysis is the obvious response to the demands of the new period. Regional growth models must be developed as counterparts to the sectoral models, and the two must be combined in a national framework. Inter-regional effects of investment must be traced with precision. Investment guidelines must be stated with such accuracy that they can be unambiguously related to a regional development path. Alternative regional policies must be tested for their effects on target variables.

It is not clear whether the methodology exists to provide answers to historically relevant questions. But the survival of the regional planning function at the national level probably depends on it.

CHILE'S REGIONAL PLANNING–PROSPECTS AND INTERPRETATION

The Chilean situation may be seen as typical not only in terms of the regional planning experience, but also for the possible significance of regional science in public affairs. It is so, because the regional question is most serious for economies, such as Chile's, that are transitional to industrialism.[22] In pre-industrial, industrial, and post-industrial economies, questions of spatial structure fail to have quite the import they carry in societies that are faced with the necessity of building a modern nation. These societies generate most of the active demand for regional planning and regional science, and if the latter wishes to survive as something more than the private sport of a few academicians independent of external support, this demand must be taken seriously.

Regional planning in Chile began as a form of innovative planning and has moved increasingly in the direction of planning for allocation.[23] In the earlier stages, the interest of Chilean planners was primarily to create those institutions that would permit the Government to advance towards a series of vaguely perceived and rather badly understood objectives, including administrative decentralization, economic deconcentration, and national integration. Progress along these lines was made through the establishment within the National Planning Office of a fully staffed Regional Planning Division and the creation of a number of planning offices at the level of administrative regions. At the same time, pressures developed that obliged Chile's planners to regard resource allocation as a problem of growing importance. Indeed, the justification for regional planning as a part of national economic planning was to a large degree found in the somewhat ambitious claim of regionalists that they knew how to arrive at a regional allocation of investments (both on an annual and on a multi-year basis) that was optimal from a national standpoint and would simultaneously promote the growth of national production, the reduction of urban unemployment, a politically defensible balance among regions, a strengthening of local economies, a slowdown of expansion in the Santiago area, and a distribution of social infrastructure investments that would be closely related to economic objectives.

Initially, Chilean planners looked upon regional science chiefly as a fund of received doctrine. There was no time for complicated and possibly inconclusive studies. Whatever data existed about regional economies had to be collected laboriously from many, often obscure, Government offices, but the compilation and analytical treatment of these data was rather simple.

Verified truth was all too frequently replaced by mere assertion. Improvisation and intuition filled in where information was unavailable. Experts were imported to bear sacred truths acquired at the fount of knowledge in distant countries overseas. Their message was at first not questioned. But what served these experts as truth was, indeed, little more than the generally accepted body of doctrine which informs nearly all regional analysis. They expounded the concept of the region as an open economy linked with other regions into a system, they proposed the simultaneous use of several regional types (administrative, development, and nodal), they insisted on the influence which the costs of overcoming economic distance had on the location of activities and on their productivity, they taught the significance of gravity fields and urbanization economies for increasing production and emphasized the key role of core regions in the development process, they introduced the notions of hierarchy and functional arrangement of cities and applied what was then known of regional growth theory.

For a time, this fund of doctrine and regional folklore served Chile's needs quite well, but the shift to allocative planning made doctrine gradually yield to technique. For neither foreign experts nor Chilean planners were really sure that what they preached had any relevance to the domestic situation. They wanted detailed studies of the Chilean space economy, its evolution, and its structural strengths and weaknesses. Is it true or not, for instance, that the productivity of agriculture declines with geographic distance from cities? This hypothesis is widely held, but the only supporting data are from the United States.[24] And if it did not hold for Chile, why not, and with what exceptions? At the same time, specific needs arose, such as for regional projections and the forecast of the local effects of government programs. Inter-regional models, perhaps of the input-output type, would be needed for this purpose. Finally, the ultimate legitimation of the regional planning function would come from the use of quantitative methods in analysis: especially mathematically complex and elegant ones. For regional planning had to pass the test not only of political usefulness but also of intellectual respectability. The time was approaching when a major push in the direction of regional economic research had to be made.[25]

Although planning and science appear thus to be drawn more closely together into a working partnership, it has to be admitted that there continue to be areas of substantial, if latent, conflict between them. These arise mainly for the following reasons.

We may start, perhaps, with the assertion that regional planners normally use only a small percentage of the techniques and propositional knowledge of regional science; there are no statistics, of course, but a fair guess would be that the proportion of regional science knowledge actively used in policy and planning lies between 10 and 20 percent. The bulk of academic literature on regional science remains irrelevant for practical purposes. This may be shocking news, though most planners would probably agree with it. Planning always operates under severe constraints of time and skills and funds. Only the simplest analytical techniques can be used: those with the least data requirements that can be handled by a technical staff whose educational level rarely exceeds a college undergraduate degree. Involved and complicated studies, while they may add prestige to the sponsoring agency, seldom yield results that are useful for day-to-day decisions. A country the size of Chile would need at least twenty regional economists at the master's degree and doctoral levels to support the ongoing planning activities with appropriate studies. None exists today, and it would take nearly a decade to train all of them.[26] Time and resource restrictions are equally serious. These difficulties are sometimes avoided, it is thought, by concentrating on research to the exclusion of other interests, so that planning becomes practically synonymous with research. The planner thus withdraws from the daily toss and tumble of decisions and focuses on what he understands to be major policy issues, preparing basic "alternatives and consequences" for those he identifies as top-level decision makers. But this image of planning is false. To be effective in influencing policy, the planner cannot remain aloof. He must press for particular solutions, involving himself in the daily work of keeping an organization going. He must be always on the lookout for ways and means to get action. Shielding the research staff from contact with immediate problems is one way of linking research and planning; no such device is successful. Priorities are constantly rearranged (the "daily blueprint"), and this alone makes adherence to long-range basic research exceedingly difficult.

A second reason why planners neglect so much of what is done in regional science is its lack of relevance to the questions they must answer. Most regional models are expressed in equilibrium terms, while planners need models for change in which the variables that can be controlled are explicitly identified and the objectives clearly stated.[27] Planners continue to look in vain for guidance in several basic tasks:

(1) rigorous identification of core regions at different levels of importance in terms of their potential contribution to the national economy;
(2) study of the effects of regional infrastructure investments on local production;
(3) measurement of urbanization economies for different categories of economic activity;
(4) definition of optimal boundaries for sets of development regions;
(5) prediction of the volume and direction of migration flows which can be expected to follow upon a given spatial distribution of investments;
(6) estimation of location responses to differential financial and infrastructure investment incentives;
(7) measurement of the impact of alternative settlement schemes on long term economic growth and political stability;
(8) measurement of the relation of agricultural to urban development under imperfect market conditions;
(9) execution of cost-benefit analyses for alternate urbanization patterns; and
(10) construction of predictive models of regional and interregional growth.

These are by no means all the important questions that remain unanswered, but they do serve to underscore the need to reorient research in the field.

The last reason for the parting of planning and science is found (as noted above) in the fact that planners are not exclusively dedicated to analysis but must spend much of their time in intellectually disciplined but nonetheless political action. A dialogue must be sustained; new legislation must be drafted; doubters and opponents must be persuaded. Above all, adequate resources must be obtained. These actions, though usually focused on only a few men, will usually require a large technical staff for back-up work not accepted as research by academicians. Regional planning is, therefore, a process that includes science but goes beyond it in seeking to transform existing patterns of spatial relationship in accord with social values.

CONCLUSION

We have shown the parting of ways for regional scientists and planners. Now, let us suggest how they might approach each other and work together on a common task. Indeed, regional scientists have been urged by one of their presidents to evaluate their work not only on the basis of its "internal merits" (perfection of models and formal methods of verification), but also from the standpoint of its "external merits" (its utility to the world at large).[28]

Historically speaking, regional science has developed rapidly in countries that were little prepared to implement related planning efforts on a national scale. In the United States, for instance, rich data sources could be mined and sophisticated models built without anyone seriously raising the question of immediate practicality. In spite of some research for government agencies, regional scientists continued to operate largely without the benefit of testing and application.[29] At the same time, the greatest potential consumers of regional science knowledge are the industrializing countries where economic changes bring about vast geographic shifts in population and activities. This imbalance between excess supply and unsatisfied demand is gradually being resolved by the transformation of regional science into a worldwide movement and by international technical assistance.

Experience in Chile suggests that regional science can make the greatest contributions where the institution-building phase of planning is well advanced and only professional acceptance remains to be gained. This lag means precious time gained for training local regional scientists and initiating certain kinds of basic research. On the other hand, it also means that during the initial phases, more regional scientists from countries where research and teaching traditions have been longer established will have to serve in the field as planning advisers. This will not only provide for more rapid feedback from practice to theory, but will also offer a splendid opportunity for comparative, cross-cultural research.

Thus, a nucleus of national scientific workers on regional problems will spring into being just when demands for sophisticated research will start to be generated.

It may be appropriate to call attention to possible relationships between the universities and planning agencies. Traditionally, regional science has been securely harbored in academia. Recently, however, planning agencies have themselves become engaged in various types of fundamental research. Whatever the reason (the planners' motivations may play an important role), this excursion into research has a cost: less time for planning. Loose but continuing contractual relationships with university institutes would resolve this problem; where these are lacking, pressure for the results of research should be used as a lever to encourage the creation of such institutes.

Such a relationship would have two important consequences. It would encourage regional scientists to expand their interests to problems of institutional, political, administrative, and technological development which arise with particular force during the introductory phases of a nation's regional planning efforts. These matters now tend to be left to sociologists, political scientists, and others. Yet in dynamic models of regional development, it will be difficult to avoid these questions. Just as economics is once more becoming political economy, so regional science would do well to become one of the so-called policy sciences.

The contractual relationship would also provide a bridge that would make it easier for academicians to turn into planners temporarily, and vice versa. Often, and especially in Europe, it is common for the two groups to remain coolly aloof. The result for both is isolation and sterility. Such status boundaries become increasingly anachronistic with the rate of scientific change. This rapid amortization of knowledge makes it imperative for practicing planners to return to school at fairly frequent intervals at the same time that it obliges regional scientists to step into the spotlight as planners under real-life conditions. In the process both thought and action will be improved.

NOTES

1. M. M. Webber, "Culture, Territoriality, and the Elastic Mile," Papers of the Regional Science Association, vol. 13, 1964, pp. 59-70. The ambiguity of the distance concept is also brought out in Gunnar Olsson, *Distance and Human Interaction; A Review and Bibliography,* (Bibliography Series No. 2). Philadelphia: Regional Science Research Institute, 1965.

2. The importance of regional economic policy in public policies generally is clearly made evident in two recent publications: *Area Redevelopment Policies in Britain and the Countries of the Common Market,* U.S. Department of Commerce, Area Redevelopment Administration, Washington, GOP, 1965; and E. S. Kirschen and L. Morissens, "The Objectives and Instruments of Economic Policy," in Bert G. Kickmann, ed., *Quantitative Planning of Economic Policy.* Washington: The Brookings Institute, 1965.

3. L. Rodwin, "Regional Science: Quo Vadis?" Papers and Proceedings of the Regional Science Association, vol. 5, 1959, pp. 3-20.

4. Walter Isard et al., *Methods of Regional Analysis: an Introduction to Regional Science.* Cambridge, Mass.: and New York, Technology Press and Wiley, 1960.

5. R. Vining, "An Outline of a Stochastic Model for the Study of the Spatial Structure and Development of a Human Population System," Papers of the Regional Science Association, vol. 13, 1964, pp. 15-40; and B.J.L. Berry, "Cities as Systems within Systems of Cities," ibid., pp. 147-164, also reprinted in John Friedmann and William Alonso, eds., *Regional Development and Planning: A Reader.* Cambridge, Mass.: The M.I.T. Press, 1964.

6. Much of this literature is expertly reviewed by Gunnar Olsson, op. cit.

7. The relevant literature is summarized in: Friedmann and Alonso, op. cit.; and in John Friedmann, *Regional Development Policy, A Case Study of Venezuela.* Cambridge, Mass.: The M.I.T. Press, 1966.

8. Walter Christaller, *Die Zentralen Orte in Süddeutschland.* Jena, Gustav Fischer, 1933. For later elaborations, see items listed in: Brian J. L. Berry and Allen Pred, *Central Place Studies,* (Bibliography Series no. 1). Philadelphia: Regional Science Research Institute, 1965, 2nd edition with Supplement through 1964.

9. John Friedmann and John Miller, "The Urban Field" Journal of the American Institute of Planners, vol. 31, no. 4, Nov. 1965, pp. 312-320.

10. The concept of hierarchy is rapidly disappearing from organization literature. See, e.g.: Saul M. Katz, *A Systems Approach to Development Administration,* (Papers in Comparative Public Administration, Special Series no. 6), American Society for Public Administration, Comparative Administration Group, 1965; B. M. Gross, "The Managers of National Economic Change," in Roscoe C. Martin, ed., *Public Administration and Democracy, Essays in Honor of Paul H. Appleby.* Syracuse: Syracuse University Press, 1965; W. G. Bennis, "A New Role for the Behavioral Sciences: Effecting Organizational Change," Administrative Sciences Quarterly vol. 8, no. 2, Sept. 1963, pp. 125-165; and A. O. Hirschman, "Models of Reform-mongering," Quarterly Journal of Economics, vol. 77, May 1963, pp. 236-257.

11. Recent evidence supports this view. See Barclay G. Jones and Jon T. Lang, *Studies in Regional Development: Population Activities, and Incomes in Chenango, Delaware, and Otsego Counties.* Ithaca, N.Y.: Center for Housing and Environmental Studies, Division of Urban Studies, Cornell University, 1965, pp. 119ff.

12. A. Altschuler, "Rationality and Influence in the Public Services," Public Administration Review, September 1965.

13. Altschuler's article, op. cit., is one of the few serious attempts in this direction. Perceptive comments on the problem are also found in Joseph La Palombara, ed., *Bureaucracy and Political Development.* Princeton: Princeton University Press, 1963, passim. However, a full-scale study of the problem remains to be written.

14. A. Marfan, "Desarrollo y fomento regionales," Industria (Monthly publication of the Sociedad de Fomento Fabril), no. 77, September 1960, pp. 53-59.

15. See Friedmann, *Regional Development Policy,* op. cit., ch. 3, for further discussion.

16. One purpose was to devise an optimal set of regions for the administration of development. This subject has been of some interest to regional scientists, but the contributions to date, although suggestive, have not as yet been made operational. See: W. Isard, "Spatial Organization and Regional Planning: Some Hypotheses for Empirical Testing," Papers and Proceedings of the First Far East Conference of the Regional Science Association. Tokyo, 1965, pp. 3-23. Also P. Ylvisaker, "Some Criteria for a Proper Areal Division of Governmental Powers," in Friedmann and Alonso, eds., op. cit., pp. 519-541.

17. Lloyd Rodwin, "Choosing Regions for Development," in Friedmann and Alonso, eds., op. cit., pp. 37-58.

18. T. A. Reiner, "Subnational and National Planning: Decision Criteria," Papers of the Regional Science Association, vol. 14, 1964.

19. A. Ganz, "Regional Planning as a Key to the Present Stage of Economic Development of Latin America: The Case of the Guayana Region, A Frontier Region," Cuadernos de la Sociedad Venezolana de Planificación, Special Issue, September 1963 (Papers presented to the First Latin American Conference on Regional Science held at CENDES, Caracas, November 1962), pp. 117-127.

20. *Manual of Economic Development Projects,* New York, United Nations, 1958, Part I, "The Substance of a Project."

21. Oskar Morgenstern, *On the Accuracy of Economic Observations.* Princeton: Princteon University Press, 1963, 2nd ed., ch. 14.

22. Friedmann, *Regional Development Policy,* op. cit., ch. 1.

23. For definitions of these two concepts of planning, see: J. Friedmann, "Planning as Innovation: the Chilean Experience," Journal of the American Institute of Planners, vol. XXXII, no. 4 (July 1966), pp. 194-203.

24. William B. Nicholls, "Industrialization, Factor Markets, and Agricultural Development," in: Friedmann and Alonso, eds., op. cit.

25. The reverse sequence—from research to action—would have been sure to founder. In the absence of an action program, there would have been no client.

26. At least 20 regional economists at the M.A. and Ph.D. level would be required. We assume 5 students entering a program of at least 2 years' duration each year, and an attrition rate of 50 percent (this includes successful students who do not work for the Government). It should be noted that 5 students embarking annually upon an advanced regional science program would be a high average figure. For instance, at the post-graduate ESCOLATINA of the Economics Faculty of Chile, there are 30 students in the first, preparatory year, and 15 in the second. These students are drawn from all of Latin America. They concentrate in two fields: economic development and agricultural economics. Chilean students represent only a fraction of the total, and it is questionable whether more than one or two Chilean economists at ESCOLATINA could be persuaded to study regional economics intensively over a two-year period. This situation holds for the nation's only graduate school of economics.

27. Jan Tinbergen, *Economic Policy: Principles and Design.* Amsterdam, North Holland: 1956. Roland Artle's recent work on the Hawaiian economy represents a significant and interesting departure from the basic Tinbergen model.

28. W. L. Garrison, "Values of Regional Science," Papers, Regional Science Association, vol. 13, 1964, pp. 7-11.

29. J. Friedmann, "Regional Development in Post-Industrial Society," Journal of the American Institute of Planners, vol. XXX, no. 2 (May 1964), pp. 84-90.

Chapter 14

AN APPROACH TO THE DEVELOPMENT OF NEW CORE REGIONS: THE GUAYANA PROGRAM IN A REGIONAL PERSPECTIVE

PREFATORY NOTE

Among the more attractive policy ideas for regional development is the activation of new "growth poles" in a country's national periphery. As used by planners, "growth pole" refers to an urban center of some size which has a basis for sustained and rapid economic expansion. In the terminology of the present volume, *core region* has been used with an essentially similar meaning. Common to all policies aimed at the activation of core regions is an industrial complex or set of dynamic manufacturing industries which, because of their multiple backward and forward linkages in an urban economy, creates conditions favorable to the location of additional industries and services in the same area.

The theoretical justification of a national core region policy is that (1) the economic potential of peripheral areas is underutilized in the absence of peripherally located, if subsidiary core regions, (2) core-region development is economically more efficient than a policy of spatially decentralized industrial development, (3) the process of innovation diffusion into the national periphery is accelerated by encouraging core-region growth, and (4) the social benefits of core-region growth will filter down more rapidly to peripheral populations than would otherwise be the case. In more general terms, core-region policies are expected to promote the social and economic integration of peripheral populations into a developing national system.

Ideally, this process passes through three major phases. In Phase I (mobilization), resources and technology are transferred from the existing national core area to the periphery. Heavy investments are made in infrastructure and directly productive activities at the site of the new core region. The incipient industrial center initially exports the bulk of its production to core-region markets outside its own boundaries. In Phase II (growth), additional resources (chiefly in the form of labor and raw materials) are transferred to the new core region from its own emerging periphery. With the expansion of income in the region, an increasing proportion of its production is marketed locally, encouraging the establishment of small manufacturing and service firms. At the same time, and because of closer and more frequent contacts of peripheral populations with the new core, the flow of urban innovations into the core's periphery is accelerated. In Phase III (consolidation), growing regional specialization occurs, as the new core area economy becomes successively more integrated with the rest of the country's urban system, net transfers of resources from the older core region decline to zero, and the new core passes the threshold of self-sustaining growth. The drainage effects of its own development are diminished as the new core region begins to shift resources (chiefly capital) into its dependent periphery, thereby initiating a cycle of expansion at new locations.

This succession of phases is by no means inevitable. The mobilization phase may break down; the growth phase may further drain an already weak periphery, exacerbating its problems of internal adjustment; and the consolidation phase may present urban planners with insurmountable problems of accelerated population and economic growth.

Despite the great popularity of core-region development as a national policy, scarcely any empirical studies exist to sustain its major theses. The best and most complete study to date has been done for the city of Guayana in southeastern Venezuela from which the present chapter has been extracted.[1] An area blessed with a great wealth of natural resources, the site of the present city had only about 4,000 inhabitants in 1950. Some twenty-odd years later, it has become Venezuela's principal center of heavy industry, has productively absorbed a population of nearly 200,000, and is at present a major exporter of intermediate products to both national and international markets. This result, amazing for a country that less than a generation ago was predominantly rural in outlook and subsisted chiefly on the export of crude oil, was the result of a national policy and a concerted long-range program of direct government intervention. The full story of this development saga is told elsewhere.[2] In this chapter, I wish to analyze a number of strategic choices that were made with the intention of tracing their effects on the development of this new core region. I believe that something may be learned from this analysis that goes beyond the particular historical circumstances surrounding the creation of Ciudad Guayana.

REGIONAL DEVELOPMENT: CONTRASTING EXPERIENCES

What can be learned from the Guayana experiment in regional resource development? Is it a unique experiment that bears no special lessons for other planners, or will it become, like the Tennessee Valley program, a model inspiring similar efforts in other parts of the world? Its similarities to the Tennessee Valley effort are in many ways striking. Here is the same bold scale, the same determination to bring under a central authority of the national government the management of natural resources within a broad geographic area primarily for economic ends. But the contrasts between the two gigantic undertakings are more interesting. A comparative analysis may lead to a clearer understanding of the Guayana's important contribution to the theory and practice of regional planning. The relevant points of divergence between the Guayana and Tennessee Valley programs may be summarized as follows:[3]

Guayana	*Tennessee Valley*
Purpose	
Economic development of the nation through the coordinated development of all resources, including but not limited to water, in a relatively new and sparsely populated region; regional consequences considered secondary to the national significance of the program.	Economic development of a depressed region through the coordinated development of primarily its water resources for a number of related uses and appropriate land management; national consequences considered secondary to the regional significance of the program.
The region	
A remote resource frontier loosely defined geographically but focused upon a new and rapidly expanding urban-industrial complex within a national economy in the process of industrializing.	An old and economically backward area defined as a river basin and as a power service area, within an already industrialized and technologically advanced national economy.
Program emphasis	
Manufacturing industry, electric power, urban development, mining, transportation, agricultural development.	Water and land resource development and conservation, agricultural and rural development, forestry, electric power, river navigation, tourism.[4]
Local participation	
A national design imposed upon the region with only minor participation of local, state, or regional interests; political support sought and obtained on a national basis.	Importance of local, state and regional participation emphasized as a matter of doctrine; major efforts to strengthen local institutions and to obtain political support within the region.
Planning style of development agency	
Planning institutionalized; long-range planning studies for both the economic and the urban-physical development of the region.	Informal planning only; absence of long-range planning studies for the economic growth of the region; such planning left to the states and to private institutions.

Guayana	Tennessee Valley
Performance criteria	
Number of new industrial plants; industrial production, exports, and total employment; regional population growth; integration of the region with national economic development.	Financial solvency and economic efficiency of development authority; physical measures of increase in regional output (electric power, river traffic, acres, reforested, etc.); electric power consumption per capita; growth of regional per capita income relative to that of the nation.

These contrasts in the approach to regional development in the Guayana and the Tennessee Valley are remarkable. They reflect differences primarily in the structural conditions for planning. The Guayana is an emerging frontier region in a country with a unitary government and an economy that is still transitional to industrialism. Administration is highly centralized; local participation in public decision making has not yet taken root; and the private sector still lacks a strong institutional base. The Tennessee Valley, on the other hand, was a backward regional economy with a long history of settlement, a large agricultural sector, and a substantial urban population divided among a number of medium-sized and smaller cities and towns. The United States has a federal government in which both states and municipalities enjoy considerable autonomy. Its economy during the 1930's was already the most powerful in the world, and its private institutions were and continue to be imbued with a strong sense of independence.

Each program of regional development may therefore be regarded as a type. The Guayana experiment is an apparently successful response to the problems of resource frontier areas in developing economies organized as unitary states; the Tennessee Valley program represents an equally successful, if much older, effort to cope with the problems of regional economic backwardness in an advanced industrial society organized as a federation of states. The lessons of the Guayana appear for this reason particularly appropriate to countries in South America, Africa, and the Middle East, which offer similar conditions and in which resource frontiers may come to play a decisive part in national development.[5]

Resource frontiers are remote from their country's principal centers of population. They come into existence as the result of a commitment, usually by private firms, to exploit important natural resources or resource complexes on a commercial basis. Typically, they are isolated from the rest of the country by natural barriers to transportation such as deserts or mountain ranges; they are, at least initially, controlled by foreign concerns that operate the mining or other important economic establishments in the region; their economies are focused upon a city or quasi-urban settlement that, however, has very limited central service functions because of the great emptiness that surrounds it; their future depends to a great extent on the behavior of foreign markets and, in the longer term, on the strategic resource reserves that can economically be exploited; and they acquire a social system that, although it is extremely transitory, is yet highly stratified.

The Guayana region met all of these conditions when the government of Venezuela decided in 1958 to transform this distant frontier area into an integral part of the national economy, creating on the Orinoco River a major center of manufacturing based on iron ore, hardwood forests, hydroelectric power, natural gas, and petroleum. This move was attractive to Venezuelans not only because of the obvious wealth of the region's natural resources but also because of a general dissatisfaction with the continued growth of the national capital, Caracas, at the expense of the provincial periphery. Thus two important national goals were invoked in support of the proposed development scheme: the diversification and strengthening of the country's export base, and the regional decentralization of productive activities and population. Ciudad Guayana came to be regarded as a counterweight to Caracas and began to compete with the capital both for investment and for professional and managerial groups:

How this problem was attacked may be studied by analyzing a number of decisions that affected the course of the Guayana's development in a major way. Admittedly the choices were not always made with complete awareness of their implications or of all the relevant alternatives.

PLAN IMPLEMENTATION: COORDINATIVE VERSUS CORPORATE PLANNING

One important question was whether the development of the Guayana's resource economy could be carried out efficiently and expeditiously through the coordinated action of many public agencies or whether the complexity and scale of the task would require an organization endowed with broad executive powers. The choice is a familiar one in regional planning.

In Venezuela the government opted for the corporate solution and created the Corporación Venezolana de Guayana (CVG), vesting in it full powers to plan, program, and implement the investments that would have to be made. What were the reasons for this far-reaching choice? Some were subtle and intangible, yet extremely important to the ultimate success of the enterprise. First, a development corporation would underscore the determination of the new government to assign top priority to the Guayana program among all other claims pressing for attention. Second, decisive coordinated action could be more readily achieved, it was thought, by concentrating in a single institution all the necessary authority and economic power. Third, resources for regional development could probably be mobilized more readily through the agency of a corporation, with its comprehensive but exclusive responsibilities, than through a dispersed, haphazardly coordinated program involving many centers of power. Fourth, only a corporation would be able to take a global view of the program and of the strategic interrelations among mining, industry, urbanization, transportation, housing, agriculture, forestry, social services, and electric energy, to name only the major sectors that were to be coordinated.

In short, the very complexity of the program suggested the necessity for a concentration of powers capable of overcoming the many limitations inherent in traditional government bureaucracy. Experience with interagency coordination in Venezuela had rarely been happy and argued strongly for a centralized institutional structure to ensure rapid progress. The national planning agency (Oficina Central de Coordinación y Planificación–Cordiplan) or the national development corporation (Corporación Venezolana de Fomento–CVF) might indeed have been given a leading role in the over-all coordination, but Cordiplan was to be carefully isolated from executive responsibilities, and the scale of the Guayana program was too ambitious to be entrusted to the CVF as one of its many activities. Above all, in view of the traditional autonomy of the ministries, no one would have been clearly accountable for the program. Finally, a regional development corporation was probably thought to be more capable of dealing energetically with the powerful foreign mining concerns that were active in the area than a program in which responsibilities would be divided and whose ultimate authority would justifiably be regarded as uncertain.

REGIONAL BOUNDARIES: FIXED OR INDETERMINATE

A question of perpetual concern to regional planners is the delimitation of the area to be planned. In the case of the Guayana, the problem was to determine precisely the area within which the CVG would have authority to make decisions. The problem was complicated by the fact that there were no clear natural boundaries. The geographic concept of the Guayana itself was ambiguous, a loose designation for all of the area south of the Orinoco, encompassing two-thirds of the nation's land area. The river basin concept, on the other hand, would have been inappropriate in view of the DVG's concern with the development of resources other than water. In the absence of definite physical criteria, it was difficult not to be arbitrary in the drawing of boundaries.

A novel solution was finally found. A region is a time-space continuum. Why, then, invest it with Euclidean properties? From the outset it was evident that the major impact of the CVG would be on a small and tightly circumscribed zone in the area surrounding the confluence of the Caroní and Orinoco rivers. This would be the focus of the CVG, and it was carefully described by a coordinate system in the legislation establishing the Corporation (though the corresponding map was never published). Within this development zone (*zona de desarrollo)* the CVG would have complete authority. But–and this was the novelty–the Corporation was legally entitled to undertake investment programs *beyond* this zone if their relevance to activities within it could be demonstrated.

As a result of this decision, the CVG now operates a coal mine on the coast, some 250 miles from the *zona de desarrollo,* is helping to finance road-building projects throughout the eastern provinces, and has initiated a large land reclamation scheme in the Orinoco Delta. This ingenious device for regional delimitation serves to constrain the Corporation's powers by establishing a criterion of relevance, but it leaves the Corporation flexible to cope with problems that could not be foreseen at the time of the initial legislation. Above all, this device helps to concentrate the attention of CVG planners at the same time that it minimizes potential conflicts with provincial governments and national ministries.

REGIONAL STRATEGY: DEVELOPING A CORE REGION

From the beginning the *zona de desarrollo* was conceived primarily as an urban region of metropolitan size. In the language of regional theory, the decision was to create on the frontier of Venezuela's effectively settled space a core region (*pôle de croissance*) whose successful development was believed essential to the attainment of national objectives. Ciudad Guayana was expected not only to stimulate the economy of the Guayana region and the eastern provinces but also to help to integrate, as a functionally specialized element, the Guayana's wealth in resources with national development. The DVG's planners were therefore confronted with a unique and very difficult task: to coordinate the physical development of a new metropolis with the economic development of an extensive resource complex.

This task was defined as an empirical problem; to avoid it would have been impossible. Initial construction of the steel mill, the first of a series of hydroelectric dams on the Caroní River, and related facilities such as local access roads and workers' housing attracted large numbers of migrants from all parts of eastern Venezuela. Within a few years, the population of Ciudad Guayana increased more than tenfold. The additional investment in urban service facilities for this growing population reinforced the attractiveness of the core region for migrants from rural areas, who kept streaming into the region at the rate of about 1,000 a month. The rate at which new jobs became available was insufficient to absorb the steadily swelling labor force, and this created pressures for expanding the economic base of the community. Thus was set into motion a rapidly upward-turning spiral that contributed to the political urgency of the enterprise and gave to urban development an importance at least equal to that of economic programs in the region.

This interaction was more visible in the Guayana region than in Caracas, where the economic development function was obscured by the urgency to find solutions to the urban problems of this rapidly expanding metropolis. Housing and physical planning in Caracas were typically handled in gross ignorance of underlying economic forces, with results so clearly unsatisfactory that high priority was given to urban decentralization as an objective of national policy.

In the Guayana, on the other hand, the physical planning of the metropolis was done in close correspondence with the economic planning for the region. This was reflected in the distribution of types and location of houses, in the selection of the appropriate structure of community facilities, and in the main outlines of the internal circulation network. It should be emphasized, however, that the physical plan for Ciudad Guayana was not wholly derived from the economic plan for the region but made a significant contribution of its own. Interregional competition, especially with Caracas, suggested the need for creating in the Guayana an urban environment that would be able to attract and hold professional and managerial groups. The aim of urban development, therefore, was not merely to provide minimum facilities for the industrial labor force but to create a socially diversified community by attending to the needs of the middle class for cultural and physical amenities of a type that exceeded the functional minimum of a purely industrial city.

Three conclusions may be drawn concerning this close, interactive relationship between urban and economic development. First, the long-term growth of the city was set into motion by concentrating initially on the export of manufactured products from the region. The export of raw materials alone did not lead to urbanization. And, without a local market of sufficient size to support by itself an industrial structure capable of generating further expansion, the Guayana's economy became vitally dependent on the series of industrial projects that would, in the first instance, serve primarily international and national markets. To assure the competitive advantage of Ciudad Guayana as a location matrix therefore became a major objective of development policy for the region. This called for a careful adjustment of physical and economic planning to each other.

Second, the full potential of urban development could be realized only by going outside the urban nucleus itself and developing the natural resources in the surrounding areas on which the city's growth as a manufacturing center would have to depend. The natural resource wealth of the region would thus be brought together and organized by the city through its several industrial complexes. The interests of city and region were viewed as complementary to each other, and the city, acting through the agency of the CVG, would have to reach out and assure for itself an adequate supply of the basic materials on which its own progress depended. Equally important was the concentration upon the city of a network of transport routes that would channel raw materials and energy (roads, railroads, pipelines, high-tension transmission lines) to its industries and carry the city's products to other parts of the nation and to the world (waterways, highways, airports).

Third, the development of this core region would have a major impact on other areas of the nation. Thus, the rural areas from which the city obtained its population would be drained of much of their manpower and would have to readjust their local economies accordingly. Complementary market-oriented industries and services would have to be established in other urban centers, particularly along the coastal belt in the eastern provinces, which would obtain from the Guayana intermediate products for final processing. Little attention was given to this phase of the regional development effort, partly because the problem was regarded as lying outside the specific responsibilities of the CVG. In the absence of a national policy for regional development and with an intensification of regional development efforts in other parts of the country, full advantage could not be taken of the Guayana program. This problem will be discussed more extensively later in this chapter.

LOCATIONAL EFFICIENCY: PROJECTS VERSUS PROGRAMS

The locational strategy of the Guayana program was clearly aimed at maximizing so-called external economies that might be obtained from the coordinated implementation of a group of related industrial projects. Since it is practically impossible to measure external economies with any degree of precision, this strategy was more an act of faith than a fully rational determination. The alternative would have been a fragmented approach in which each project would be separately evaluated in relation to similar potential projects in other locations. What would have been the optimal location for a steel mill? Which would have been more economical: a number of small thermal plants using natural gas and located in the vicinity of the large consuming centers, or Guri Dam, with its power potential of several million kilowatts? What port improvements in what locations would have the largest payoffs for Venezuela's economy as a whole? It might well be true that some of the investments that eventually came to be located in Ciudad Guayana would have been better located elsewhere, had each determination been made separately and independently of all other projects. But a dispersal of the individual components that made up the Guayana program would have left the region more or less as it had been during the early fifties: a typical resource frontier dominated by foreign interests, an exclave mining economy on the periphery of the nation. The mobilization of resources for projects unrelated to a regional program would have been more difficult. Internal political forces might have been set into motion that, ultimately, could have fought each other to a standoff. A new core region might not have come into being, and possibly fewer resources might have been incorporated into the stream of national production.

In the Guayana program we have a case where the locational efficiencies of its separate projects can be measured only after the fact, for the maximum benefit from externalities can obviously be obtained only when the program has attained maturity and all of its component parts are in place. The Guyana scheme must therefore be viewed as primarily a program for establishing the conditions for efficient industrial location. The ultimate test will be the vitality and capacity for further growth of Ciudad Guayana and its region once public investments decline to levels comparable to those in other Venezuelan cities.

PLANNING STRATEGY: IMBALANCE VERSUS BALANCE

The general planning strategy for the Guayana was closely related to the decision on locational strategy, with its emphasis on external and agglomeration economies. The two important alternatives were imbalance or balance in the development of the different sectors of the regional economy. The case for imbalance has been effectively argued by Albert Hirschman and that for balance by Paul Rosenstein-Rodan in his theory of the "Big Push."[6] A strategy of imbalance would permit a development agency to concentrate its resources on a few key sectors or projects without taking into account complementary investments. The resulting imbalances would, according to Hirschman's theory, create political pressures for action on other fronts and eventually lead to a reallocation of resources on an emergency basis. Thus the pendulum would swing back and forth from crisis to crisis, all the while maintaining a high degree of forward momentum.

It is doubtful whether such a strategy would have worked in the Guayana, where the CVG controlled all essential elements of the program. In the absence of "countervailing powers," imbalances might very well have remained imbalances as the Corporation strove for dramatic breakthroughs in a few key sectors—say, the steel mill and Guri Dam. The CVG was in fact created precisely for the purpose of doing "balanced" planning, since it was feared that the imbalances created by alternative organizational arrangements would *not* be self-correcting. It was primarily for this reason that the CVG instituted a formal planning process whose aim was the simultaneous advance on a number of roughly balanced fronts. (The TVA, acting in a richer institutional environment, could dispense with formal planning and internal balances.) By rapidly building a balanced physical and economic environment, the CVG hoped to engender the dynamism that would assure the sustained growth of the region, create an efficient location matrix, and exude the magic of success so necessary to an enterprise on the scale of the Guayana. On the basis of this experience, it may be suggested that the theory of the "Big Push" probably has greater application to the activation of core regions in transitional economies than to national economic development with its more complex institutional setting.

CONTROL: ORDER VERSUS GROWTH

The approach to planning through a series of internal balances automatically gives preference to the aesthetics of order over the dynamics of rapid growth. If every component of a balanced program were to be put in its right place and its timing determined with precision, then indeed the whole arrangement might be subjected to an aesthetic criterion that would render the plan exceedingly rigid and resistant to modification in the course of its implementation. The problem may be reduced to a question of control: to what extent should spontaneous forces be controlled in the interest of harmony and order? During the early years of the CVG the desire for controlled order was very great; it reflected the typical planning bias in favor of equilibrium arrangements intended to maximize community welfare in the long run. But soon the impossibility of reconciling order with growth became apparent, as the several elements of the program refused to fall into place in accordance with the plan. The desire for total control had therefore to be reduced to a desire for *strategic* controls. As a result, planning became less and less precise as a means for imposing a preconceived scheme but gained importance as a general framework for making decisions. Neither the actual physical form of the city nor the temporal sequence of industrial investments could be determined beforehand by the CVG's planners; their planning had to reflect the spontaneously evolving reality of the region's economy. If balanced or comprehensive planning contributed to the institutionalization of the planning function within the CVG and argued for the full exercise of formal powers of control, the exigencies of spontaneous growth steadily gnawed away at planned formalisms. The CVG's planners were unable to control decisions from the top; they had to contribute to decisions with the specialized knowledge they acquired from their studies and long-range projections.

INTERREGIONAL RELATIONS: IMPROVISATION VERSUS NATIONAL POLICY

One unanticipated consequence of the Guayana program was the intense interest it generated throughout Venezuela in regional development planning. There was an awakening of regional consciousness and of a desire on the part of provincial interests to participate more actively in the great task of national transformation. The CVG was so obviously successful in concentrating financial resources for the development of regional resources that it was natural for other parts of the country to wish to imitate this experience and to put pressure on the central government for the creation of development corporations with equally extensive powers. It was obvious that this was done without sufficiently differentiating among regional problems. While the CVG appeared to be an eminently successful means for integrating a remote resource frontier into the national economy, the problems of backward, downward-transitional areas such as the Andean region might call for a totally different approach. The national planning office, Cordiplan, resisted these pressures as much as it could, because it feared that the extension of the regional planning concept to other areas would divert energies from the main tasks of national development. On the other hand, it lacked a clear vision of the regional economy of Venezuela, of the country's urban structure in terms of functional specialization, and of the close interrelationships among the problems of regional development. It failed to see, for example, that the backwardness of the coastal belt in the eastern provinces might be aggravated by the Guayana program farther to the south or that this program might be used in combination with other efforts to overcome the traditional economic problems of that region. What was lacking, then, was a national policy for regional development that might be used to guide the newly found enthusiasm for regional development into constructive paths and to forestall such obviously mistaken measures as the creation of the Corporación de los Andes on the pattern of the CVG.

The difficulties in implementing a national policy for regional development would indeed be numerous if, at the same time, one wished to avoid the multiplication of the development corporation concept as a means for overcoming them. For at any one time, a country the size of Venezuela can ill afford more than one concentrated and centrally administered regional development effort. The difficulties inherent in the normal processes of coordination led to the creation of the CVG; other regions, however, would have to be content with less imposing organizational arrangements. But even the first steps in this direction could not be taken until Cordiplan acquired the intimate knowledge of Venezuela's regional economies and the relations among them that would be needed to provide central guidance in this area of public policy. The failure to develop such expert knowledge left a critical gap between a national investment policy elaborated along sectoral lines and the implementation of specific projects. It also left Cordiplan incapable of exercising the leadership necessary to prevent regional development initiatives from working against the basic purposes of national development. The issue is still unresolved, but it appears that Cordiplan is taking a closer interest in the problem than heretofore and may be gradually acquiring the expertise needed to fill this vacuum of public policy.

CONCLUSIONS

The preceding analysis demonstrates, I believe, the relevance of the Guayana experiment to other nations. As a means for transforming a resource frontier into an integral part of the national economy and for activating a potential core region as part of this effort, the experiment must be considered a success that holds important lessons for regional planning theory as well as practice.

It would have been relatively easy for the program to go wrong. The critical points of decision were many, and the possibilities for mistakes—in the absence of a clear theory—were obviously great. One need only imagine what would have become of this experiment if alternative choices had been made:

(1) if loose coordination had been substituted for corporate planning
(2) if an essentially arbitrary spatial system of references had been permanently fixed in legislation, with the consequent diffusion of energies and the chance inclusion (or exclusion) of strategic resources
(3) if instead of concentrating on the activation of Ciudad Guayana as a core region, a number of focal points had been chosen simultaneously or, worse, if urbanization had not been brought into functional relation with resource development
(4) if instead of a regional development program, individual investment projects had been evaluated without taking into account mutual complementarities and external and agglomeration economies
(5) if a strategy of imbalance had been followed in the hope that imbalances would ultimately be self-correcting through the mechanism of political pressures or autonomous action on the part of private entrepreneurs and public institutions
(6) if planned order had been preferred to a more flexible framework of planning adapted to the exigencies of a rapidly changing situation.

It does not matter that some of these alternatives would have been inconsistent with others. One or two wrong choices would have been enough to reduce dramatically the effectiveness of the program.

In the Guayana experience, one idea seems to run as a theme through all the decisions. This is the idea of *integration:* the integration of the Guayana's empty spaces into the effectively settled space of the country through a functionally specialized core region, the integration of the urban economy of Ciudad Guayana, and the integration of urban physical development with planning for the city's economic base. *It would be quite appropriate to call this the master idea of regional planning.* Where it is ignored, or where ineffectual measures are taken to implement it, regional development planning in transitional societies cannot be expected to be successful.

NOTES

1. Lloyd Rodwin and Associates, *Planning Urban Growth and Regional Development: The Experience of the Guayana Program of Venezuela.* Cambridge, Mass.: The M.I.T. Press, 1969.

2. John Friedmann, *Regional Development Policy.* A Case Study of Venezuela. Cambridge, Mass.: The M.I.T. Press, 1966, chapters 7 and 8; and John Friedmann, *Venezuela: From Doctrine to Dialogue.* Syracuse: Syracuse University Press, 1965.

3. For a general review of the TVA experience see Roscoe C. Martin, ed., *TVA: The First Twenty Years.* Knoxville, Tenn.: The University of Tennessee Press, 1956. Also relevant in this context is John Friedmann, *The Spatial Structure of Economic Development in the Tennessee Valley.* Research Paper No. 39. The University of Chicago, Department of Geography, March 1955.

4. To the extent that the TVA has completed its task of multiple-purpose river development and has turned its attention primarily to the question of producing and selling electric power, it has become interested in industrial development. This interest, however, has remained essentially a passive one, and the Authority has continued to rely on state, local, and private groups to further the growth of manufacturing industries in the region. Similarly, the TVA has stimulated the regional coal mining industry during recent years as it has begun the construction and operation of a number of huge thermoelectric plants. But at no time did the TVA consider becoming actively involved in the development problems of the coal mining industry.

5. For further discussion of the differences between regional planning in transitional and in industrialized economies see John Friedmann, "Poor Regions and Poor Nations: Perspectives on the Problem of Appalachia," Southern Economics Journal, Vol. XXXII, No. 4 (April 1966), pp. 465-473.

6. Albert O. Hirschman, *The Strategy of Economic Development.* New Haven, Conn.: Yale University Press, 1958, and Paul Rosenstein-Rodan, "Notes on the Theory of the 'Big Push,'" in Howard S. Ellis and Henry C. Wallich, eds., *Economic Development for Latin America.* New York: St. Martin's Press, 1961, pp. 57-81.

Chapter 15

INTENTION AND REALITY: THE AMERICAN PLANNER OVERSEAS

City planning in America is to a great extent a homemade profession. Its evolution has run parallel to the course of the American city and, in the detail of its organization and practice, it reflects the character of our national institutions. In spite of this history, the profession has not remained completely sealed off from the outside world. Thanks to the pioneering efforts of a few dedicated individuals,[1] foreign students have been accepted in increasing numbers at leading planning schools, some research has been started, and foreign assistance agencies have slowly come to realize that accelerated urbanization in the newly industrializing countries requires expert attention.[2]

These new directions notwithstanding, the overall impact of overseas development on the American planning profession has been relatively modest. As late as 1967, students from developing countries accounted for less than 10 percent of the total enrollment in graduate planning schools; the percentage of articles devoted to the subject in the leading journal of the profession was less than 5 percent during a recent three-year period; and short-term consulting assignments abroad have probably involved no more than 150 professionals on brief, sporadic missions and perhaps less than thirty planners on a more or less continuing basis.

Under the leadership of American planners, a few ambitious ventures in overseas development were started in Lima, Ciudad Guayana, Calcutta, Santiago, and Ljubljana, but these programs employed fewer city planners than professionals from other fields. In Chile, for example, only four out of fourteen advisers stationed in Santiago were degree-holding planners; the rest were architects, economists, engineers, computer scientists, anthropologists, and sociologists. From this evidence one is forced to conclude that overseas experience has, on the whole, remained tangential to the main thrust of the profession.

Nevertheless, some American planners do get to work abroad for varying periods of time, and an even larger number of foreign students come to study planning at American universities. The questions, then, I wish to pose are these: How effective are American-trained planners as potential agents of change in the newly developing countries? Is their preparation appropriate to their assignment? What difficulties are they likely to encounter?

One of the major premises of a recent conference was that "willy-nilly U.S.- (or European-) trained professionals working in low-income countries are called to assume the role of agents of economic, social, and political change to a much greater extent than they normally do in advanced countries. This role involves them in activities that are (1) normally considered as part of the profession in advanced countries; and (2) in other activities that are not normally considered as part of the profession."[3] The present chapter is intended to explore this proposition in detail.

AGENTS OF CHANGE?

Let us consider the foreign planner who returns to his country after the usual two years necessary for obtaining his professional degree in the United States. It may be foolhardy to venture sweeping generalizations about two-thirds of the world's population, but the following description of the social environment to which our planner returns is no less true than an assertion that two-thirds of the world's population are poor and technologically backward.[4]

The social environment cannot be taken for granted. Like the air we breathe, it surrounds us on all sides. We become conscious of the presence of air when we have difficulty breathing; we become conscious of the social environment when we have difficulty planning. American planners are aware of selected aspects of their own social environment; planners in foreign countries are likely to become aware of different environmental aspects. These differences are frequently so great, that the American style of planning, applied to foreign situations, is rendered virtually useless.

To begin with, the foreign planner trained in the United States cannot expect to find a great deal of political or even administrative support upon returning to his country. Because a civic spirit in the European-American tradition is missing, local populations rarely advocate better cities.[5] The great mass of urbanites, composed of recent rural immigrants to the city, lack effective spokesmen and depend on the spontaneous good will of central bureaucrats for improvements in their condition; other groups, though clearly better off, will generally pursue more narrow interests than the environmental quality of the cities they inhabit. The quality of urban life is therefore left for the government to decide and, more specifically, the handful of specialists who occupy leading positions in the national administration. Municipalities have very little to contribute. Without effective power, they are usually incompetent to maintain more than purely housekeeping activities.

Other than those mentioned, institutional arrangements are likely to be weak. Legislation regulating urban growth tends to be obsolete and fragmentary. Though specialized agencies concerned with public housing may exist, the newer, more encompassing ideas of urban development will generally go begging for an institutional framework capable of giving them some substance. At the national level, economic planners have not even taken the trouble to learn the vocabulary for engaging city planners in a meaningful dialogue.

Questions of vocabulary are secondary, however, when even simple facts of urban life remain unknown. With very few exceptions, countries in the developing world do not sustain research into urban questions, and the scraps of data that come readily to hand are for the most part drawn from sources that are old and unreliable. Because there are no national institutions devoted to urban research, nor scholars trained in the subject, nor funds to support such research, there is no way to remedy abysmal ignorance. Given this background, the planner operates on hunches, guesses, or plain misinformation. Insofar as this is true, his word is not demonstrably better than that of any layman; anyone inclined to challenge his conclusions may do so with relative impunity.

The planner in a developing country moves in a highly politicized environment in which apparently objective statements are frequently suspected of having an ulterior motivation. Indeed, in the context of a faction-ridden closed bureaucracy, dispassionate analysis is the most difficult of achievements. Uncontested public values do not exist, and opinions cannot be separated from the persons holding them.

The bureaucracy through which he has to work is creakily inefficient, more concerned with maintaining itself than with producing results. In the absence of a career civil service, the occupants of bureaucratic office are insecure in their positions. The planner may start in his profession; he may end up doing jobs for which he was never prepared.

Professional isolation is yet another aspect of the planner's environment abroad. The number of colleagues with backgrounds similar to his will be too few to start a journal or professional society. Contacts with his former professors and friends in the United States will lapse after some time. Mail may be lost; interests may change; books and foreign journals cost dollars that he does not have. As his brief American experience sinks back into the misty recesses of memory, his youthful aspirations will tend to drop to levels more immediately attainable in his society.

Yet even apart from the environment in which he must survive, would his foreign training have been serviceable? Is the intellectual equipment he brings back from two years' study overseas in any way useful in his work? The reality model implicit in his education was that of the living American city and its containing frame of laws and institutions. Can this model serve him for understanding the city-forming processes in his own country? The techniques of planning and analysis he was taught presupposed the existence of massive information systems and a culture accustomed to management by experts. But do information systems with similar capacities exist where he must make his mark as a professional?[6]

Without attempting to justify my answer, I shall reply to all these questions with a flat no. The problems of urban development in industrializing countries are altogether different from those in the United States. The reality map must be another one. The analytical techniques appropriate to post-industrial society are too refined for application abroad. The values underlying policy in one social situation are unacceptable in another. In short, the planner returning from America has been prepared for the wrong job.

Agent of change? He will lack even the power that stems from scientific knowledge. And if he should finally, against all reasonable odds, succeed in transplanting a bit of foreign culture, it is likely to be a formula with no organic bonds to his own society. Half a generation ago, the master plan concept was introduced into Latin America, but in that particular environment it has been even less successful than in the United States where it is rapidly being abandoned in favor of other techniques for guiding urban change.[7]

The American-born planner going abroad will not only experience many of the same frustrations as his foreign counterpart, but will additionally be handicapped in other ways. Rarely will he speak the language of the country to which he has been sent, nor will he be at home with its culture. In contrast to the native-born planner, who has to spend a lifetime in the same surroundings, he will try to stamp out quick and tangible results. Where the native-born planner looks to a stable support group of family, friends, and associates related to each other through patterns of reciprocal commitments, the American planning consultant must not only satisfy the immediate demands of his employer—generally an international bureaucracy—but also stay in favor with his professional colleagues at home. The consultant, therefore, has no time to nurture new institutions through their initial growing pains; he has no time to mobilize political support for his ideas; he has no time to persuade those who are hostile or indifferent to his ideas.

The consultant's professional outlook was formed in the parochial schools of U.S. planning. He knows very little of the cultural and institutional environment in which his work has to be done. In a recent period of little less than two years, an American consulting firm completed 35 master plans for Libyan cities and "oases." The job was done but what results can be expected from this exercise in which most of the work was actually done at the firm's temporary headquarters in Rome? Planners are still expected to be experts in preparing plans. Their job is rarely defined as that of building institutions capable of carrying the work of planning forward after they have gone.

But this, of course, is what is needed. A plan results in a more or less balanced allocation of resources, but institutional development is an instance of *innovative planning* for major changes in a complex system.[8] As the Libyan example demonstrates, a rudimentary plan can be drafted in a matter of months, but successful institutional development requires years of patient effort. The Guayana project in Venezuela involving professors from both MIT and Harvard, and the Chilean program for urban and regional development supported by the Ford Foundation, are examples of continuing involvement over many years. These projects are exceptions to the usual consulting arrangement. Their results will have to be measured in terms of the institutional capabilities created to carry on the job of dealing with a variety of ever-changing problems in the future. This calls for subtlety of measurement and new criteria of institutional performance.[9]

THE TAO OF PLANNING

Innovative planners must learn to practice what I should like to call the Tao of Planning. And the Tao says: "Truly, a great cutter does not cut." This counsel is even more appropriate applied to foreign technical advisers.

Every national culture has its own sense of time. Ours is the sequential rhythm of linear time. John G. Gunnell writes:[10]

> "Development as a mode of thought involves the emergence of an orientation toward time and change which has found its paradigmatic expression in modern Western society. In general this means a focus where society as well as the life of the individual is perceived as moving along a unilinear plane from the past, through the present and toward the future, and where the future is understood not only as a dimension of existence and a boundary of life, but as an object of intentional action which can be anticipated, appropriated, controlled, and made to conform to goals devised in the present. Here the future is something to be planned and actualized by consciously innovative and creative action."

We know very little about time conceptions in the newly industrializing societies. We sense that they are different from our own, though we also believe that if a society adopts development as its overriding objective, it must adopt the Western conception.[11] But this may turn out to be simply an ethnocentric illusion. When a foreign expert acts according to this time conception, he will almost surely experience the nagging frustrations of delay and misunderstanding and suffer from an inability to communicate effectively with others. Many of these frustrations might be avoided if the planner learns to follow the rhythm native to the country where he works.

"Tao invariably does nothing, and yet there is nothing that is not done." This famous quotation from the *Tao-te Ching* sounds as paradoxical to Western man as does the following: "The most yielding of things outruns the most unyielding." Many foreign advisers wish to replace, as quickly as possible, what they consider to be outmoded ways of conduct with their own inventions. But their environment acts rather like a rubber wall: it yields to the initial impact only to spring back to its original position as soon as pressure is relaxed. To achieve a permanent change, there must be time—the right time to let initiatives float up to the surface, the right time for consensus to emerge, the right time for resistance to wear down.[12] Innovations must be fitted organically into the structure of things; artificially implanted, they cannot be sustained for very long.

The technical expert must also know the limits of his knowledge. He must be humble enough to learn from others, to know when to draw back, and to listen with attention. If he listens long enough and carefully, his own thoughts may be given back to him as the ideas of others. As the *Tao-te Ching* expresses it: "To give life, but not to own, to achieve but not to cherish, to lead but not to be master—this is the Mystic Virtue."

The purpose of innovative planning is to accomplish a structural transformation, but according to Lao Tzu, "all things go through their own transformations." If the innovative effort is concentrated at the critical intersections so that a small initial change produces large adjustments in other parts of the system, innovation is possible on a societal scale. Even a minor shift in the position of the fulcrum will tip the seesaw to one side.

Tao is the wisdom of nonaction, but innovative planning calls for building new institutional devices.[13] To this end, leadership must be selected and developed; long-term financing must be secured; client groups must be carefully nurtured; the varying impacts of innovation must be calculated for those groups which have the power to destroy or to impede the innovation; and the objectives towards which the action is pointed must be accepted as legitimate by a wider public.

American planners are taught neither the skills of institutional innovation nor the difficult art of giving advice. The dominant professional ideology tends to reject these skills because, in the American setting, innovations are accepted as a matter of course: the greater need is seen to lie in holding the multiple innovations of the society in balance. With a cultural revolution under way in America, we have for the first time created a climate propitious for the acceptance of innovative planning; conflict is now more readily accepted as a condition of social advance. But professional ideology is slow to adjust. It may be years before institutional change is accepted as one of the central tasks of urban planning.[14]

FOCAL POINTS OF CHANGE

The innovative planner must get to know the sensitive pressure points of his society. He must discover precisely where to apply creative energy within the social system. To do this, he must acquire a deep and comprehensive understanding of the structure and internal dynamics of his society. Since important social ends are bound to vary between developing and post-industrial societies, the major focal points for intervention are also likely to be different in each. I shall argue that, in the case of the newly developing countries, urbanization must be considered a strategic focus for public policy. I shall further suggest that the study of innovative planning should concentrate on clarifying relationships among key variables in urbanization and on projecting possible outcomes for societal change of different policy "settings" of these variables.

Over thirty years ago, Karl Mannheim suggested that planners search for and study what he called the *principia media* of society.[15] In the contemporary language of social science, we would call them positive feedback processes that lead to changes in societal structure. Elements of social systems are organized by these processes as iron filings are organized within a moving magnetic field. The changes set up in this field, whether the result of self-generated internal dynamics or planned intervention, will eventually produce new structural arrangements in society at large. Mannheim believed that any process of social change results from contending principia media, some of which are emergent, others dominant, and still others declining. Depending on their relative strengths, these forces are capable of producing quite different outcomes for society. The task for planning then, he said, is to discover the *principia media* operating in specific historical contexts and, by encouraging some while supressing others, to lend a helping hand to history. For Mannheim, planning was a form of guided social change.

If this idea is followed through, we are led to conclude that innovative planning should concentrate on the intersections of those processes which, from a purely instrumental point of view, are likely to result in a desired pattern of change in the society. In developing countries, one of the emergent *principia media* appears to be the complex of processes we call urbanization.[16] Accepting this view, we may regard the city as a major constructive force for national development rather than a congeries of problems which is still the prevailing view.[17] As William Mangin has said in reference to squatter settlements, the problem is itself the solution.[18]

Urbanization occurs as an inevitable accompaniment of national development. It is also a process capable of being guided through planned intervention. This formulation allows one to see a functional linkage between planning for urban and national development. Partially independent of one another, urban development can occur without commensurate expansion of the national economy, while policies for economic development may be promoted in ways that are independent of urbanization. But, to the extent that urban development affects the behavior of the national economy, the two processes are contingent upon each other. The first requirement is therefore to determine national policies for urban development and to learn the use of instruments for carrying them out.[19]

This interpretation of the city as an agent of social change is different from the prevailing concept in American planning which sees urbanization as resulting chiefly in local problems–blight, smog, crime, and congestion. "Regardless of the future," writes sociologist William McCord, "it seems beyond dispute that the developing nations are already overurbanized, burdened with festering cities beyond their capacity to support."[20] And he goes on to say:

> "However false the tinsel of these cities, their image of luxury continues to inveigle the peasant into immigration. At first, when the ex-villager enters the city, he must be overwhelmed by the glory of New Delhi's Secretariat, Mexico's Reforma Boulevard, Accra's government buildings, or the Paris-like magnificence of Cairo along the Nile. But soon he has to face the physical realities of city life–the *barriada,* or *favela,* or *bidonville* which are the rotten cores within the apple. Unless he has unusual training, he quickly discovers that no one wants his labor."

For this observer, at least, the city occupies the "outer edge of civilization."[21] But the facts are different. One of the major problems of urban-regional development in the newly industrializing countries is how to generate and capture innovative impulses at selected points within the spatial system, and to encourage migration towards *core regions and social development poles* rather than to turn them back from the "outer edge of civilization" toward the heartland of a rural utopia.[22]

The other major problem is how to maintain the equilibrium of a national system undergoing accelerated urbanization in order to achieve the progressive integration of the national economy in its territorial dimension through a balanced system of cities.

Both functions of urban development planning–using cities as innovating, transforming elements in a national society and achieving its social, economic, and political integration across regional space–drop out of view in American planning practice, because these problems have been largely solved in the United States. In the underdeveloped areas of the world, however, the process of urbanization is still a major force in national development and, as such, suggests a need for formulating policies capable of guiding this critical process as a whole towards socially desired ends.[23]

SOME SUGGESTIONS

Given (1) a continuing though small involvement of American planning schools in problems of urban development overseas, especially in connection with the preparation of planners from the newly industrializing countries; (2) the convergence of certain areas of interest and knowledge between metropolitan America and these countries, such as social planning and problems of low-income populations; and (3) the creation of national centers of planning education and research overseas, it is possible to suggest possible new directions for American planning schools. Work on developing countries might be concentrated in three or four of our top-flight universities. These programs would develop curricula emphasizing course work important to those planners who, during most of their professional lives, will be working in the newly industrializing countries of the world. They would include such subjects as regional development planning, housing in developing societies, national urban policy, economic development theory, and political development. Students would also be offered work in the history, culture, and languages of major world regions.

Research would be closely related to these subjects. Collaborative research with national centers abroad, under conditions which encourage mutual learning, should be initiated in those subject areas where a convergence of interests is manifest. Four specific topics suggest themselves: (1) the management of urban poverty; (2) the integration of marginal migrant groups into urban society; (3) the building of new urban institutions; and (4) the relationship of urbanization to national development. This research would be greatly helped if it were done on a comparative basis within a carefully defined theoretical framework.

These same university programs might wish to think of entering long-term arrangements of professional exchange with sister schools abroad. Such a program, involving university centers, might include mutual help in curriculum development, exchange of teachers and students, joint research, document exchange, and transfer of academic credits.

One of the major responsibilities of a participating American urban development center would be to translate what has been learned from these joint endeavors into the language of planning practice at home—rethinking its theoretical assumptions, reformulating some of the traditional problems, suggesting new methodologies of planning, and proposing new approaches to complex problem-solving. While the suggested overseas involvement of American professional schools will remain small, its potential significance for the United States is great. Joint programs will have equally important consequences for the practice of planning in the developing world. Since the latter is the usual argument, I have stressed the reverse effect which, if properly handled, could be substantial.

NOTES

1. Some of the planners active in this effort have included: Frederick J. Adams, Ernest Weissmann, Catherine Bauer, Francis Violich, Anatole Solow, Harvey Perloff, Richard L. Meier, Charles Abrams, John Turner, and Lloyd Rodwin.

2. Evidence of budding interest will be found in United Nations, "Training for Town

and Country Planning," Housing, Building and Planning, United Nations Bulletin No. 11, New York: United Nations, 1957. More recently, the magazine Ekistiks has played an important role in popularizing the role of urban planning in the developing countries. The part played by the United Nations in promoting urban and regional development in the underdeveloped regions of the world is ably described by the person who most contributed to this effort–Ernest Weissmann, "The Role of the United Nations in Urban Research and Planning," in Leo F. Schnore and Henry Fagin (eds.), *Urban Research and Policy Planning.* Beverly Hills, Calif.: Sage Publications, Inc., 1967, chap. 19.

3. Conference on "The Role of the Professional as an Agent of Economic, Social, and Political Change in Low-Income Countries," Berkeley, May 24-26, 1968.

4. Albert Waterston, *Development Planning: Lessons of Experience.* Baltimore: The Johns Hopkins Press, 1965; John Friedmann, "The Institutional Context," in Bertram M. Gross (ed.), *Action Under Planning: The Guidance of Economic Development.* New York: McGraw-Hill, 1967, chap. 2. The importance of environment for city planning is emphasized by Richard Bolan, "Emerging Views of Planning," Journal of the American Institute of Planners, XXXIII, No. 4 (July 1967), 233-245.

5. The civic spirit associated with local communities has grown out of a Western European tradition which has found scant resonance in other parts of the world where central bureaucracies have generally been responsible for maintaining a minimum amount of order in cities. For a typical nonwestern tradition of urban government, see the brilliant study by Marvin Lapidus, *Muslim Cities in the Later Middle Ages.* Cambridge, Mass.: Harvard University Press, 1967.

6. Ralph Gakenheimer has suggested additional problems in the transfer of American planning experience to other countries. The foreign planning student in the United States is unlikely to penetrate very deeply into the cultural and institutional matrix of American society and will consequently learn American planning approaches by rote and will not have a full understanding of their implications. Since he probably fails to grasp the unique aspects of American society that conditioned the particular approaches to planning in the first place, he will not be in a position to reformulate and adapt them to the culture of his own country. His lack of acquaintance with the pedagogic style of American universities and his struggles with the English language may enhance this particular difficulty (personal communication). There is, moreover, an antinomy inherent in this situation–to the extent to which the foreign planning student overcomes these difficulties, he becomes progressively estranged from his own society and will be tempted to remain in the United States, a country in which he will feel increasingly at home. The problem of migration of scientific personnel from the developing countries to the United States and/or Europe has lately received a good deal of attention. One recent study is by Luis Giorgi, "Extent, Nature, and Causes of the Loss of Scientists and Engineers Through Migration to More Advanced Countries." A Report to the Conference on the Application of Science and Technology to the Development of Latin America, Santiago, Chile, September 13-22, 1965, published in Final Report (New York: UNESCO, 1966). On the question of professional alienation (which, however, has deeper roots than the causes mentioned above), see, Francisco M. Suarez, Julio F. Aurelio, and Luis A Rigal, "Alienación professional en contextos transicionales," Revista Lationoamericana de Sociologia, II, No. 3 (November 1966), 378-395.

7. Hans Blumenfeld, "The Role of Design," Journal of the American Institute of Planners, XXXIII, No. 5 (September 1967), 304-310. Other articles in the same issue of this journal are also relevant here.

8. John Friedmann, "A Conceptual Model for the Analysis of Planning Behavior," Administrative Science Quarterly, XII, No. 2 (September 1967), 225-252.

9. Professor John W. Hanson suggests the following criteria for successful institution building (see Milton J. Esman, *The Institution-Building Concepts: An Interim Appraisal.* Pittsburgh: Inter-University Research Program in Institution-Building, Graduate School of Public and International Affairs, University of Pittsburgh, 1967, pp. 28-29):

1. *Use of services:* The extent to which programs are willingly used or requested by publics the organization is designed to serve.

2. *Survival:* The maintenance (and presumably also the expansion) of values and distinctive identity. These apply both to the institution itself and to its norms and include the concept of "innovation transferral"–innovations pioneered by the organization which have been transferred to other organizations which might more appropriately perform them.

3. *Support:* The extent to which capital resources . . . are providing for initiating, maintaining, and expanding programs, and the extent to which other organizations mobilize influence behind the subject organization's leadership and problems.

4. *Respect and Approval:* The extent to which the programs and the personnel of the organization are judged to be serving accepted or emergent goals well or as being qualified to serve such goals well.

5. *Normative Spread:* The extent to which action and belief patterns incorporated in the organization have become normative both within the organization and for other social units or related to its sphere of operation.

6. *Autonomy:* The degree of freedom of an organization to set and implement goals.

7. *Innovative Thrust:* The capacity of the organization to continue to innovate.

10. John G. Gunnell, "Time and the Concept of Development," CAG Occasional Papers, Bloomington, Indiana (April 1966), pp. 35-36. Other relevant papers are: Alfred Diamant, "The Temporal Dimension in Models of Administration and Organization," CAG Occasional Papers. Bloomington, Indiana (April 1966); and especially Hahn-Been Lee, "Developmentalist Time and Leadership in Developing Countries," CAG Occasional Papers. Bloomington, Indiana (February 1967).

11. This viewpoint is defended by Warren F. Ilchman, "New Time in Old Clocks: Productivity, Development, and Comparative Public Administration," CAG Occasional Papers. Bloomington, Indiana (February 1968).

12. A survey inquiring about professional support of this thesis was taken among approximately 750 foreign and national experts working in the area of induced social change in developing countries. The results indicate that 78 percent of the foreign experts and 80 percent of the nationals "believe that proceeding at a slow pace of action generally has the greater chance of success in the face of traditional ways; only 15 percent of the foreign and 17 percent of the national experts favour fast action. This overwhelming support of a slow tempo appears in every region and sphere of activity." As one expert in Africa commented: "It was found the pace of development could not be forced. Things moved somewhat slower than one is accustomed to in Europe and any attempt to accelerate this resulted in resentment, enemies, and frustration. To go slowly then is more than a slogan. It represents considered advice." From Herbert H. Hyman, Gene N. Levine, and Charles R. Wright, *Inducing Social Change in Developing Countries: An International Survey of Expert Advice.* Geneva: United Nations Research Institute for Social Development, (1967), pp. 70-73 and 109.

For a general discussion of alternatives in establishing a pace of change (and priorities in development programming), see Wilbert E. Moore, "Problems in Timing, Balance, and Priorities in Development Measures," Economic Development and Cultural Change, II, No. 4 (April 1954), 239-248. A very competent handling of the various roles of the foreign expert in an unfamiliar cultural environment is provided by Sune Carlson, *Development Economics and Administration.* Bonniers: Svenska Bok for Laget, 1964. Particularly descriptive is what Carlson calls the "second month depression." During the first month of an assignment, the expert finds everything new and interesting as he explores the situation with little expectation of immediate results. Gradually, however, he begins to make an assessment of the tasks to be accomplished and starts to push his work only to encounter a given set of conditions that prevent the attainment of his time-programmed tasks. "The transport and office facilities," says Carlson, "are not what he expected, he does not get the information he needs, he has no one to work with and there is no one who wants to listen to him. *Only time will solve these problems"* (Italics mine).

13. Milton J. Esman, *The Institution-Building Concepts,* p. 36ff.

14. This judgment may be too pessimistic. A Seminar on Innovative Planning will be offered in the Department of City and Regional Planning of the University of Toronto during the spring term of 1968 by Professor Gerald Hodge. On the other hand, I have been challenged on the assertion that skills in innovation and advice can be acquired by academic study. It is my impression, however, that the relevant literature is accumulating. In addition to Milton Esman's study cited above, and the work referred to therein, Bertram Gross, *The Managing of Organizations.* New York: The Free Press of Glencoe, 1964, may be considered basic. Although not directly pertinent to the issue, it does provide the intellectual foundations for an analysis of innovative efforts (and advice). Other relevant works include: W. A. Jöhr and H. W. Singer, *The Role of the Economist as Official Advisor.* London: Allen and Unwin, 1954; and Lyman Bryce, "Notes on a Theory of Advice," Political Science Quarterly, LXVI (1951), 321-339. An interesting case study of an attempted institutional reform that failed is described by Roderick T. Groves, "Administrative Reform and the Politics of Reform," Public Administration Review, XXVII, No. 5 (December 1967), 436-445. Case studies of this sort need to be built up to develop both a descriptive and normative theory of innovative planning.

15. Karl Mannheim, *Man and Society in an Age of Reconstruction.* New York: Harcourt, Brace, and Co., 1949, pp. 174-196.

16. David Popenoe, "On the Meaning of 'Urban' in Urban Studies," Urban Affairs Quarterly, I, No. 1 (September 1965), 17-34; and John Friedmann, "Two Concepts of Urbanization: A Comment," Urban Affairs Quarterly, I, No. 4 (June 1966), 78-84.

17. See Chapter 2.

18. William Mangin, "Latin American Squatter Settlements: A Problem and a Solution," Latin American Research Review, II, No. 3 (1967), 65-98.

19. See Chapter 8.

20. William McCord, "Portrait of Transitional Man," in Irving Louis Horowitz (ed.), *The New Sociology: Essays in Social Science and Social Theory in Honor of C. Wright Mills.* New York: Oxford University Press, 1964, p. 430.

21. Ibid., p. 437.

22. See Chapter 9.

23. See Chapter 5.

Chapter 16

EDUCATION FOR REGIONAL PLANNING IN DEVELOPING COUNTRIES

More than a decade ago, Harvey Perloff published a path-breaking essay on education for regional planners.[1] The essay did more than formulate a curriculum; it contributed significantly to the definition of regional planning as a field of study.

Perloff wrote primarily for planners in an American setting; he thought of professional careers in the United States. In the intervening years, however, regional planning has received growing attention in the developing countries, not only in connection with programs of resource development, such as the efforts in the Damodar, Cauca, and Volta River Basins, but also as a framework for national development planning. A number of countries now have systems in which regional planning appears both at national and regional levels. In these countries, the national economy is treated not only in global and sectoral terms, which is traditional, but also in terms of spatial analysis.

The rapid growth of interest in regional planning has created a demand for certain kinds of professional skills and has put pressure on educational institutions to provide them. This chapter proposes the elements out of which a curriculum capable of supplying the necessary skills may be devised.

Perloff wrote: "Experience . . . suggests that regional planning . . . has its best opportunity for substantial and continuous contribution when it is tied to ongoing, clearly focused operations."[2] Therefore, if we wish to build a curriculum for preparing planners to work abroad, it is necessary to identify "the ongoing, clearly focused operations" of contemporary planning practice in the developing countries.

REGIONAL PLANNING ACTIVITIES

In what sorts of activities do regional planners typically engage? In answering this question, I shall present a model which encompasses planning at both the national and regional levels. Although the model assumes a unitary form of government, it is readily modified to take into account the more complex structure of federal states. In this event, a third level of decision-making would have to be inserted between region and nation; the character of the major planning activities, however, would remain the same. The principal steps in the formulation and implementation of regional plans as a part of national planning are described below in accordance with the sequence identified in Figure 1.

1. *Formulation of regional development, migration and settlement policies.* Preparation of guidelines and criteria for the regional allocation of resources; definition of planning and administrative regions; priorities for regional investment; desired geographical distribution of major urban functions (administration, manufacturing, education, etc.); policies for the development of industrial satellite cities in metropolitan regions (national level).

2. *Preparation of regional strategies, plans and programs.* Formulation of objectives for regional economic development; identification of investment priorities by sectors and sub-regions; recommendations on institutional measures for implementing the regional plan; preparation of multi-year development programs consistent with the main lines of policy established (regional level).

3. *Identification and preliminary evaluation of regional investment projects.* Identification of promising projects for regional investment as part of an integrated program sequence; preliminary evaluation of these projects in accord with feasibility and desirability criteria, regional program objectives, and cost/benefit analyses (regional level).

4. *Integration of development strategies, plans and programs for individual regions into a national set.* Achieving internal coherence and consistency of all regional development strategies, plans and programs with each other and their integration into a national set.

5. *Integration of the national set of regional strategies, plans, and programs with global and sectoral plans.* Achieving internal consistency and coherence of all regional strategies, plans, and programs with national objectives for production, investment, balance of trade, and employment, as well as with the breakdown of these objectives into targets for activity sectors and sub-sectors (national level).

6. *Preparation of annual program budget guidelines.* Preparing a preliminary regional allocation of the projected investment budget for the nation, together with indications of desired program emphasis as a basis for detailed programming and capital budgeting at the regional level (national level).

7. *Preparation of regional program budget.* Preparing the annual program budget for the region on the basis of guidelines received and the multi-year regional program evolved in Step 2 (regional level).

8. Integration of regional program budgets into a national set and integration of this set with the program budget for the nation. Achieving internal coherence and consistency of all regional program budgets with each other and with the program budget of the nation conceived along sectoral lines (national level).

9. Coordination of regional investment and action programs. Facilitating the coordination of sectoral agencies in the implementation of the regional program budget (regional level).

10. Promotion of regional development. Promoting the economic development of the region, particularly in the industrial sector, by identifying private investment opportunities, helping to create a consciousness of regional development problems, and providing technical and information services to potential investors (regional level).

11. Regional planning analysis. Continuing study of progress, problems, and possibilities in regional economic development and of the regional impact of proposed investments on economic welfare, employment, urban expansion, regional integration, etc. (national and regional level).

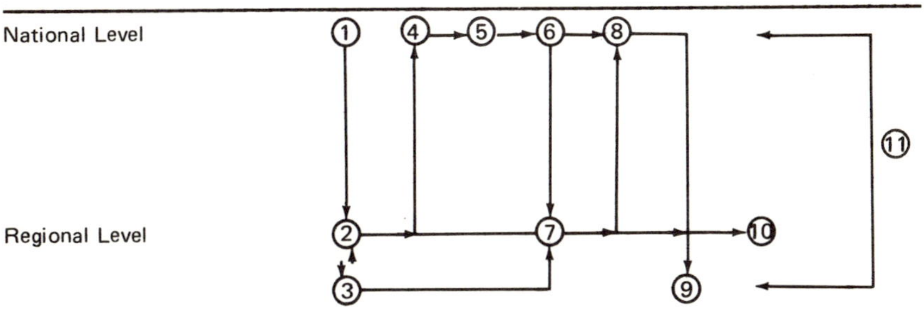

Key

1. Formulation of regional development, migration and settlement policies.
2. Preparation of regional strategies, plans and programs.
3. Identification and preliminary evaluation of regional investment projects.
4. Integration of development strategies, plans and programs for individual regions into a national set.
5. Integration of the national set of regional strategies, plans and programs with global and sectoral national plans.
6. Preparation of annual program budget guidelines.
7. Preparation of regional program budget.
8. Integration of regional program budgets into a national set and integration of this set with the program budget for the nation.
9. Coordination of regional investment and action programs.
10. Promotion of regional development.
11. Regional planning analysis.

Figure 1: Simplified Regional Planning Activity Flow Chart

REGIONAL PLANNING ANALYSIS

The specifically technical contribution to the planning process (Step 11) will now be described in greater detail. As may be seen from Figure 1, analysis enters in all of the operations at national and regional levels. The following types of regional planning analysis are generally needed.

1. Analysis of spatial organization. Statistical and map analysis of geographical distribution patterns (spatial structure) of all the relevant variables for regional analysis; delimitation of regions; city functions, hierarchies, and urban influence areas; application of gravity models to regional analysis; etc.

2. Location analysis. Analysis of location factors across regional or inter-regional space; finding optimum locations for individual firms and enterprises; location analysis of central services; industrial complex analysis; identification of industrial development opportunities; agricultural location patterns, etc.

3. Urban complex analysis. Analysis of the physical structure and evolution of the principal urban areas; urban-metropolitan transport problems; social and economic infrastructure needs of urban areas; housing problems; urban land use, etc.

4. Regional land-use analysis. Geographic analysis of land-use requirements implicit in programs for regional development by type, location, and total area; study of special land-use requirements in connection with programs for erosion control, tourism, recreation, wildlife conservation, etc.; preparation of a long-term regional land-use plan; etc.

5. Analysis of demographic characteristics and behavior. Statistical study of demographic structure and its evolution by regions; urbanization patterns; migration flows; socio-economic studies of the decision to migrate; labor force participation rates; dependency ratios; etc.

6. Human resource analysis. Studies for manpower requirements of regional programs; studies of the educational and technical skill levels of regional populations in light of future requirements; institutional conditions for meeting these requirements; special problems in the full development of human resource potentials; policies and programs, etc.

7. Natural resource analysis. Economic analysis of existing natural resource complexes in different regions and in the context of evolving international, national, and regional markets; continuing study of technical innovations in relation to regional resource use, etc.

8. Economic activity analysis. Study of the structure, evolution and characteristics of the different sectors and sub-sectors of regional economies in terms of firms, employment, production, productive capacity, efficiency, organization, and inter-industry relation; analysis of inter-regional transport and electric energy networks; etc.

9. Regional trade pattern analysis. Inter-regional input-output analysis;

inter-regional flow patterns (products, services, information); market analysis of regional products; etc.

10. Regional accounts and development models. Preparation of a regional product accounts framework and corresponding time series; indicators of regional economic performance; regional economic forecasting models; quantitative development planning models permitting integration of regional with national global and sectoral targets; etc.

11. Project analysis. Feasibility and cost/benefit studies of major investment projects in a regional context.

12. Regional program budget analysis. Analysis of regional program budgets, optimal allocations and time sequences; investment and operating capital requirements; possible sources of financing; etc.

REGIONAL PLANNING SKILLS

The discussion so far may be summarized by saying that the purpose of planning education is to make available the specific analytical skills which appear as the technical inputs to the planning activities identified in Figure 1.

It is possible to divide these skills into two parts. Professionals possessing basic skills in the analysis of spatial organization, location of firms, regional trade patterns, regional accounts, and program budgeting for regional development will be called *regional planners.* Persons skilled in all other forms of relevant analysis will be called *regional planning specialists.*

Common to regional planners is a basis in regional economics and quantitative geography. These fields may be grouped under the general title of spatial analysis.

Spatial analysis is concerned with the quantitative study of spatial relationships and their transformation.[3] This study is basic to regional planning, first, because space, like time, provides a convenient framework for integrating development activities; second, because the spatial organization of an economy is a fundamental variable in structuring economic opportunities; and third, because the location of people and activities is the principal means available to regional planners for changing the spatial organization of an economy and affecting the objectives of regional development in a decisive way. It is by acting on location behavior that regional planners are able to influence the economic growth characteristics of different regions, the pattern of human settlement, and the main flows of migration.

Regional planners have essentially integrative skills; regional planning specialists, on the other hand, make their principal contribution along lines that are functional to regional analysis (e.g., resource analysis). Both types of professionals, however, operate within a planning framework. In addition to their technical expertise, therefore, they must both be capable of thinking in planning terms and of working effectively together.

	Regional Planners	*Regional Planning Specialists*
FIRST YEAR[a] *Parent Discipline*	For students other than economists: micro- and macro-economic theory; theory of economic development; statistics; and modelling.	One year make-up work if student is from other than required parent disciplines indicated in parentheses below.
SECOND YEAR *Basic Discipline or Specialization*	*Spatial Analysis*[b] 1. Introduction to Theoretical Geography 2. Advanced Theoretical Geography 3. Location Theory 4. Quantitative Methods in Spatial Analysis 5. Theory of Urban-Regional Development 6. Regional Accounts and Development Models 7. Transportation and Economic Development 8. Regional Analysis Workshop 9. Preparation for Comprehensive Exams	*Specialized Forms of Regional Analysis* A. *Natural Resource Analysis* (Economic Geography) B. *Regional Land Use Analysis* (Geography and/or Physical Planning) C. *Urban Complex Analysis* (Physical Planning, Geography) D. *Demographic Analysis* (Sociology) E. *Human Resource Analysis* (Sociology and/or Economics) F. *Economic Activity Analysis* (Economics and/or Industrial Engineering) G. *Local and Regional Development Administration* (Political Science and/or Public Administration)
THIRD YEAR (for all students) *Regional Policy Planning*[c]	1. Introduction to Spatial Analysis (for Regional Planning Specialists only) 2. Theory of Planning 3. Urbanization and Economic Development 4. Regional Policy and Planning 5. Political Aspects of Regional Development 6. Project and Program Evaluation 7. Instruments of Regional Development Planning 8. Regional Program Budgeting 9. Regional Policy Workshop 10. Thesis	
FOURTH YEAR (for all students)	Study tours and/or internships in regional planning agencies (3 to 6 months)	

a. Entering students are expected to have acquired a first university degree.
b. Minimum of five courses required, terminating in comprehensive examination. For course contents, see Appendix E.
c. Minimum of five courses plus thesis required. For course contents, see Appendix F.

Figure 2: Proposed Curriculum for Regional Planners and Planning Specialists

To think as a planner means (1) to organize one's studies within a normative framework of goals and objectives; (2) to conduct analysis with a view to the future; (3) to relate technical to normative considerations in an integrative way, identifying relationships within a total system; and (4) to emphasize incremental processes of reaching desirable end-states.

To work effectively within an institutionalized planning process means to have an awareness of the limits of technical competence, the relationships between expertise and political power, the way conditions under which planning is carried out may influence the form and outcome of planning, and the influence of forms of plan implementation on the formulation of the plan itself. A knowledge of planning theory is therefore an essential part of the intellectual equipment of both regional planners and specialists.

A PROPOSED CURRICULUM

Given these preliminary considerations, it is now possible to suggest the elements of a curriculum that will give advanced professional training to regional planners and specialists. The basic scheme is shown in Figure 2, with course contents summarized in Appendices E and F at back of book.

Different courses of study are proposed for each major type of planning professional. The main sequence, however, is concerned with the preparation of regional planners as such, and it is only this sequence which has been specified in some detail as to course contents. Both regional planners and specialists join towards the end of their studies in a final year devoted to policy planning.

Entering students are presumed to have a university degree. Furthermore, if they wish to pursue a career in regional planning, they should have majored in economics during their undergraduate studies. Economics, in this case, is called the *parent discipline.* If students have majored in some other subject, they will be required to spend one year in the intensive study of economic theory, statistics, and modelling. In the second year, then, they are prepared to enter the spatial analysis sequence, or *basic discipline,* which terminates at a level equal to that required for the master's degree in the United States. During this year, students will acquire the basic skills for the analysis of spatial phenomena.

Having concluded this part of their education, students are ready to enter the third year which is built around regional planning proper, that is, around the application of the basic skills acquired to problems of policy formulation and planning. This year will be the major interdisciplinary experience of the students; they will join here with students from other academic backgrounds and specializations. Although this third year is common to both regional planners and specialists, the latter will be expected to take an introductory course in spatial analysis that will help them to understand some of the basic concepts and terminology in use. The sequence terminates with a thesis at the professional degree level.[4]

The formal course of study concludes here. It may be desirable, however, to follow up with a period of study-observation at foreign regional planning institutions and/or an internship lasting for several months in a major regional planning agency.

The curriculum for regional planning specialists may also be divided into parent and basic disciplines. The general requirements are apparent from Figure 2.

ALTERNATIVES TO THE STANDARD CURRICULUM

The curriculum discussed so far may be thought of as comprising the standard elements of a program of study for regional planning. It is quite conceivable—indeed probable—that the concrete situation in different countries will require an adaptation of the standard model to the prevailing conditions.

For instance, planning is so closely related to elements in the culture and political life of a nation, that foreign study of planning subjects (the hypothetical third year in the curriculum) is unlikely to prepare the student adequately for his career. These courses should therefore be taught in the country or multi-country region where the student expects to be working. Spatial analysis, on the other hand, is currently taught in only a few universities in the United States and Europe. It may therefore be advisable—and it is entirely possible—to reverse the sequence suggested, so that students would first enter as specialists what appears in Figure 2 as the third or regional planning year, with the best graduates from this course being sent abroad for advanced study in spatial analysis.

Another possibility would be to extend the period of specialized study in the basic disciplines by perhaps 6 months by offering simultaneously and as part of the basic curriculum, interdisciplinary courses in planning theory and method (Appendix F) for students interested in regional planning as a "minor" field within their larger specialization. This "minor" might be offered either within a school (department, center, institute) of planning or as a number of separate courses in different university centers, coordinated by a central committee and open to all students registered in the program.

The standard curriculum outlined in Figure 2 may also be abbreviated to fit a more rapid time schedule. In many countries, the demand for regional planners may develop so rapidly that there will be insufficient time to insist on the full three-year course of study (or two years in the case of students who come from the relevant parent disciplines), even leaving aside the suggested study tours and/or internships which may require another three to six months. In this case, it is entirely feasible to train only regional planning specialists, at least in the beginning, and to insist on nothing more than the common third year of regional planning on the assumption that entering students will already possess some degree of the relevant basic specializations. This one year can be further reduced to six months of intensive work, if the thesis requirement is waived. Going further in the direction of saving time, these six months can, in principle, be collapsed into a super-intensive sequence lasting for only three months in which the rudiments of regional planning would be learned. Any course shorter than this would have to be extremely selective in its subject matter and would include only a few specialized topics for professionals already engaged in some form of regional planning practice. Such courses are very important for maintaining and increasing professional competence and should be given to all planning staff on a regular basis.

POSTSCRIPT

The ideas contained in the chapter were presented in 1967 at an International Workshop on the occasion of the Third Lustrum of the Institute of Social Studies at The Hague. In the course of the ensuing discussion, a number of questions were raised which, I believe, will be of interest to the reader. I shall comment briefly on them in the hope that this dialogue will help to clarify a number of important issues suggested by my proposal.

1. Regional planning often occurs in the absence of national planning. In this case, what are the educational requirements for specialists in the preparation of comprehensive area plans?

A. The proposed model curriculum is, indeed, based on the assumption that regional planning will be integrated with plans for national development. The professional requirements, however, are not likely to differ substantially where this fails to be true. Just as the national space is divided into regions, so is regional space divided into sub-regions. The skills involved in handling the planning problems on these two scales are essentially the same.

2. You have described the contribution of planning technicians. But planning as an instance of guided social change is broader and must result in programs and in actions. Where, in your model, is the planning strategist and plan administrator to be found?

A. I do not see strategy and administration as special *regional* skills. What is new in regional planning is our ability to place scientific and technical knowledge at the service of a larger community. Planners seek to bring a technical competence to bear on social problems. Regional strategy and administration, on the other hand, are based on intuition, experience, and managerial ability much more than on a trained, analytical capacity.

3. Your curriculum seems excessively theoretical. Students should be immersed in reality. They should be taught how to prepare regional plans.

A. Anyone can prepare plans without theory. The making of plans poses less of a problem than planning analysis capable of withstanding sophisticated criticism. The planner has no special knowledge of the public interest. But he can defend alternative proposals on technical grounds, he can reduce the area of choice by showing what is feasible and what is not, he can sharpen the issues for political decision. And this special ability is nourished by theory. Nevertheless, some of the courses can be taught as problem workshops, permitting students to relate theory to policy in the solution of significant problems. In these workshops, the presentation of results to a public will be as important as the analysis and planning synthesis itself.

4. In your curriculum, insistence on the need for comprehensiveness in planning is missing. Is this not an important shortcoming of your proposal?

A. Comprehensive planning is a vague term. I prefer to speak of systems and the analysis of systems. There are many kinds of systems and they are interacting in complex ways across regional space. The proposed model curriculum is intended to familiarize the student with these systems and to give him some tools for their analysis.

5. Is team work not a crucial element in successful regional planning? Is this not a valuable habit that should be acquired early in the course of one's studies?

A. Yes, team work is fundamental. The proposed planning curriculum, however, offers many opportunities to encourage this particular mode of problem-solving.

6. The distinction between regional planners and regional planning specialists does not appear to be felicitous. Why distinguish between them at all?

A. The distinction is mainly heuristic; I would hope that it does not carry over into a professional image. One might also, and perhaps more appropriately, speak of basic and contributory skills in regional planning. In the end, titles do not matter very much.

7. Is there not a danger that only received theory will be taught? There is no mention of the importance to link teaching with research, so that new knowledge may be quickly passed on to the students.

A. I am in wholehearted agreement with this position. Planning education without research will rapidly become obsolete, repeating empty formulas. Research in all the relevant subject matters must be incorporated into a full educational program from the start. Both students and faculty should participate in the advancement of scientific knowledge.

NOTES

1. Harvey S. Perloff, *Education for Planning: City, State, and Regional.* Baltimore: Johns Hopkins Press, 1957, Part II.

2. Ibid., p. 91.

3. Walter Isard and Thomas Reiner, "Regional Science: Retrospect and Prospect," Regional Science Association Papers, Vol. 16 (1966), pp. 1-16.

4. If this full sequence were followed, the degree offered at the end of the third year might be a doctorate. This would not be a research degree, however, such as the Ph.D., but an advanced professional degree.

APPENDICES

Appendix A

ANNOTATED BIBLIOGRAPHY*

(Chapter Four)

A. LATIN AMERICA

Balán, Jorge, "Migrant-Native Socioeconomic Differences in Latin American Cities: A Structural Analysis," in *Latin American Research Review,* Vol. IV, No. 1 (Spring, 1969), pp. 3-29.

Suggests an analytical framework for predicting migrant streams to account for inter-city variations in the socio-economic position of migrants. Though the attempt at constructing the model is not entirely successful, a refined model based on this first attempt may prove to be a helpful tool in pulling together the behavioral-descriptive studies of migration that we now have.

Beuchler, Hans C., "The Reorganization of Counties in the Bolivian Highlands: An Analysis of Rural-Urban Networks and Hierarchies," in Elizabeth Eddy, ed., *Urban Anthropology: Research Perspectives and Strategies.* Athens, Ga.: Southern Anthropology Society, 1968, pp. 48-57.

A descriptive analysis of the formation and transformation of *cantones,* and of the complex intertwining of relationships involved in this process.

*No effort has been made to update this bibliography, which covers exclusively materials reviewed in Chapter 4 and includes writings available to the authors through June 1970.

Browning, Harley L., "Urbanization and Modernization in Latin America: The Demographic Perspective," in Glenn H. Beyer, ed., *The Urban Explosion in Latin America.* Ithaca, New York: Cornell University Press, 1967, pp. 71-92.

A short explanatory essay on the positive roles played by demographic variables (population size and density, mortality, fertility, and migration) in the process of modernization.

Browning, Harley L. and Waltrout Feindt, "Selectivity of Migrants to a Metropolis in a Developing Country: A Mexican Case Study," in Demography, Vol. 6, No. 4 (November, 1969), pp. 347-357.

A study of migrant-native differences. Suggest possible causes and effects of the degree of selectivity and pattern (over time) of selectivity that was found. Zachariah and Sovani in studies of India suggest similar findings on the degree of selectivity, though they and others have not discussed selectivity over time.

de Jesus, Carolina Maria, *Child of the Dark,* trans. David St. Clair. New York: E. P. Dutton and Co., Inc., 1962.

This diary of an uneducated slum Negress describes contemporary Brazilian society—the people of the favela, Brazil's politicians, the lives of the slum dwellers, the economic organization of the slum—as seen by someone at the very bottom of the social structure. Particularly interesting is Carolina's description of the economic activities in the slum, a part of the bazaar economy described by Geertz and Armstrong and McGee in South Asia.

Flinn William L., *Rural to Urban Migration: A Columbian Case.* Reprint No. 19. Madison: University of Wisconsin Land Tenure Center, July, 1966.

This study lists the motives given by migrants for moving from a rural to an urban area. The findings lend credence to the suggestion given by Herrick in Chile and duToit and Little in Africa that many migrants move only a short distance and that aid from a friend or relative is an important factor in moving and adjusting to the urban milieu.

Frank, André Gunder, *Capitalism and Underdevelopment in Latin America: Historical Studies of Chile and Brazil.* New York and London: Monthly Review Press, 1967.

Series of historical essays written at different times and for varying purposes. Each in its own way is intended to clarify how the structure and development of capitalism has generated and maintained underdevelopment. A Marxist interpretation.

Friedmann, John, *Regional Development Policy: A Case Study of Venezuela.* Cambridge, Mass.: The M.I.T. Press, 1966.

Core-periphery thesis illustrated with Venezuelan data. Sees need for regional development theory by which one can clarify reordering of spatial relations that occur under conditions of growth and explain the changing influence of spatial patterns on system-wide growth.

Friedmann, John, *Urban and Regional Development In Chile: A Case Study of Innovative Planning.* The Ford Foundation Urban Regional Development Advisory Program in Chile, Santiago, Chile, June, 1969.

Report of ambitions and scope of Ford Foundation Program in Chile which was to explore the possibilities of coordinative development planning of subnational levels–regions and cities–within the broad context of national policy.

Gauthier, Howard L., "Transportation and the Growth of the São Paulo Economy," in Journal of Regional Science, Vol. 8, No. 1 (1968), pp. 77-94.

Investigates the relationship between changes in accessibility to highway networks and the growth of urban centers. Taaffe, Morrill and Gould drawing on evidence from Ghana and Nigeria have written a more descriptive and readable paper on this same subject.

Germani, Gino, "The Concept of Social Integration," in Glen H. Beyer, ed., *The Urban Explosion in Latin America.* Ithaca, New York: Cornell University Press, 1967, pp. 175-189.

Discusses mechanisms by which the city acts as a social integrator, a process which will eventually lead to the modernization of society.

Graham, Douglas, "Divergent and Convergent Regional Economic Growth and Internal Migration in Brazil–1940-1960," in *Economic Development and Cultural Change,* Vol. 18, No. 3 (April, 1970) p. 362-382.

Concentrates on phenomenon of internal migration, its changing volume and differential incidence among states over time, and its role in the trends of regional growth from 1940 to 1960, concluding with a brief summary of major factors that led to an increased volume of internal migration in the 1950's. This type of analysis has not been done for Africa and Asia, though Soja and McNulty touch on it in their analysis of the spatial organization of modernization in Nigeria and Ghana.

Herrick, Bruce, *Urban Migration and Economic Development in Chile.* Cambridge, Mass.: The M.I.T. Press, 1965.

This excellent book attempts to coordinate the record of Chile's economic development with an account of the concomitant internal migration. In particular, the shifts in urban population and changes in the structure of the labor force are explored in order to understand the role of migration in the process of development.

Kahl, Joseph A., *The Measurement of Modernism: A Study of Values in Brazil and Mexico.* Austin: The University of Texas Press, 1968.

Technical monograph. Creates measures for and describes those values which have implications for a man's current work behavior and for his long-range career planning. Relates change in values to the transformation of man and society from "traditional" to "modern."

Lewis, Oscar, "Urbanization Without Breakdown: A Case Study," in Dwight B. Heath and Richard N. Adams, eds., *Contemporary Cultures and Societies of Latin America.* New York: Random House, 1965, pp. 424-437. (Reprinted from The Scientific Monthly, Vol. LXXV (1952), pp. 31-41.)

One of the first empirical studies to deal with the social and psychological consequences of rural-urban migration. Discredits earlier theoretical findings which had been influenced by Wirth's Chicago studies and Redfield's rural-urban continuum. Most studies on migrants in Africa and Asia suggest, as does Lewis, that adjustment to the urban milieu is within control of the individual.

MacDonald, Beatrice D. and John S. MacDonald, "Motives and Objectives of Migration: Selective Migration and Preferences Towards Rural and Urban Life in Guayana," in Ekistics, Vol. 28, No. 128 (November, 1969), pp. 321-327.

Study of rural-urban differentials, and motives and objectives given by rural people for migrating or not migrating.

Mack, Raymond W., "Race, Power and Class in an Urban Plantocracy: Barbados," in Scott Greer, et al., eds., *The New Urbanization.* New York: St. Martin's Press, 1968, pp. 72-8.

Describes and analyzes the way the polity may be rapidly expanded to include among the rulers those who had been the ruled, necessitating a description of the radical changes that have taken place in the defining of race as former bases of racial exclusiveness, such as education and political power, have given way. Though they do not analyze the issue in the depth that Mack does, Banton, Emerson, Epstein, among others, indicate that race or ethnic identity plays a crucial role in African nation-building.

Mamalakis, Markos J., "The Theory of Sectoral Clashes," in *Latin American Research Review,* Vol. IV, No. 3 (Fall, 1969), pp. 9-46.

Presents series of models to show how sectoral alignments in Chile have changed from the 19th century to the present. Attempts to analyze the phenomenon of competition between major economic sectors and the results of establishing government as an active partner.

Mangin, William P., "Mental Health and Migration to Cities: A Peruvian Case," in Dwight B. Heath and Richard N. Adams, eds., *Contemporary Cultures and Societies of Latin America.* New York: Random House, 1965, pp. 546-555.

Explores the psychological reasons why low-status migrants suffer relatively little mental stress and exhibit little disorganized behavior as a result of their move from a peripheral to a core area.

Mangin, William P., "The Role of Regional Associations in the Adaptation of Rural Migrants to Cities in Peru," in Dwight B. Heath and Richard N. Adams, eds., *Contemporary Cultures and Societies of Latin America.* New York: Random House, 1965, pp. 311-323.

Examines the role of the informal regional association or social club in assisting many migrants to adjust to Lima. Their importance in African cities has been discussed by Little, Banton, Gutkind and the Hannas.

Merkx, Gilbert W., "Sectoral Clashes and Political Change: The Argentine Experience," in *Latin American Research Review,* Vol. IV, No. 3 (Fall, 1969), pp. 89-116.

Presents an interpretation of Argentine social history based on Mamalakis' theory of sectoral clashes. In particular, focuses on the economic and political aspects of transactions between sectors and among income groups within sectors.

Olien, Michael D., "Levels of Urban Relationships in a Complex Society: A Costa Rican Case," in Elizabeth Eddy, ed., *Urban Anthropology: Research Perspectives and Strategies.* Athens, Ga.: Southern Anthropology Society, 1968, pp. 83-92.

Description of the intra- and inter-community network of relationships in the town of Puerto Limon.

Peattie, Lisa Redfield, *The View from the Barrio.* Ann Arbor: University of Michigan Press, 1968.

Describes evolution of social processes in a barrio of a rapidly growing planned city in the interior of Venezuela. Very good description of core-periphery relationships.

Pedersen, Poul Ove, "Innovation Diffusion Within and Between National Urban Systems," Geographical Analysis, Vol. II (1970), pp. 203-254.

Investigates the process of innovation diffusion, and its constraints, within a single country, Chile, and between the South American countries, suggesting a rather orderly pattern that governs the diffusion process. Relates diffusion process to levels of economic and urban network development. Similar studies have not as yet been undertaken in Africa and Asia.

Pedersen, Poul Ove and Walter Stöhr, "Economic Integration and the Spatial Development of South America," in American Behavioral Scientist, Vol. XII, No. 4 (May-June, 1969), pp. 2-12.

Attempts to determine the spatial impact of policies promoting national and continental integration.

Preston, David A., "Rural Emigration in Andean America," in *Human Organization,* Vol. 28, No. 4 (Winter, 1969), pp. 279-286.

Paper reviews the scale and direction of migration in Andean America and outlines a variety of problems associated with rural emigration. Offers evidence that permanent migration is preceded by temporary migration in Bolivia and Ecuador, a situation described by duToit and Elkan in Africa and Lambert in India.

Robock, Stefan H., *The Rural Push for Urbanization in Latin America: The Case for Northeast Brazil.* Occasional Paper No. 1. East Lansing: Michigan State University Latin American Studies Center, 1968.

Discussion of rural "push" factors which are accelerating the rate of urbanization in Brazil. Sovani, in Asia, has attempted to cast doubt on the push-pull dichotomy which Robock is discussing.

Rycroft, W. Stanley and Myrtle M. Clemmer, *A Study of Urbanization in Latin America.* New York: Commission of Ecumenical Mission and Relations, The United Presbyterian Church in the U.S.A., 1963.

Compilation of data and statistics. Attempt to describe and analyze urbanization in relation to the social and economic development of Latin America. Slightly dated.

United Nations, "Recent Changes in Urban and Rural Settlement Patterns in Latin America," in *Urbanization: Development Policies and Planning, International Social Development Review,* Vol. 1. New York, 1968, pp. 55-62.

Cursory survey of the recent trends in urban and rural population redistribution and resettlement patterns in Latin America.

United Nations, "Rural Social Change," in *Social Change and Social Development Policy in Latin America.* New York: Economic and Social Council of the Economic Commission for Latin America, February 1969, pp. 118-131.

Superficial analysis of the changing rural situation in Latin America.

Willems, Emilio, "Urban Classes and Acculturation in Latin America," in Elizabeth Eddy, ed., *Urban Anthropology: Research Perspectives and Strategies.* Athens, Ga.: Southern Anthropology Society, pp. 75-82.

Investigates manner by which different cultural groups are absorbed into the urban class structure, hypothesizing that neither the cultural groups nor the class structure remain unchanged in the process.

B. AFRICA

Abu-Lughod, Ibrahim, "The Mass Media and Egyptian Village Life," in Social Forces, Vol. 42, No. 1 (October, 1963), pp. 97-104.

Investigates the role of the mass media (radio and newspapers) as a shaper of public opinion and as an agent of diffusion.

Abu-Lughod, Janet, "Urban-Rural Differences as a Function of the Demographic Transition: Egyptian Data and an Analytical Model," in The American Journal of Sociology, Vol. 69, No. 5 (March, 1964), pp. 476-490.

Uses data from Egypt to refute almost all generalizations about urban-rural demographic differences that have been developed from the study of industrialized countries. However, then uses data from Egypt to construct a temporal sequence model of modernization, through which descrepancies revealed from the cross-cultural comparison can be reconciled and even predicted.

Abu-Lughod, Janet, "Urbanization in Egypt: Present State and Future Prospects," in Economic Development and Cultural Change, Vol. XIII, No. 3 (April, 1965), pp. 313-343.

Attempts both a diagnosis and a prognosis of Egyptian urbanization, investigating current extent of overconcentration, the geographic distribution of urban population, and the trends that may intensify or mitigate the problem of overconcentration in the future.

Aldous, Joan, "Urbanization, The Extended Family and Kinship Ties in West Africa," in Social Forces, Vol. 41, No. 1 (October, 1962), pp. 6-12.

Study of how African kinship ties are affected by urbanization. Evidence indicates, as does that offered by Lewis in Mexico and Mukherjee in India, that the extended family is present in the urban milieu and that it performs a crucial function.

Banton, Michael, "Social Alignment and Identity in a West African City," in H. Kuper, ed., *Urbanization and Migration in West Africa.* Berkeley, Calif.: University of California Press, 1965, pp. 131-147.

Examines the kinds of relationships that emerged among immigrants to Freetown, Sierra Leone. Particularly important were the voluntary associations, based on tribal ties, very similar to those described by Mangin in Peru.

Deutsch, K. W., "The Growth of Nations: Some Recurrent Patterns of Political and Social Integration," in World Politics, Vol. IV (1953), pp. 168-195.

Discusses the impact of urbanization on attitudes and aspirations and its roundabout affect on political organization and action.

duToit, Brian M., "Cultural Continuity and African Urbanization," in Elizabeth Eddy, ed., *Urban Anthropology: Research Perspectives and Strategies.* Athens, Ga.: Southern Anthropology Society, 1968, pp. 58-74.

Discusses kinds of physical mobility that mark the African scene and the changes or adaptations that occur when the African rural migrant leaves the village to become an urban laborer or resident. Mangin, Preston and the MacDonalds in Latin America and Mukherjee and Lambert in Asia have also discussed the above phenomenon.

Elkan, Walter, "Circular Migration and the Growth of Towns in East Africa," in Ekistics, Vol. 25, No. 149 (April, 1968), pp. 238-241.

An essay on one facet of movement into towns of East Africa, the pattern of circular migration. No attempt is made to describe or quantify labor migration as a whole. Olien, in his study of Puerto Limon, Costa Rica, and Lambert in India indicate that this pattern is quite common for those areas.

Emerson, Rupert, "Nation-Building in Africa," in K. W. Deutsch and W. J. Foltz, eds., *Nation Building.* New York: Atherton, 1966, pp. 95-116.

Superficial analysis of several factors hindering the development of sub-Saharan African nations.

Epstein, A. L., "Urbanization and Social Change in Africa," in Gerald Breese, Ed., *The City in Newly Developing Countries: Readings on Urbanism and Urbanization.* Englewood Cliffs, N.J.: Prentice-Hall, Inc., 1969, pp. 246-263.

Examines some of the variables that shape the structure of social relations in modern African towns. At times, his analysis sounds like that of Willems' study of urban classes and acculturation in Latin America.

Gould, Peter, "A Note on Research into the Diffusion of Development," in Journal of Modern African Studies, Vol. 2 (1964), pp. 123-125.

Article apparently served as an impetus for the start of research into the spatial patterns of modernization and urbanization in African countries. Gould pointed out that political, social and economic development takes place in space as well as time, and challenged students of social change to take space into account when researching, which Soja, McNulty and Gould himself did do in later studies.

Gould, Peter, "Tanzania 1920-63: The Spatial Impress of the Modernization Process," in World Politics, Vol. 22, No. 1 (January, 1970), pp. 149-170.

Study of Tanzania's evolving spatial structure. Very similar to Soja's study of Nigeria and McNulty's of Ghana. Unfortunately, this type of study has not been done in Latin America or Asia.

Green, L. P. and T.J.D. Fair, *Development in Africa: A Study in Regional Analysis with Special Reference to Southern Africa,* Part I. Johannesburg: Witwatersrand University Press, 1969.

Regional analysis of the pattern of economic and social development. Though published recently, most data and statistics are dated in the 1950's.

Grove, David and Laslo Huszar, *The Towns of Ghana: The Role of Service Centers in Regional Planning.* Accra: Ghana Universities Press, 1964.

Study of the changing relationships between urban centers and the hinterlands of Ghana within a general analytical framework of the relationship of city development to national development. Discusses the policy implications of such relationships. In philosophical orientation, study is similar to those of Chile and Venezuela done by John Friedmann.

Gutkind, Peter C.W., "The African Urban Milieu: A Force in Rapid Change," in Civilisations, Vol. XII, No. 2 (1962), pp. 167-191.

Examination of contemporary African urban problems and characteristics–social, economic, political and physical–all of which have policy implications for the developing nations. A survey of the type done in Latin America by the United Nations.

Gutkind, Peter C. W., "African Urbanism, Mobility and Social Network" in Gerald Breese, ed., *The City in Newly Developing Countries: Readings on Urbanism and Urbanization.* Englewood Cliffs, N.J.: Prentice-Hall, Inc., 1969, pp. 289-400.

Indicates some of the characteristics of urban social networks and suggests how these networks are shaped by extent and type of rural-urban mobility. Gutkind indicates in this study of Kampala, Uganda, as does Gallin in his study in Taipei, that the urban networks are important determinants of a migrant's success in adjusting to and being accepted into the urban sociopolitical structure.

Hanna, William J. and Judith L. Hanna, "The Integrative Role of Urban Africa's Middleplaces and Middlemen," in Civilisations, Vol. XVII (1967), pp. 12-29.

Examines the problem of vertical integration in states which incorporate vastly contrasting sub-cultures, focusing on the integrative role of middleplaces and middlemen. In Latin America, the regional association and the *cholo* perform the same roles as the middleplaces and middlemen. (See William P. Mangin's article on regional associations.)

Hanna, William J. and Judity L. Hanna, "Polyethnicity and Political Integration in Umuahia and Mbale," in Robert Daland, ed., *Comparative Urban Research: The Administration and Politics of Cities.* Beverly Hills, Calif.: Sage Publications, Inc., 1969, pp. 162-202.

Examines the ethnic character of African cities and the interactions between and within ethnic groups. Stress the merging of modern and traditional patterns.

Horvath, Ronald, "Around Addis Ababa: The Study of the Impact of a City on Its Surroundings." Unpublished Ph.D. Dissertation, University of California, Los Angeles, 1966.

Discusses role of the city in the transformation of the area adjacent to it.

Little, Kenneth, "Organization of Voluntary Associations in West Africa," in Civilisations, Vol. IX, No. 3 (1959), pp. 283-297.

Describes the organizational elements that are common to most all types of voluntary associations–fraternity, social interaction, mutual benefit and aid. These organizational elements are also common to the voluntary associations in Latin America and are described for that region by William Mangin.

Little, Kenneth, "Some Contemporary Trends in African Urbanization." Lecture delivered at Northwestern University on April 20, 1965 and published by Northwestern University Press.

Discusses the voluntary association in terms of its contribution to the process of urbanization, specifically its function as an adaptive mechanism and as a provider of social and economic security for the newly arrived urban migrant.

Mabogunje, Akin, "The Economic Implications of the Pattern of Urbanization in Nigeria," in Nigerian Journal of Economic and Social Studies, Vol. VII (March, 1965), pp. 9-30.

Examines Nigeria's attempt to build up infrastructure. Brings up the dilemma faced by most developing countries, that is, the choice between national integration and short-term increments of economic growth.

Mabogunje, Akin, "Urbanization in Nigeria: A Constraint on Economic Development," in Economic Development and Cultural Change, Vol. XIII, No. 4, Part I (July, 1965), pp. 413-438.

Traces the development of and analyzes Nigerian system of cities. As in earlier article, emphasis is on the policy implications for national development of Nigeria's accelerated urbanization.

Mayer, Philip, "Migrancy and the Study of Africans in Towns," in American Anthropologist, Vol. 64, No. 3, Part I (June, 1962), pp. 576-592.

Good summary of migrancy research that had been done in Africa up to 1962. Also examines extra-town ties of town dwellers and assesses them in relation to within-town ties.

McNulty, Michael L., "Urban Structure and Development: The Urban System of Ghana," in Journal of Developing Areas, Vol. III (January, 1969), pp. 159-176.

Analysis of the urban system of Ghana. Provides a brief examination of concomitant growth of the urban system and of space economy, identifies some of the major sources of variation and urban differentiation, and relates these structural elements to the roles of the various urban centers in the organization of the economy. Excellent study. Technique could be profitably applied to Latin America and Asia.

Safier, Michael, "The Urban System in Africa; Patterns of Disjunction and Integration," in African Urban Notes, Vol. 2, No. 3 (May, 1967), pp. 4-11.

African development has not been endogenous or generic, but grafted onto and over another societal structure. This has resulted in a pattern of uneven development and spatial organization. Unfortunately, article was published without footnotes.

Soja, Edward, *The Geography of Modernization in Kenya.* Syracuse: Syracuse University Press, 1968.

Analyzes the geography of modernization in Kenya and examines the implications of the spatial perspective for the analysis of modernization. Very important and useful study, as are those by Gould and McNulty. Latin American and Oriental specialists would do well to apply these research methods and techniques to their continents.

Southall, Aidan, "Kampala-Mengo," in H. Miner, ed., *The City in Modern Africa.* London: Pall Mall, 1967, pp. 297-332.

Historical and spatial analysis of long established traditional Ugandan city that has developed into a multi-racial modern city. Particularly interesting are the new social relationships that are being worked out, relationships that are neither African nor Western

Taaffe, Edward J., Richard L. Morill, and Peter R. Gould, "Transport Expansion in Underdeveloped Countries: A Comparative Analysis," in The Geographical Review, Vol. LIII, No. 4 (October, 1963), pp. 503-529.

Examines the process of change initiated by improvement in internal accessibility, using as examples Ghana and Nigeria. A more superficial study on this subject has been done by Howard Dauthier, drawing on evidence from the São Paulo region of Brazil.

C. SOUTH ASIA

Armstrong, W. R. and T. G. McGee, "Revolutionary Change and the Third World City," in Civilisations, Vol. 18, No. 3 (1968), pp. 353-378.

Description of the bazaar economy—the whole pattern of peddling and processing activities characteristic of so many Third World cities. Stresses its social-cultural, as well as economic role. In Latin America, this phenomenon has been described by Carolina de Jesus and Lisa Peattie.

Berry, Brian J.L., "City Size and Economic Development: Conceptual Synthesis and Policy Problems, with Special Reference to South and South-East Asia." Unpublished MS, 1970.

A systematic study of current policy debate in developing countries, with special attention given to the factor of primacy in national development. This concern is expressed by Mabogunje in his studies of Nigeria.

Eames, Edwin, "Urban Migration and Joint Family in a North Indian Village," in The Journal of Developing Areas, Vol. 1, No. 2 (January, 1967), pp. 163-178.

Presents data collected in a North Indian village which indicate that the joint family is a facilitating mechanism in the process of rural-urban migration and which also suggest that the process of rural-urban migration increases the viability of this unit in the village.

Gallin, Bernard, "Political Attitudes and Political Participation of Rural-to-Urban Migrants in Taipei." Paper presented at Conference on Urban Society and Political Development in China. New York: Social Science Research Council, 1969.

Discusses effect that urban social-political structure has on migrants' integration into that structure and result of partial integration.

Geertz, Clifford, *Peddlers and Princes: Social Change and Economic Modernization in Two Indonesian Towns.* Chicago: University of Chicago Press, 1963.

Comparative study of the economic system and modernizing agents in two Indonesian towns, characterizing aspects of the urban social system of the poor, particularly the bazaar economy. Armstrong and McGee have elaborated on Geertz's description for Asia and Lisa Peattie and Carolina de Jesus describe this interesting phenomenon in Latin America.

Guyot, James F., "Creeping Urbanism and Political Development in Malaysia," in Robert Daland, ed., *Comparative Urban Research: The Administration and Policies of Cities.* Beverly Hills, Calif.: Sage Publications, Inc. 1969, pp. 124-161.

Empirical study and speculative forecast of creeping urbanism in rural development program of the Malaysian government and of contributions this development could make to general trend toward more competitive politics among the Malays.

Jakobson, Leo and Ved Prakash, "Urbanization and Regional Planning in India," in Urban Affairs Quarterly, Vol. 2, No. 3 (March, 1967), pp. 36-65.

Presents their analysis of urbanization pattern in India. Propose policy of integrated linear system of urban centers in which each node interacts with others along efficient, rapid, multipurpose transportation corridors.

Lambert, Richard D., "The Impact of Urban Society Upon Village Life," in Roy Turner, ed., *India's Urban Future.* Berkeley, Calif.: University of California Press, 1961, pp. 117-141.

Summarizes the findings of Indian urban studies up to 1961 and weighs the evidence of the impact of urbanization on villages in India. Horvath, also interested in the impact of the city on its surrounding area, is more systematic in his study of Addis Ababa, Ethiopia, than is Lambert in this study of India.

McGee, T. G., "Catalysts or Cancers? The Role of Cities in Asian Society," in *Urbanization and National Development,* edited by Leo Jakobson and Ved Prakash (Vol. I, South and Southeast Asia Urban Affairs Annuals), Beverly Hills, Calif.: Sage Publications, Inc., 1971, pp. 157-181.

Attempts to construct a model of the system of cities in Asia. Discusses the functions and roles cities play in process of development. Like Germani in Latin America and Gutkind in Africa, McGee feels that the city plays a very important role in restructuring and modernizing society.

Mukherjee, Ramkrishna, "Urbanization and Social Transformation in India," in *International Journal of Comparative Sociology,* Vol. 4, No. 2 (September, 1963), pp. 179-210.

Study of effects of urban development on the Indian social system, examining in particular migration patterns, the relative prevalence of different family types, and the prevalence of the caste system.

Murphy, Rhoads, "Urbanization in Asia," in Gerald Breese, ed., *The City in Newly Developing Countries: Readings on Urbanism and Urbanization.* Englewood Cliffs, N.J.: Prentice-Hall, Inc., 1969, pp. 58-75.

Identifies two urban patterns in Monsoon Asia, the traditional Asian city and the colonial city, demonstrating the tenacity of long established networks.

Sendut, Hamzah, "Patterns of Urbanization in Malaya," in Journal of Tropical Geography (University of Malaya), Vol. 16 (October, 1962), pp. 114-130.

Identifies numerous factors which have influenced the pattern of urbanization in Malaya.

Sovani, N.V., "Some Notes on In-Migrants in Indian Cities," in *Urbanization and Urban India.* New York: Asia Publishing House, 1966, pp. 142-155.

Examines the social composition of migrant populations. Many of his findings are similar to those of Bruce Herrick in Chile.

Vakil, C.N. and Krishna Roy, "Growth of Cities and their Role in the Development of India," in Civilisations, Vol. 15, No. 3 (1965), pp. 326-359.

Identify chief characteristics of India's urban centers in different states and suggest some (rather simplistic) policies to reverse these existing trends.

Van Alstyne, Arthur J., "Urban Influences on Rural Communities: A Taiwanese Example," in The Chung Chi Journal (Hong Kong), Vol. 7, No. 2 (May, 1968), pp. 197-245.

Attempts to determine relationship between diffusion of innovations and proximity and accessibility to urban centers. Due to simplistic assumptions, the validity of his conclusions is questionable.

Zachariah, K.C., "Bombay Migration Studies: A Pilot Analysis of Migration to an Asian Metropolis," in Gerald Breese, ed., *The City in Newly Developing Countries: Readings on Urbanism and Urbanization.* Englewood Cliffs, N.J.: Prentice-Hall, Inc., 1969, pp. 360-375.

An analysis of the 1960 population census, with special emphasis given to data concerning migration. In purpose, very much like compilation of statistics done by Rycroft and Clemmer for Latin America.

Appendix B

A NOTE ON THRESHOLD SIZES FOR URBAN DEVELOPMENT

(Chapter Ten)

Speculations concerning policies for spatial development rest, in considerable measure, on the assumption that local area economies must attain a minimum market size before their development can become self-sustaining. There is no proof, however, that urban threshold sizes in this sense exist. Although true for individual services, the applicability of the concept of threshold to the capacity of urban systems for sustaining economic growth has not as yet been tested. The available evidence is somewhat casual and indirect, and may be summarized as follows:

(1) As the internal market of a local area economy increases in size, the ratio of its export to local service industries will tend to decline. A larger percentage of total growth will therefore be generated internally, and the economy will become less vulnerable to changes in the composition of external demands.

(2) With increasing market size, the industrial and service specialization of local areas will tend to increase, creating external economies that will reinforce the propensity of the system to grow.

(3) With increasing market size, the capacity of the local population to organize itself for economic growth will tend to increase.

(4) Higher level services will tend to locate in areas of increasing market size and help extend the effective market area to surrounding regions.

If we assume, along with authors such as Brian Berry, that the population threshold for American cities lies in the neighborhood of 250,000, the equivalent figure for Chile with a much lower per capita income than the U.S., might be found somewhere between 1.0 and 1.5 million. If it were possible, therefore, to accelerate the growth of existing urban subsystems up to and beyond the appropriate threshold size, a process of self-sustaining growth might be initiated. In general, large cities—or large and highly integrated urban systems—behave more autonomously than small centers. It should not be thought, however, that the transition to self-sustaining growth will be a sudden one as the critical size is approached. Only a gradual acceleration of the capacity for self-sustaining growth would be expected to occur.

Within the Capital Region of Chile, there is currently only one urban subsystem other than Santiago that offers a remote possibility for reaching this size within the next twenty years. This is Valparaíso, whose existing labor market area incorporates slightly more than 700,000 people. However, the area is not yet fully integrated economically, and its present size is perhaps only one-half of what the threshold hypothesis would require under Chilean conditions.

Appendix C

MULTIPLIER EFFECTS OF ALTERNATIVE INDUSTRIAL LOCATION POLICIES

(Chapter Ten)

The spatial development policy outlined in Chapter 10 includes a strong recommendation for concentrating industries in a few strategic locations (Santiago, Quintero, Rancagua), thereby avoiding the random spread of industrial plants among a large number of dispersed points along the agro-industrial corridors of the Capital Region or in its "intermetropolitan periphery."

This recommendation is based on the following analysis of local multiplier effects resulting from industrial development under two contrasting hypotheses for location: concentration and dispersion.

Types of Effect	*Concentration*	*Dispersal*
Demographic	Accelerated migration towards the sub-center in reaction to other effects listed below; substantially reduced migration to Santiago	Small increment of population near industrial locations; few secondary effects (see below); and continued rapid growth of Santiago
Economic	Accelerated mutiplier effects, especially in tertiary and quaternary sectors resulting from increase in disposable income, greater integration of industry with local community, and threshold effects for higher-order services	Principal multiplier effects in Santiago; internalization of essential services by each industry, with resulting higher production costs; nonintegration of industry with the local community

Types of Effect	*Concentration*	*Dispersal*
Labor Market	Greater individual employment ability (alternative sources of employment in local center); greater individual and collective bargaining power; improved social-economic situation	Employment mobility attainable only through commuting or migration, both of which impose high costs on individual; greater rigidity of occupational structure; lower bargaining power and greater dependency on employer
	Because of greater integration of industry with the community, accelerated effects in aspirations of the traditional population sector through demonstration; management staff will live increasingly near the enterprise itself	Industrial "enclaves" with continued maintenance of dualistic social structures (modern-traditional); the industry is self-contained and socially invisible; management staff will continue to make its home in Santiago
Psycho-social	Increase in the internal capacity to promote and guide the local process of economic and social development; relative decrease in central powers	Continued dependency on Santiago and increase in central powers
	Gradual increase in the possibilities for upward social mobility in the local center	Upward social mobility possible only in Santiago
Location	Greater attraction of local community for future industrial and service locations (feedback effect)	Except where strong external economies are created, the location of an industrial plant will have a neutral and even negative effect on future locations
Spatial	High level of integration based on subregional complementarities and partial autonomy of principal centers	Low-level integration with continued high predominance of Santiago as regional center; increasing dependency of subcenters, with usual "drainage" effects

Appendix D

OUTLINE OF A PROPOSAL FOR A METROPOLITAN DEVELOPMENT CORPORATION

(Chapter Ten)

BASIC ASSUMPTIONS

Major metropolitan areas located in the periphery of Santiago—specifically Valparaíso and Concepción—have the possibility of evolving an economic life of their own that is not exclusively dependent on the central agencies of management and control in the nation's capital. The future growth of their economies will in large part depend on their ability to initiate and maintain a continuous process of development (as opposed to a random, discontinuous process) and to provide the necessary institutional guidance for this process.

As an appropriate institutional mechanism for this purpose a Metropolitan Development Corporation is proposed. The general characteristics of the corporation would be:

- a strictly technical orientation
- responsible political control, with active participation of the affected population
- the fusion of planning and action
- a project orientation within the context of general area development
- responsibility for the economic, social, and physical development of an area
- autonomy, but central accountability and coordination
- a publicly chartered, non-profit corporation

The following proposal is an attempt to illustrate the idea of a Metropolitan Development Corporation with specific detail. It is not our intention to suggest this as the most appropriate institutional arrangement, one that would, in any event, have to be created by special law of the National Congress. But the detail will, we hope, make the proposal sufficiently concrete to allow it to be considered as a serious alternative to the activation of some of the major policy recommendations contained in the main body of our study.

A METROPOLITAN DEVELOPMENT CORPORATION (MDC)

1. Structure and Functions

1.1. The MDC is a publicly chartered, autonomous non-profit Corporation.

1.2. Its area of responsibility extends to the entire labor market area of the principal metropolitan center. However, adjacent areas may be incorporated to the extent they are important in furthering metropolitan development.

1.3. The Directorate of the MDC is a collective body, consisting of three Directors who are appointed with the advice and consent of the Policies Council by the President of the Republic. Each Director is in charge of one major functional area of the MDC's activities: planning and budgeting, projects and loans (Figure 1).

1.4. The Policies Council of the MDC is composed of the three Directors, the Intendente of the Province (or if more than one Province is included in its area of responsibility, the Intendentes of all affected Provinces), the

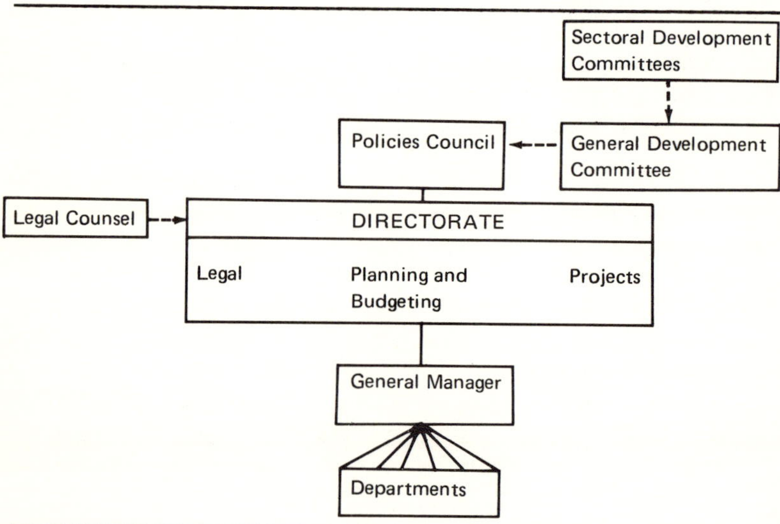

Figure 1: Structure of Metropolitan Development Corporation

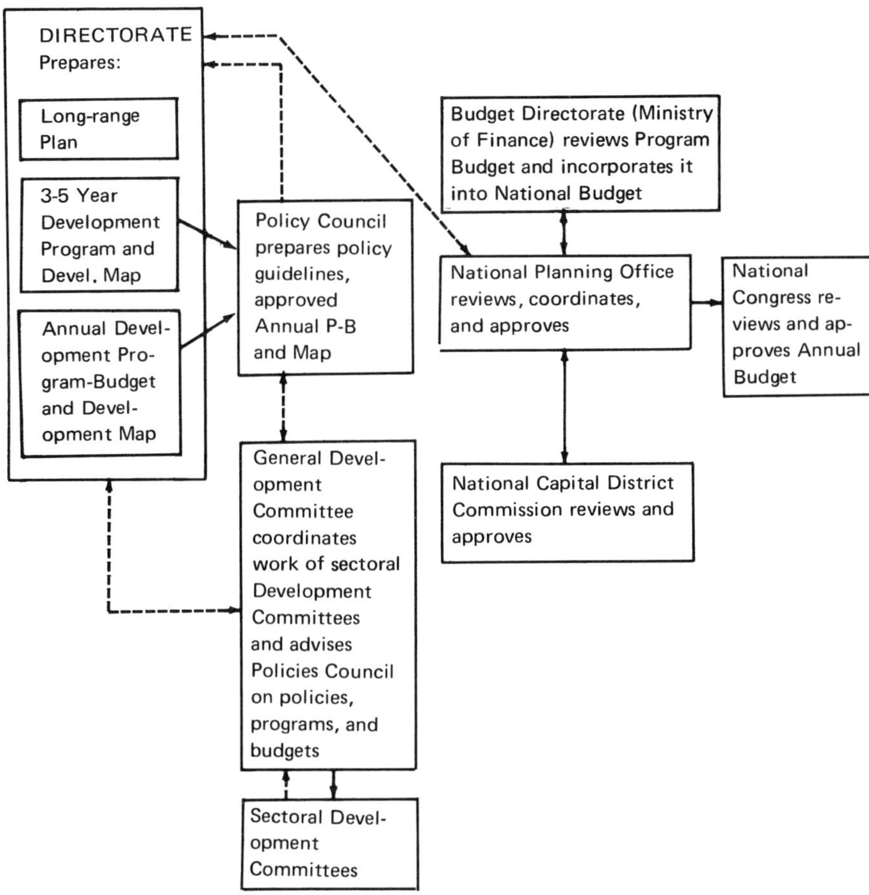

Figure 2: Metropolitan Development Corporation: Planning Process

Mayors of the principal cities, and a certain number of representatives from the major productive sectors of the economy, the Neighborhood Council Association, and the Universities.

1.5. The MDC would be empowered to:
- make loans to private enterprise and municipalities at low rates of interest through a revolving development loan fund
- prepare long-range plans, medium-range programs, and annual program budgets
- contract for the construction of development projects with both the private and the public sectors
- manage and operate productive enterprise where necessary or convenient, such as a bus line or a sewage treatment plan. These

enterprises may be set up as autonomous, non-profit corporations, subsidiary to the MDC.

1.6. A General Manager of the MDC is appointed by the Directorate and is directly responsible to the Directorate.

2. Planning Process

2.1. A long-rang area development strategy for 20-25 years. This would serve as an internal reference document for the MDC's own use (Figure 2).

2.2. A 3-5 year Development Program of Investments, together with a Development Map, showing the spatial implications of the programmed investments. This document would be revised every year and submitted together with the annual Program Budget. The Development Program and Program Budget would be drawn up on basis of policy guidelines received from the Policies Council; it would be submitted to the Policies Council for approval of the annual Program Budget only.

2.3. The Annual Program Budget is approved by the Council together with appropriate portion of the Development Map, and is submitted to the National Planning Office.

The Program Budget will be divided into two major parts: one, imperative, relating to the portion of the Program which is to be carried out by the MDC itself; another, indicative, representing suggestions for other Ministries.

The National Planning Office coordinates the Program Budget with the National Capital District Commission and the Budget Directorate. Any changes are discussed with the MDC and must receive its final approval. The MDC Program Budget is then included as a special item in the National Budget, and is submitted to the Congress. However, the Congress can vote only on the total amount to be budgeted for the MDC, and not on any particular line item. Internal Program Budget adjustments are left to the Corporation.

Once approved, the MDC receives the entire annual amount of its budget through the Treasury. Normal standards of accountability for public funds apply.

2.4. The Policies Council requires an internal reporting and evaluation function on a quarterly and annual basis.

3. Participation

In addition to the normal kind of coordination that would be required between MDC and Ministries and Municipalities, each of the sectors represented in the Policies Council (Municipalities, Juntas de Vecinos, and productive sectors) would set up special committees to consider the Medium-range Development Program and the Annual Program Budget. These would be organized as sectoral Development Committees. A General Development Committee, including all sectors, would serve as the principal advisory body to the Policies Council.

4. Financing

Among the sources of funds for the MDC would be the following:

4.1. Development Loan Fund (revolving)–initial capitalization
4.2. Annual Program Budget (from National Budget)
4.3. Loan Agreements with international agencies
4.4. Special Development Tax, as a token of local commitment
4.5. Interest on loans
4.6. Income from facilities and services managed by the MDC
4.7. Special regional funds (e.g., ley de cobre)
4.8. Special municipal funds

5. Special Note

The MDC would be directly responsible for development projects of a non-recurring, non-routine nature only. Routine maintenance activities would be left as the primary function of the municipalities in each area.

Expenditure ceilings for each area should not be applied arbitrarily. However, the law creating the MDC may specify a certain percentage (e.g., 60%) of the total Program Budget for whose implementation it would be directly responsible. These would be the key projects; the rest would be left to the Ministries.

Appendix E

COURSES IN REGIONAL ANALYSIS

(Chapter Sixteen)

1. *Introduction to theoretical geography*
   Models of locational structure; movement, networks, nodes, hierarchies, surfaces; general field theory; central place theory; characteristics of urban systems; mathematical techniques for regional delimitation.

2. *Advanced theoretical geography*
   Space perception and location behavior; diffusion models; stochastic models of location and movement; interaction models; simulation models in geography; migration and settlement.

3. *Location theory*
   Theory of location of the firm; location rent and the spatial structure of metropolitan and other regions; intra-metropolitan transportation and location models; industrial complex analysis; linear programming models.

4. *Quantitative methods in spatial analysis*
   Spatial measures of location, density, central tendency, and dispersion; factor analysis and regression and variance analysis applied to regional and spatial phenomena; sampling for areal measurement; problems of area statistics; elements of game theory and general equilibrium analysis.

5. *Theory of urban-regional development*
   Empirical regularities in the process of regional growth and decline; analysis of locational shifts; a location model of economic growth; interregional

inter-industry relations; internal determinants of regional growth; multiplier effects; regional distribution of income and movement of capital and labor; equilibrium and disequilibrium models of regional growth.

6. *Regional accounts and development models*
   Systems of social accounts; accounts and policy decisions; towards an integrated system of social accounts; data for regional accounts; integration of regional with national accounts; employment accounts; social indicators for regional development; linear programming models for regional development.

7. *Transportation and economic development*
   Mobility and development; transport requirements for development; choice of transport technology; regional effects of changes in relative accessibility to supply areas and markets; pricing transport services; financing transport investment; transportation and regional economic integration.

8. *Regional analysis workshop*
   A practicum required of all students and taken during the second semester. Students learn to apply the analytical techniques they have studied and are introduced to the computer.

9. *Preparation for Comprehensive Examination*
   At the master's degree level.

Appendix F

COURSES IN REGIONAL PLANNING

(Chapter Sixteen)

1. *Introduction to regional analysis*
   An introductory course for regional planning specialists. Central place theory; characteristics of urban systems; types of regions; elements of location theory; theory of regional development; major hypotheses; the accounts framework for regional development; linear programming models for regional development.

2. *Theory of planning*
   History of planning thought; limits of rationality; the institutional context of planning decisions; allocative and innovative planning; planning under conditions of crisis; politics and planning; problems of implementation: command and indicative forms; the sociology of participation; the uses of planning; comparative study of planning and typology of planning styles.

3. *Urbanization and economic development*
   The ecological and social meanings of urbanization; the influence of cities on economic, social, political development; the theory of hyperurbanization; the theory of polarized development; development of a system of cities under alternative conditions; diffusion of the elements of urban culture; spatial diffusion theories and the urbanization of rural areas; a communications model of urbanization; problems of urban development; strategies of deliberate urbanization.

4. *Regional policy and planning*
   The center-periphery model of regional development; types and definition of planning regions: core regions, depressed areas, resource frontiers, special problem areas; objective of a national policy for regional development; alternative development strategies for different planning regions; implications of these policies for migration and settlement; the integration of regional with national development policies and planning; international aspects.

5. *Political aspects of regional development*
   Legal-administrative organization relevant to regional development planning; the problem of administrative and political decentralizations; models for the areal division of powers; the sociology of participation; the social structure of power in different regions and localities; attitudes towards and possibilities for change; political behavior in relation to regional development; analysis of voting results; political movements; class-based and national politics.

6. *Project and program evaluation*
   Formulation of investment projects; preliminary evaluation criteria: feasibility and desirability; principles of cost-benefit analysis for economic and social projects; major problems in the application of cost-benefit studies; discount rates and opportunity costs, risk and uncertainty, intangible effects, multiplier effects; cost-benefit analysis of simple and complex projects; alternative criteria for project selection; case studies of specific projects in a regional context: urban development, manufacturing industry, transportation investments, water resource development.

7. *Instruments of regional development planning*
   The organization of regional planning in unitary and federal governments—central and regional levels; political aspects of regional planning; the problem of administrative decentralization: efficiency and participation criteria; problems of implementing regional planning: the capital budget, the private sectors, development banks, regional development corporations, financial instruments, direct controls; information systems; communication of the plans and creating a development ethos; staging the planning-development process.

8. *Regional program budgets*
   Theory of public expenditure decisions; major program components; determining optimal program sequences; programming cycles and the problem of optimal allocation; comprehensive versus focused programming; the program budget; relation of program budget to regional development objectives; the program budget as a control device; stability of program budgets; forecasting and uncertainty; simulation models for programming.

9. *Regional policy workshop*
   A practicum required of all students and taken during the second semester. The focus of the workshop is on policy analysis for a regional development problem under constraints similar to those encountered in the real world.

10. *Thesis in regional planning*
    Independent research on some aspect of regional development planning, building on the student's basic discipline. Doctoral level.

INDEX

INDEX

ABOUT THE AUTHOR

John Friedmann was born and raised in Vienna, Austria. After his military service with U.S. intelligence in Germany, he attended the University of Chicago from where he received his Ph.D. in planning, economics, and geography in 1955. In addition to working for two years on the policies planning staff of the Tennessee Valley Authority, he spent several years as government consultant and advisor in Brazil, Korea, Venezuela, and Chile. In recognition of his services, he was decorated by the Chilean Government with the Order of Bernardo O'Higgins and received an honorary doctorate from the Catholic University of Chile. From 1961 to 1965, he served as associate professor of regional planning at MIT; in 1969, he joined UCLA's new School of Architecture and Urban Planning as professor and Head of the Urban Planning Program. He has written and published extensively on questions of urbanization, regional development planning, and planning theory. In his spare time, he likes to write poetry and play the cello.